ABSTRACT ALGEBRA

PURE AND APPLIED MATHEMATICS

A Program of Monographs, Textbooks, and Lecture Notes

MONOGRAPHS AND TEXTBOOKS IN PURE AND APPLIED MATHEMATICS

1. *K. Yano,* Integral Formulas in Riemannian Geometry (1970) *(out of print)*
2. *S. Kobayashi,* Hyperbolic Manifolds and Holomorphic Mappings (1970) *(out of print)*
3. *V. S. Vladimirov,* Equations of Mathematical Physics (A. Jeffrey, editor; A. Littlewood, translator) (1970) *(out of print)*
4. *B. N. Pshenichnyi,* Necessary Conditions for an Extremum (L. Neustadt, translation editor; K. Makowski, translator) (1971)
5. *L. Narici, E. Beckenstein, and G. Bachman,* Functional Analysis and Valuation Theory (1971)
6. *D. S. Passman,* Infinite Group Rings (1971)
7. *L. Dornhoff,* Group Representation Theory (in two parts). Part A: Ordinary Representation Theory. Part B: Modular Representation Theory (1971, 1972)
8. *W. Boothby and G. L. Weiss (eds.),* Symmetric Spaces: Short Courses Presented at Washington University (1972)
9. *Y. Matsushima,* Differentiable Manifolds (E. T. Kobayashi, translator) (1972)
10. *L. E. Ward, Jr.,* Topology: An Outline for a First Course (1972) *(out of print)*
11. *A. Babakhanian,* Cohomological Methods in Group Theory (1972)
12. *R. Gilmer,* Multiplicative Ideal Theory (1972)
13. *J. Yeh,* Stochastic Processes and the Wiener Integral (1973) *(out of print)*
14. *J. Barros-Neto,* Introduction to the Theory of Distributions (1973) *(out of print)*
15. *R. Larsen,* Functional Analysis: An Introduction (1973) *(out of print)*
16. *K. Yano and S. Ishihara,* Tangent and Cotangent Bundles: Differential Geometry (1973) *(out of print)*
17. *C. Procesi,* Rings with Polynomial Identities (1973)
18. *R. Hermann,* Geometry, Physics, and Systems (1973)
19. *N. R. Wallach,* Harmonic Analysis on Homogeneous Spaces (1973) *(out of print)*
20. *J. Dieudonné,* Introduction to the Theory of Formal Groups (1973)
21. *I. Vaisman,* Cohomology and Differential Forms (1973)
22. *B.-Y. Chen,* Geometry of Submanifolds (1973)
23. *M. Marcus,* Finite Dimensional Multilinear Algebra (in two parts) (1973, 1975)
24. *R. Larsen,* Banach Algebras: An Introduction (1973)
25. *R. O. Kujala and A. L. Vitter (eds.),* Value Distribution Theory: Part A; Part B: Deficit and Bezout Estimates by Wilhelm Stoll (1973)
26. *K. B. Stolarsky,* Algebraic Numbers and Diophantine Approximation (1974)
27. *A. R. Magid,* The Separable Galois Theory of Commutative Rings (1974)
28. *B. R. McDonald,* Finite Rings with Identity (1974)
29. *J. Satake,* Linear Algebra (S. Koh, T. A. Akiba, and S. Ihara, translators) (1975)

30. *J. S. Golan,* Localization of Noncommutative Rings (1975)
31. *G. Klambauer,* Mathematical Analysis (1975)
32. *M. K. Agoston,* Algebraic Topology: A First Course (1976)
33. *K. R. Goodearl,* Ring Theory: Nonsingular Rings and Modules (1976)
34. *L. E. Mansfield,* Linear Algebra with Geometric Applications: Selected Topics (1976)
35. *N. J. Pullman,* Matrix Theory and Its Applications (1976)
36. *B. R. McDonald,* Geometric Algebra Over Local Rings (1976)
37. *C. W. Groetsch,* Generalized Inverses of Linear Operators: Representation and Approximation (1977)
38. *J. E. Kuczkowski and J. L. Gersting,* Abstract Algebra: A First Look (1977)
39. *C. O. Christenson and W. L. Voxman,* Aspects of Topology (1977)
40. *M. Nagata,* Field Theory (1977)
41. *R. L. Long,* Algebraic Number Theory (1977)
42. *W. F. Pfeffer,* Integrals and Measures (1977)
43. *R. L. Wheeden and A. Zygmund,* Measure and Integral: An Introduction to Real Analysis (1977)
44. *J. H. Curtiss,* Introduction to Functions of a Complex Variable (1978)
45. *K. Hrbacek and T. Jech,* Introduction to Set Theory (1978)
46. *W. S. Massey,* Homology and Cohomology Theory (1978)
47. *M. Marcus,* Introduction to Modern Algebra (1978)
48. *E. C. Young,* Vector and Tensor Analysis (1978)
49. *S. B. Nadler, Jr.,* Hyperspaces of Sets (1978)
50. *S. K. Segal,* Topics in Group Rings (1978)
51. *A. C. M. van Rooij,* Non-Archimedean Functional Analysis (1978)
54. *L. Corwin and R. Szczarba,* Calculus in Vector Spaces (1979)
53. *C. Sadosky,* Interpolation of Operators and Singular Integrals: An Introduction to Harmonic Analysis (1979)
54. *J. Cronin,* Differential Equations: Introduction and Quantitative Theory (1980)
55. *C. W. Groetsch,* Elements of Applicable Functional Analysis (1980)
56. *I. Vaisman,* Foundations of Three-Dimensional Euclidean Geometry (1980)
57. *H. I. Freedman,* Deterministic Mathematical Models in Population Ecology (1980)
58. *S. B. Chae,* Lebesgue Integration (1980)
59. *C. S. Rees, S. M. Shah, and C. V. Stanojević,* Theory and Applications of Fourier Analysis (1981)
60. *L. Nachbin,* Introduction to Functional Analysis: Banach Spaces and Differential Calculus (R. M. Aron, translator) (1981)
61. *G. Orzech and M. Orzech,* Plane Algebraic Curves: An Introduction Via Valuations (1981)
62. *R. Johnsonbaugh and W. E. Pfaffenberger,* Foundations of Mathematical Analysis (1981)
63. *W. L. Voxman and R. H. Goetschel,* Advanced Calculus: An Introduction to Modern Analysis (1981)
64. *L. J. Corwin and R. H. Szcarba,* Multivariable Calculus (1982)
65. *V. I. Istrătescu,* Introduction to Linear Operator Theory (1981)
66. *R. D. Järvinen,* Finite and Infinite Dimensional Linear Spaces: A Comparative Study in Algebraic and Analytic Settings (1981)

67. *J. K. Beem and P. E. Ehrlich,* Global Lorentzian Geometry (1981)
68. *D. L. Armacost,* The Structure of Locally Compact Abelian Groups (1981)
69. *J. W. Brewer and M. K. Smith, eds.,* Emmy Noether: A Tribute to Her Life and Work (1981)
70. *K. H. Kim,* Boolean Matrix Theory and Applications (1982)
71. *T. W. Wieting,* The Mathematical Theory of Chromatic Plane Ornaments (1982)
72. *D. B. Gauld,* Differential Topology: An Introduction (1982)
73. *R. L. Faber,* Foundations of Euclidean and Non-Euclidean Geometry (1983)
74. *M. Carmeli,* Statistical Theory and Random Matrices (1983)
75. *J. H. Carruth, J. A. Hildebrant, and R. J. Koch,* The Theory of Topological Semigroups (1983)
76. *R. L. Faber,* Differential Geometry and Relativity Theory: An Introduction (1983)
77. *S. Barnett,* Polynomials and Linear Control Systems (1983)
78. *G. Karpilovsky,* Commutative Group Algebras (1983)
79. *F. Van Oystaeyen and A. Verschoren,* Relative Invariants of Rings: The Commutative Theory (1983)
80. *I. Vaisman,* A First Course in Differential Geometry (1984)
81. *G. W. Swan,* Applications of Optimal Control Theory in Biomedicine (1984)
82. *T. Petrie and J. D. Randall,* Transformation Groups on Manifolds (1984)
83. *K. Goebel and S. Reich,* Uniform Convexity, Hyperbolic Geometry, and Nonexpansive Mappings (1984)
84. *T. Albu and C. Năstăsescu,* Relative Finiteness in Module Theory (1984)
85. *K. Hrbacek and T. Jech,* Introduction to Set Theory, Second Edition, Revised and Expanded (1984)
86. *F. Van Oystaeyen and A. Verschoren,* Relative Invariants of Rings: The Noncommutative Theory (1984)
87. *B. R. McDonald,* Linear Algebra Over Commutative Rings (1984)
88. *M. Namba,* Geometry of Projective Algebraic Curves (1984)
89. *G. F. Webb,* Theory of Nonlinear Age-Dependent Population Dynamics (1985)
90. *M. R. Bremner, R. V. Moody, and J. Patera,* Tables of Dominant Weight Multiplicities for Representations of Simple Lie Algebras (1985)
91. *A. E. Fekete,* Real Linear Algebra (1985)
92. *S. B. Chae,* Holomorphy and Calculus in Normed Spaces (1985)
93. *A. J. Jerri,* Introduction to Integral Equations with Applications (1985)
94. *G. Karpilovsky,* Projective Representations of Finite Groups (1985)
95. *L. Narici and E. Beckenstein,* Topological Vector Spaces (1985)
96. *J. Weeks,* The Shape of Space: How to Visualize Surfaces and Three-Dimensional Manifolds (1985)
97. *P. R. Gribik and K. O. Kortanek,* Extremal Methods of Operations Research (1985)
98. *J.-A. Chao and W. A. Woyczynski, eds.,* Probability Theory and Harmonic Analysis (1986)
99. *G. D. Crown, M. H. Fenrick, and R. J. Valenza,* Abstract Algebra (1986)

Other Volumes in Preparation

ABSTRACT ALGEBRA

Gary D. Crown
Maureen H. Fenrick
Wichita State University
Wichita, Kansas

Robert J. Valenza
Gould NavCom Systems
El Monte, California

MARCEL DEKKER, INC. **New York and Basel**

Library of Congress Cataloging-in-Publication Data

Crown, Gary D. [date]
Abstract algebra.

(Monographs and textbooks in pure and applied mathematics ; 99)
Bibliography: p.
Includes index.
1. Algebra, Abstract. I. Fenrick, Maureen H., [date] II. Valenza, Robert J., [date] III. Title. IV. Series: Monographs and textbooks in pure and applied mathematics ; v. 99.
QA162.C76 1986 512'.02 85-25439
ISBN 0-8247-7456-6

MARCEL DEKKER, INC.

270 Madison Avenue, New York, New York 10016

Current printing (last digit):
10 9 8 7 6 5 4 3 2 1

PRINTED IN THE UNITED STATES OF AMERICA

Preface

Ontology recapitulates philology.
-- J. G. Miller

This text covers the fundamental concepts of abstract algebra at a level appropriate to an upper-division undergraduate or first-year graduate course. Our purpose throughout is to promote genuine mathematical growth through a clear, rigorous exposition supplemented by a wealth of examples drawn from other, familiar areas of mathematics. We hope to induce in our students not only technical mastery of the subject, but also correct expository habits and a certain cultural broadening. Along the way, they should also begin to grasp the tremendous unifying power of abstract algebra -- a major rationale for its existence.

We made and have maintained an early decision that this book should be self-justifying. That is, we did not choose to write an encyclopedia of abstract algebra to be ground through now and only truly understood in light of later courses. For this reason, much space is devoted to solid, fundamental examples, and correspondingly less space is available for advanced topics. We based our decision on the clear truth that only a minority of algebra students go on to later courses which in turn justify the naturalness and universality of the fundamental algebraic structures. For the majority, the point of an encyclopedic course without concrete referents is almost certainly lost.

The exercises fall into three varied categories: First, there are instantiations of propositions and definitions. From these the deeper meanings of the austere formalisms emerge. Second, we provide the routine combinatoric drills which have become the traditional mainstay of this type of text. Third, there are extended sequential exercises which develop important supplementary topics and compel the student to apply ideas from various chapters (admittedly only a nod in the direction of reality). With these resources we feel that the development of the material is successfully motivated without the introduction of obtrusive and wordy digressions into the main exposition.

There is enormous value to an abstract algebra course beyond the mathematics it conveys or prepares us for. Those who have studied how experts in a given field solve problems tell us that the expert expends a great deal more time in the abstraction phase than the novice does. Thus his or her expertise lies not in accessing more of the available facts, but rather in recognizing which of the facts are germane to the problem at hand and limiting consideration to only those. By abstraction he or she brings the problem out of its original context and into a simplified problem space where it is then dealt with decisively. We know of no training which might better cultivate this expert methodology than abstract algebra, provided that the central definitions and theorems are understood for what they are -- the distillation of centuries of diverse mathematical experience. Thus this discipline can serve the engineer and scientist in ways that completely transcend mere mathematical technique.

We thank the editors at Marcel Dekker, Inc., and their reviewers for the patience and expertise they have shown throughout the evolution of our manuscript. And with sincere affection we thank our friend and silent partner in all of this, Suzanne Frantz, whose incomparably polished production of the original text launched us on our way.

Gary D. Crown
Maureen H. Fenrick
Robert J. Valenza

Contents

1
Preliminaries

1. SET OPERATIONS AND FUNCTIONS

In this section we will define, discuss, and give examples of the basic set operations and functions. The approach taken here is naive rather than axiomatic. We suggest the student fill in all missing details in the examples and work through all the exercises. The student should also try to construct several examples to illustrate each new concept introduced.

(1.1) NOTATION AND TERMINOLOGY We will take as primitive, or undefined, the concepts of set, element, and the relationship 'is an element of.' Usually, upper case letters will denote sets and lower case letters will denote elements. If A is a set then we will write $a \in A$ whenever a is an element of A. As usual in mathematics we will negate a relationship with a vertical line. Thus, if a is not an element of the set A, we write $a \notin A$.

Two sets A and B are equal if they have the same elements. A set A is said to be a subset of a set B if $a \in A$ implies $a \in B$. In this case we write $A \subseteq B$ and read 'A is a subset of B.' Evidently, $A = B$ if and only if $A \subseteq B$ and $B \subseteq A$.

In this book we will use two commonly used notations for denoting sets. First, if the set is finite, we will sometimes simply list the elements of the set between braces. Thus the set whose only elements are a, b, and c will be denoted

$$\{a, b, c\}$$

The second notation which we will use for denoting sets is

$$\{x: R(x)\}$$

and we will read 'the set of all x such that R(x).' In this case R(x) is some definitive statement about x, for example, x is a real number greater than 2.

The following notation is standard and will be used throughout this book.

$$\mathbf{N} = \{x: x \text{ is a natural number}\}$$
$$\mathbf{Z} = \{x: x \text{ is an integer}\}$$
$$\mathbf{Q} = \{x: x \text{ is a rational number}\}$$
$$\mathbf{R} = \{x: x \text{ is a real number}\}$$
$$\mathbf{C} = \{x: x \text{ is a complex number}\}$$

Evidently we have the following chain of set inclusions.

$$\mathbf{N} \subseteq \mathbf{Z} \subseteq \mathbf{Q} \subseteq \mathbf{R} \subseteq \mathbf{C}$$

If A is any subset of $\mathbf{R}$, then we will denote

$$A^+ = \{x: x \in A \text{ and } x > 0\}$$

Thus, $\mathbf{R}^+$ is the set of all positive real numbers, $\mathbf{Q}^+$ is the set of all positive rational numbers, $\mathbf{Z}^+ = \mathbf{N}$, etc.

(1.2) DEFINITION If A and B are sets then we define

(a) $A \cup B = \{x: x \in A \text{ or } x \in B\}$

(b) $A \cap B = \{x: x \in A \text{ and } x \in B\}$

(c) $A - B = \{x: x \in A \text{ and } x \notin B\}$

(d) $A \times B = \{(a, b): a \in A \text{ and } b \in B\}$

The operation defined in (a) is called the union of A and B and is read 'A union B.' The operation defined in (b) is called the intersection of A and B and is read 'A intersect B.' The operation defined in (c) is called the relative difference of B in A and is read 'A minus B.' The operation defined in (d) is called the product of A and B and is read 'A cross B.'

As is typical in mathematics, we will employ the use of the inclusive 'or'. That is, $x \in A$ or $x \in B$ means either x is in A or x is in B or both. It is sometimes convenient to picture operations on sets with Venn diagrams. The operations in (a), (b), and (c) are diagrammed below.

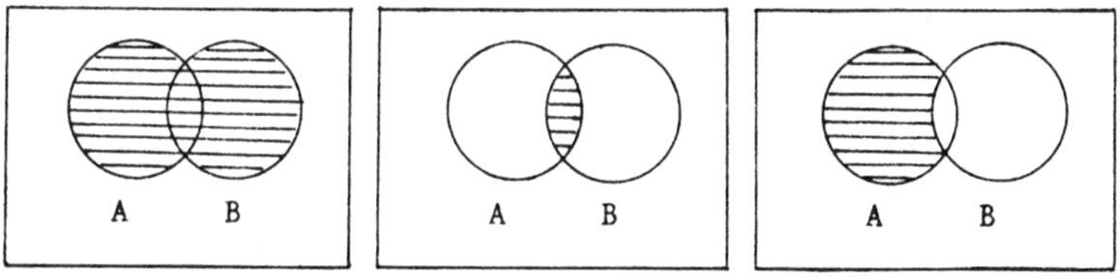

$A \cup B$ $\quad$ $A \cap B$ $\quad$ $A - B$

In each case the points inside the circles represent the elements of either A or B as specified. The shaded area represents the elements of the set diagrammed as specified below the rectangle.

As an example, we let $A = \{1, 2\}$ and $B = \{2, 3\}$. Then

$A \cup B = \{1, 2, 3\}$

$A \cap B = \{2\}$

$A - B = \{1\}$

$A \times B = \{(1, 2), (1, 3), (2, 2), (2, 3)\}$

Rigorous proofs of the properties listed in the following proposition are beyond the scope of this book. If desired, the student can take the properties listed as axioms for the set operations. With Venn diagrams, heuristic arguments can be given to illustrate some of the properties. We leave this as an exercise to the student.

(1.3) PROPOSITION Let A, B, and C be sets. Then we have the following facts.

(a) Commutative Properties

$A \cup B = B \cup A$

$A \cap B = B \cap A$

(b) Associative Properties

$(A \cup B) \cup C = A \cup (B \cup C)$

$(A \cap B) \cap C = A \cap (B \cap C)$

(c) Distributive Properties

$A \cap (B \cup C) = (A \cap B) \cup (A \cap C)$

$A \cup (B \cap C) = (A \cup B) \cap (A \cup C)$

$A \times (B \cap C) = (A \times B) \cap (A \times C)$

$A \times (B \cup C) = (A \times B) \cup (A \times C)$

(d) DeMorgan's Properties

$A - (B \cup C) = (A - B) \cap (A - C)$

$A - (B \cap C) = (A - B) \cup (A - C)$

(e) Absorption Properties

$A \cup B = B$ if and only if $A \subseteq B$

$A \cap B = A$ if and only if $A \subseteq B$

(f) Idempotent Properties

$A \cup A = A$

$A \cap A = A$

(g) Containment

$A \cap B \subseteq A$

$A \subseteq A \cup B$

(h) Complementation

$A \cap (B - A) = \emptyset$

$B = (B - A) \cup (A \cap B)$

(1.4) EXAMPLES We will now present examples which illustrate some of the previous facts. The student is advised to study these examples completely and to supply all missing details. The student should also construct other examples to illustrate the ideas in the text. We feel that a greater understanding of the subject is only obtained through a careful study of many examples.

(1.4.1) The empty set. The empty set is the set with no elements and will be denoted by $\emptyset$. The following facts about the empty set are clear.

(a) For any set A, $\emptyset \subseteq A$.

(b) For any set A, $\emptyset \cap A = \emptyset$.

(c) For any set A, $\emptyset \cup A = A$.

(d) For any set A, $\emptyset \times A = \emptyset$.

(e) For any set A, $A - \emptyset = A$.

(1.4.2) The power set of a set. If S is any set, then we will define the power set of S by

$$\mathbf{P}(S) = \{A: A \subseteq S\}$$

In this setting, we will use a special notation for the relative difference of S and A. In particular, if $A \subseteq S$ then we will denote $C(A) = S - A$. Keeping this notation in mind and recalling (1.3), we may now state some facts about $\mathbf{P}(S)$. For every A, B in $\mathbf{P}(S)$ we have.

(a) $C(A \cup B) = C(A) \cap C(B)$

(b) $C(A \cap B) = C(A) \cup C(B)$

(c) $A \subseteq B$ if and only if $C(B) \subseteq C(A)$

(d) $A \cup B \in \mathbf{P}(S)$

(e) $A \cap B \in \mathbf{P}(S)$

(f) $A - B \in \mathbf{P}(S)$

A convenient method for picturing the inclusion relation in $\mathbf{P}(S)$, or any other ordered collection, is the Hasse diagram. As an example we include here the Hasse diagram of $\mathbf{P}(S)$, where $S = \{a, b, c\}$.

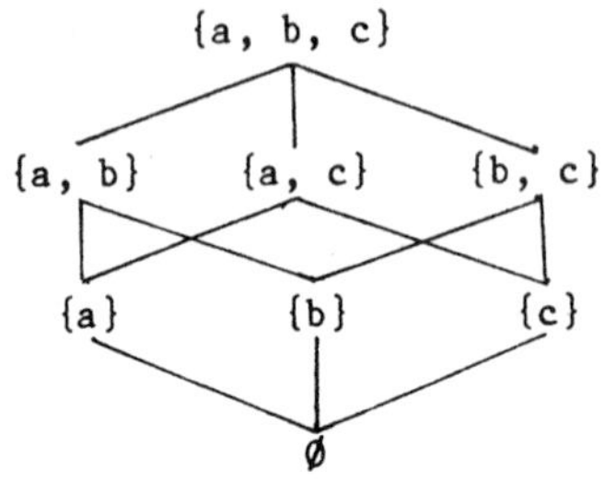

In this diagram, a polygonal line segement drawn upwards from a set A to a set B indicates that $A \subseteq B$.

(1.4.3) The cardinality of a finite set. We will accept the idea of a finite set as intuitive and, at this point, we will not attempt a rigorous definition. If A is a finite set, Card(A), read 'cardinality of A,' will denote the number of elements in A. The following facts about the cardinality of finite sets are basic:

(a) $\text{Card}(A \cup B) = \text{Card}(A) + \text{Card}(B) - \text{Card}(A \cap B)$

(b) if $A \cap B = \emptyset$, then $\text{Card}(A \cup B) = \text{Card}(A) + \text{Card}(B)$

(c) $\text{Card}(A \times B) = \text{Card}(A)\text{Card}(B)$

(1.4.4) The symmetric difference. If A and B are two sets, then the symmetric difference of A and B is defined as follows.

$$A \triangledown B = (A - B) \cup (B - A)$$

If A, B, and C are sets then the following facts follow from (1.3).

(a) $A \triangledown B = B \triangledown A$

(b) $A \triangledown A = \emptyset$

(c) $\emptyset \triangledown A = A$

(d) $(A \triangledown B) \triangledown C = A \triangledown (B \triangledown C)$

(e) $A \cap (B \triangledown C) = (A \cap B) \triangledown (A \cap C)$

Complete arguments can be given, using (1.3), to establish these facts. We will leave this as an exercise to the student.

(1.5) DEFINITION Let A and B be sets, then a function from A to B is a triple (A, f, B) which satisfies the following axioms:

(a) $f \subseteq A \times B$

(b) If $a \in A$, then there is a $b \in B$ such that $(a, b) \in f$.

(c) If $(a, b) \in f$ and $(a, b') \in f$, then $b = b'$.'

In this case, we will frequently say that f is a function from A to B and we write $f: A \to B$, or sometimes we write $A \xrightarrow{f} B$. If $(a, b) \in f$, then we will write $b = f(a)$. We will refer to b as the action of f on a. The symbol f(a) will be read 'f of a.'

The set A is called the domain of the function. The set B is called the codomain of the function. If f is a function from A to B, then we will specify f diagramatically as follows:

$$A \xrightarrow{f} B$$
$$x \mapsto f(x)$$

If f is a function from A to B then f is said to be injective if for every $x, y \in A$, $f(x) = f(y)$ implies $x = y$. We will say that f is surjective if for every $b \in B$ there is an $a \in A$ such that $f(a) = b$. We will say that f is bijective if f is both injective and surjective. If f is a function from A to B then we define im(f), read 'image of f,' by

$$\mathrm{im}(f) = \{b: b \in B \text{ and there is an } a \in A \text{ such that } f(a) = b\}$$

Clearly, f is surjective if and only if $\mathrm{im}(f) = B$.

If $f: A \to B$ and $g: C \to D$ are functions then they will be called equal if and only if the following conditions are satisfied.

(1) $A = C$;

(2) $B = D$;

(3) For every $a \in A$ we have $f(a) = g(a)$.

In case that the domain A and codomain B have been specified, we will often identify f with the function (A, f, B). Thus we will say 'the function f' instead of 'the function (A, f, B).' With this in mind, if f and g are functions with the same domain and codomain then we will write $f = g$ whenever the two functions are equal.

(1.6) DEFINITION Let f be a function from A to B and let g be a function from B to C. Then we define the composition $g \circ f$ as follows.

(a) $g \circ f$ is a function from A to C.

(b) $(g \circ f)(a) = g(f(a))$ for every $a \in A$.

The symbol $g \circ f$ will be read 'g circle f.'

(1.7) PROPOSITION If $f: A \to B$, $g: B \to C$ and $h: C \to D$ are functions, then

$$h \circ (g \circ f) = (h \circ g) \circ f$$

PROOF Since the domains and codomains agree, we need only check the action. Let $a \in A$. Then, from the definition of composition, we have

$$\begin{aligned} h \circ (g \circ f)(a) &= h((g \circ f)(a)) \\ &= h(g(f(a))) \\ &= (h \circ g)(f(a)) \\ &= ((h \circ g) \circ f)(a) \end{aligned}$$

It follows that $h \circ (g \circ f) = (h \circ g) \circ f$.

(1.8) PROPOSITION Let $f: A \to B$, $g: B \to C$ be functions.

(a) If f and g are injective, then $g \circ f$ is injective.

(b) If f and g are surjective, then $g \circ f$ is surjective.

(c) If f and g are bijective, then $g \circ f$ is bijective.

(d) If $g \circ f$ is injective, then f is injective.

(e) If $g \circ f$ is surjective, then g is surjective.

PROOF We will prove (a) and (e). The remaining proofs will be left as exercises. To prove (a), we let x and y be elements of A such that $(g \circ f)(x) = (g \circ f)(y)$. Then by definition of composition we have $g(f(x)) = g(f(y))$. Since g is injective, we have $f(x) = f(y)$. But f is also injective, hence $x = y$. It follows that $g \circ f$ is injective.

We will now establish (e). Assume that $g \circ f$ is surjective and that $c \in C$, then there is an $a \in A$ such that $(g \circ f)(a) = c$. But by definition of composition, $g(f(a)) = (g \circ f)(a) = c$. Thus we have found an element $f(a)$ in B such that $g(f(a)) = c$. Therefore, g is surjective.

The following proposition shows that there is a kind of duality between the ideas of injective and surjective mappings. This duality will be a topic for later discussion.

(1.9) PROPOSITION Let $f: A \to B$ be a function. Then

(a) f is surjective if and only if for every set C and for every pair of functions g, h from B to C we have $g \circ f = h \circ f$ implies $g = h$.

(b) f is injective if and only if for every set C and for every pair of functions g, h from C to A we have $f \circ g = f \circ h$ implies $g = h$.

PROOF We will prove (a) and leave the proof of (b) as an exercise. First assume that f is surjective, C is a set and g, h are functions from B to C such that $g \circ f = h \circ f$. Since the domain and codomain of g and h agree, we need only check the action. If $b \in B$, then since f is surjective, there is an $a \in A$ such that $f(a) = b$. Since $g \circ f = h \circ f$, we have $g(f(a)) = h(f(a))$. But then $g(b) = h(b)$. Since the actions of g and h agree, we have $g = h$.

Conversely, assume that f is a function from A to B such that for every set C and for every pair of functions g, h from B to C we have $g \circ f = h \circ f$ implies $g = h$. Suppose that there is a $b \in B$ such that for every $a \in A$ we have that $f(a) \neq b$. Let $C = \{1, 2\}$ and define two functions g, h from B to C as follows: $g(x) = 1$ for every $x \in B$ and

$$h(x) = \begin{cases} 1 & \text{if } x \in B \text{ and } x \neq b \\ 2 & \text{if } x = b \end{cases}$$

It is easily checked that $g \circ f = h \circ f$ and yet $g \neq h$. Hence, we have a contradiction and our original assumption about b must be false. Thus, f is surjective.

(1.10) EXAMPLES We will now include some examples of functions and compositions. In some cases, the examples included here are as important as the material in the text. Thus, as always, the student is urged to study the examples diligently.

(1.10.1) Let f, g, and h be defined as follows.

$$f: \mathbf{R} \to \mathbf{R} \qquad g: \mathbf{R}^+ \to \mathbf{R}^+ \qquad h: \mathbf{R} \to \mathbf{R}^+ \cup \{0\}$$
$$f(x) = x^2 \qquad g(x) = x^2 \qquad h(x) = x^2$$

Then f, g and h are all functions with the same action yet they are all distinct.

(1.10.2) Let f and g be defined as follows.

$$f: \mathbf{R} \to \mathbf{R} \qquad g: \mathbf{R} \to \mathbf{R}$$
$$f(x) = x^2 + 3x - 1 \qquad g(x) = 2x + 3$$

Then $f \circ g$ and $g \circ f$ are functions with domain and codomain **R** and

$$(f \circ g)(x) = 4x^2 + 18x + 17$$
$$(g \circ f)(x) = 2x^2 + 6x + 1$$

Note that the actions of $f \circ g$ and $g \circ f$ are not equal.

(1.10.3) The Identity Function on a Set. Let A be a set and define a function, called the identity function on A, as follows

$$1_A: A \to A$$
$$a \mapsto a$$

Then 1_A is a bijection from A to A. Moreover, if B and C are sets and $f: A \to B$ and $g: C \to A$ are functions, then $f \circ 1_A = f$ and $1_A \circ g = g$.

(1.10.4) Permutations. Let

$$I_n = \{1, 2, \ldots, n\}$$

Let S_n denote the set of all bijections from I_n to I_n. The elements of S_n will be called permutations. We will temporarily employ the following notation for denoting permutations.

$$f = \begin{pmatrix} 1 & 2 & 3 & 4 & 5 & 6 \\ 2 & 3 & 1 & 4 & 6 & 5 \end{pmatrix}$$

In this case, f is an element of S_6. The action of f on any element is listed just below that element. Thus, $f(1) = 2$, $f(4) = 4$, and

$f(6) = 5$. If we define g in S_6 by

$$g = \begin{pmatrix} 1 & 2 & 3 & 4 & 5 & 6 \\ 3 & 2 & 5 & 4 & 1 & 6 \end{pmatrix}$$

then we can compute $g \circ f$ to get

$$g \circ f = \begin{pmatrix} 1 & 2 & 3 & 4 & 5 & 6 \\ 2 & 5 & 3 & 4 & 6 & 1 \end{pmatrix}$$

Also, we can compute $f \circ g$ and get

$$f \circ g = \begin{pmatrix} 1 & 2 & 3 & 4 & 5 & 6 \\ 1 & 3 & 6 & 4 & 2 & 5 \end{pmatrix}$$

Note that $f \circ g$ and $g \circ f$ are not equal.

(1.10.5) Restriction of functions. Let $f: A \to B$ be a function and let $A' \subseteq A$. Then we can define a function $f|_{A'}$ from A' to B, which we call f restricted to A', by

$\quad f|_{A'}(x) = f(x)$ for every $x \in A'$

(1.10.6) Diagrams. The following schematic of sets, functions and composition of functions is called a diagram.

(1)

$$\begin{array}{ccc} A & \xrightarrow{f} & B \\ & h \searrow \swarrow g & \\ & C & \end{array}$$

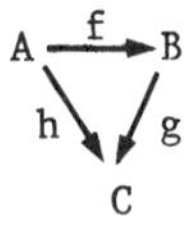

Other examples of diagrams are given below.

(2)

$$\begin{array}{ccc} A & \xrightarrow{f} & B \\ h \downarrow & & \downarrow g \\ C & \xrightarrow{k} & D \end{array}$$

(3)

$$\begin{array}{ccc} A & \xrightarrow{f} & B \\ h \uparrow & & \downarrow g \\ C & \xrightarrow{k} & D \end{array}$$

(4)

$$\begin{array}{ccc} A & \xrightarrow{f} & B \\ k \downarrow & & \downarrow g \\ C & & D \\ & p \searrow \swarrow h & \\ & E & \end{array}$$

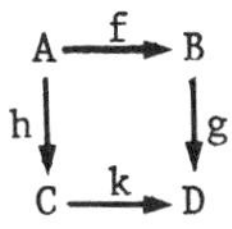

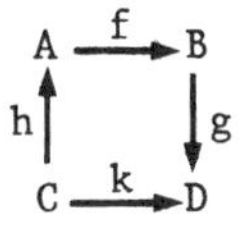

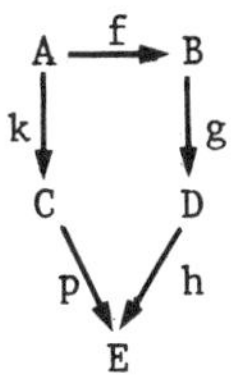

(5)

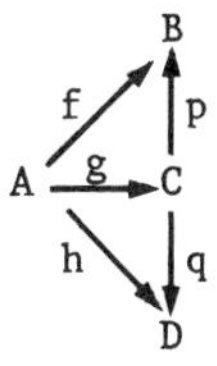

A diagram will be called commutative if any two functions, or composition of functions, with the same domain and codomain are equal. Thus,

(1) is commutative if $g \circ f = h$.

(2) is commutative if $g \circ f = k \circ h$.

(3) is commutative if $g \circ f \circ h = k$.

(4) is commutative if $h \circ g \circ f = p \circ k$.

(5) is commutative if $p \circ g = f$ and $q \circ g = h$.

(1.11) DEFINITION Let $f: A \to B$ and $g: A \to C$ be functions. Then f is said to factor through g if there is a function $h: C \to B$ which makes the following diagram commutative

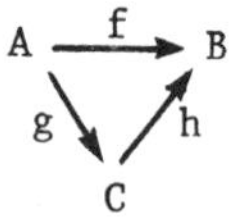

that is, such that $f = h \circ g$. We will say that f factors uniquely through g if the function h is unique.

(1.12) THEOREM The Fundamental Factorization Theorem. Let $f: A \to B$ and $g: A \to C$ be functions such that g is surjective and

(F) For each $x, y \in A$ $g(x) = g(y)$ implies $f(x) = f(y)$

Then f factors uniquely through g.

PROOF Let $c \in C$. Then, since g is surjective, there is an $a \in A$ such that $g(a) = c$. Define $h(c) = f(a)$. Condition (F) assures us that the action of h on c is independent of the selection of a. Therefore, h is well defined. Furthermore, if $a \in A$, then $h(g(a)) = f(a)$. Hence, $h \circ g = f$. The uniqueness of h follows from the surjectivity of g and proposition (1.9).

(1.13) NOTATION AND TERMINOLOGY An indexed family of sets is a surjective function

$$I \xrightarrow{f} \mathbf{S}$$

where I is a set which we will call the index set and **S** is a set of sets. That is every element of **S** is a set. If $i \in I$, then we will write

$$i \mapsto A_i$$

and we will employ the notation

$$\{A_i: i \in I\}$$

for the family of sets. In case the index set is finite, we will sometimes write

$$\{A_1, A_2, \ldots, A_n\}$$

(1.14) DEFINITION If $\{A_i: i \in I\}$ is an indexed family of sets, then we define the union and intersection of the family as follows.

(a) $\bigcup_{i\in I} A_i = \{x: x \in A_i \text{ for some } i \in I\}$

(b) $\bigcap_{i\in I} A_i = \{x: x \in A_i \text{ for every } i \in I\}$

Note that in the case $I = \{1, 2\}$, we have $\bigcup_{i\in I} A_i = A_1 \cup A_2$ and $\bigcap_{i\in I} A_i = A_1 \cap A_2$. We can now generalize some of the facts previously given in (1.3).

(1.15) PROPOSITION Let $\{A_i: i \in I\}$ be an indexed family of sets and let B be any set, then we have the following properties.

(a) Distributive Properties

$$B \cap (\bigcup_{i\in I} A_i) = \bigcup_{i\in I} (B \cap A_i)$$

$$B \cup (\bigcap_{i\in I} A_i) = \bigcap_{i\in I} (B \cup A_i)$$

(b) DeMorgan's Properties

$$B - (\bigcup_{i\in I} A_i) = \bigcap_{i\in I} (B - A_i)$$

$$B - (\bigcap_{i\in I} A_i) = \bigcup_{i\in I} (B - A_i)$$

More general forms of the commutative and associative properties can also be given. Their precise statements will be left as an exercise to the student.

EXERCISES

(1.1) Let A and B be sets with $A \subseteq B$. Let C be any set such that $A \cup C = B$ and $A \cap C = \emptyset$. Use (1.3) to show that $C = B - A$.

(1.2) Let A, B, and C be sets. Show that

$$A \cap (B - C) = (A \cap B) - (A \cap B \cap C)$$

(1.3) Let A, B, and C be sets. Show that

$$A \cap (B - C) = (A \cap B) - C$$

(1.4) Let S be a set such that $A \subseteq S$ and $B \subseteq S$. Draw a Venn diagram to show that

$$A - B = A \cap C(B)$$

(1.5) Use (1.3) and example (1.4.2) to show the following facts.

(a) $A - (A \cap B) = A - B$

(b) $(A \cup B) - C = (A - C) \cup (B - C)$

(c) $(A \cap B) - C = (A - C) \cap (B - C)$

(d) $A \triangledown B = (A \cup B) - (A \cap B)$

(1.6) Let A, B and C be sets. Use (1.3) and example (1.4.2) to show that

(a) $A \cap (B \triangledown C) = (A \cap B) \triangledown (A \cap C)$

(b) $A \triangledown (B \triangledown C) = (A \triangledown B) \triangledown C$

(c) $(A \triangledown B) \cup (B \triangledown C) \cup (A \triangledown C) = (A \cup B \cup C) - (A \cap B \cap C)$

(d) $[(A \cap B) - C] \cup (A \cap C) = A \cap (B \cup C)$

(e) $(A - B) \cup (B - A) \cup (A \cap B) = A \cup B$

(1.7) Draw a Venn diagram of both sides of the following equality.

$$C(A \cap B) = C(A) \cup C(B)$$

(1.8) Prove (1.9) (b).

(1.9) Prove (1.8) (b), (c), and (d).

(1.10) Define f: $\mathbf{R} \to \mathbf{R}$ by $f(x) = x^3 + x + 1$. Show that f is a bijection.

(1.11) Define f: $\mathbf{R} \to \mathbf{R}$ by $f(x) = x^3 - x^2 - x + 2$. Show that f is a surjection, but f is not injective.

(1.12) Modify the notation for permutations to include all functions from a finite set A to another finite set B. Find all functions from {1, 2, 3} to the set {1, 2}.

(1.13) If A has m elements and B has n elements, then how many functions are there from A to B?

(1.14) If f: $A \to B$ is a bijection, then f^{-1}: $B \to A$ is defined by $f^{-1}(x) = y$ if and only if $f(y) = x$. If $f \in S_6$ is given by

$$f = \begin{pmatrix} 1 & 2 & 3 & 4 & 5 & 6 \\ 2 & 1 & 3 & 6 & 4 & 5 \end{pmatrix}$$

then find f^{-1}.

(1.15) How many elements are there in S_n?

(1.16) How many elements in S_n send 1 to 2 and send n to n - 1?

(1.17) If $f, g \in S_5$ are defined by

$$f = \begin{pmatrix} 1 & 2 & 3 & 4 & 5 \\ 2 & 1 & 3 & 5 & 4 \end{pmatrix} \qquad g = \begin{pmatrix} 1 & 2 & 3 & 4 & 5 \\ 2 & 1 & 4 & 5 & 3 \end{pmatrix}$$

then find $f \circ g^{-1}$.

(1.18) If A and B are sets, define $A < B$ to mean that there is an injective function from A to B, but there is no injective function from B to A. Prove that for any sets A, B, and C, if $A < B$ and $B < C$, then $A < C$.

(1.19) Let A and B be sets and define

$p_1: A \times B \to A$ $p_2: A \times B \to B$

$(a, b) \mapsto a$ $(a, b) \mapsto b$

Let X be any set and let f: X → A, g: X → B be functions. Show that there is a function h: X → A × B making the following diagram commutative.

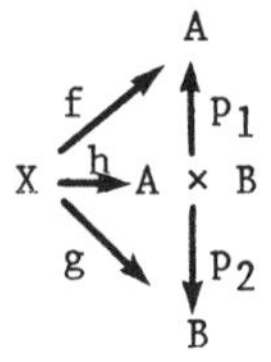

(1.20) Let f: A → B and g: C → D be functions. Define

$f \times g: A \times C \to B \times D$

by $(f \times g)(a, b) = (f(a), g(b))$. Show that $f \times g$ is the unique function which makes the following diagram commutative.

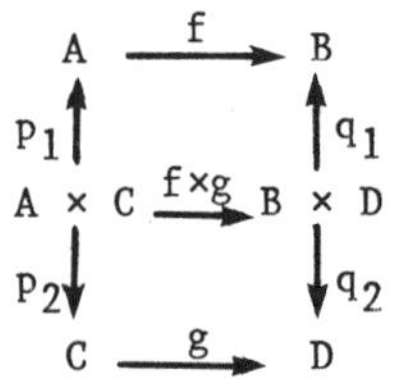

(1.21) Let f: A → B be a bijection. Use the Fundamental Factorization Theorem to deduce that there is a unique function, which we will call f inverse,

$f^{-1}: B \to A$

such that $f \circ f^{-1} = 1_B$ and $f^{-1} \circ f = 1_A$.

(1.22) Let S be any set. Show that there is no surjective function from S to **P**(S). Hint: Assume that f: S → **P**(S) is a function and define

$B = \{s: s \in S \text{ and } s \notin f(s)\}$.

Show that $B \notin \text{im}(f)$.

(1.23) A set S is called finite if every injective function $f: S \to S$ is also surjective. If S is not finite, then we will say that S is infinite. Show that

(a) $\mathbf{R}$ is infinite.

(b) If S is finite and $T \subseteq S$, then T is finite.

(c) If S is finite and $f: S \to S$ is surjective, then f is injective.

(1.24) Show that there is a bijection

(a) from $A \times B$ to $B \times A$.

(b) from $(A \times B) \times C$ to $A \times (B \times C)$.

(1.25) Let A and B be sets and let $f: A \to B$ be a function. If $A_1 \subseteq A$ and $B_1 \subseteq B$, then define the direct image of A_1 by f by

$$f(A_1) = \{b: b \in B \text{ and } b = f(a) \text{ for some } a \in A_1\}$$

and define the inverse image of B_1 by f by

$$f^{-1}(B_1) = \{a: a \in A \text{ and } f(a) \in B_1\}$$

Let A_1 and A_2 be subsets of A. Let B_1 and B_2 be subsets of B. Prove the following statements.

(a) If $A_1 \subseteq A_2$, then $f(A_1) \subseteq f(A_2)$.

(b) If $B_1 \subseteq B_2$, then $f^{-1}(B_1) \subseteq f^{-1}(B_2)$.

(c) $f(A_1 \cup A_2) = f(A_1) \cup f(A_2)$.

(d) $f(A_1 \cap A_2) \subseteq f(A_1) \cap f(A_2)$.

(e) $f^{-1}(B_1 \cup B_2) = f^{-1}(B_1) \cup f^{-1}(B_2)$.

(f) $f^{-1}(B_1 \cap B_2) = f^{-1}(B_1) \cap f^{-1}(B_2)$.

2. PARTITIONS AND EQUIVALENCE RELATIONS

In this section we will introduce the very basic concepts of partition and equivalence relation. We will see that there is essentially a one to one correspondence between equivalence relations on a set and partitions of the same set. We begin with the definition of a partition on a set A.

(2.1) DEFINITION Let A be a set and let $\{A_i: i \in I\}$ be an indexed family of sets. Then $\{A_i: i \in I\}$ is a partition of A if the following axioms are satisfied.

(a) If $i, j \in I$ such that $i \neq j$, then $A_i \cap A_j = \emptyset$.

(b) $\bigcup_{i \in I} A_i = A$.

We are now able to state a fundamental counting principle.

(2.2) PROPOSITION If $\{A_1, A_2, \ldots, A_n\}$ is a partition of a finite set A, then

$$\text{Card}(A) = \text{Card}(A_1) + \text{Card}(A_2) + \cdots + \text{Card}(A_n)$$

PROOF The proof of this fact uses induction which will be studied in section 4. Hence we will leave the proof as a future exercise to the student.

(2.3) NOTATION AND TERMINOLOGY Let A and B be sets. Then a relation from A to B is any subset R of $A \times B$. If R is a relation from A to B and if $(a, b) \in R$, then we will write aRb. If R is a relation from A to A we will say that R is a binary relation on A.

(2.4) DEFINITION Let R be a binary relation on A, then R will be called an equivalence relation if the following are satisfied.

(a) R is reflexive, that is for every $a \in A$, we have aRa.

(b) R is symmetric, that is if aRb then bRa.

(c) R is transitive, that is if aRb and bRc, then aRc.

If R is an equivalence relation on A, then we will denote

$$[a] = \{x: x \in A \text{ and } aRx\}$$

This set is called the equivalence class of a modulo R. If more than one equivalence relation is being discussed, we will connote the dependence of the notation on the relation by $[a]_R$. Finally note that if $a \in A$, then $a \in [a]$, since R is reflexive.

(2.5) PROPOSITION If A is a nonempty set and R is an equivalence relation on A, then for every $a,b \in A$ we have

$$[a] = [b] \text{ if and only if } aRb$$

PROOF Assume first that aRb and let $x \in [a]$. Then aRx, by definition of [a]. Thus, by the symmetric property, we have xRa. Hence, by transitivity of R, xRb. Finally, by the symmetric property, bRx and thus $x \in [b]$. It follows that $[a] \subseteq [b]$. Similarly, $[b] \subseteq [a]$.

Conversely assume that $[a] = [b]$. Then $b \in [b] \subseteq [a]$, therefore, aRb as required.

(2.6) PROPOSITION If A is a set and R is an equivalence relation on A, then for every $a,b \in A$ either $[a] = [b]$ or $[a] \cap [b] = \emptyset$.
PROOF Assume that $[a] \cap [b] \neq \emptyset$. Let c be a common element, then aRc and bRc. Since R is symmetric, we have cRb. By transitivity of R, we get aRb. Thus, by the previous proposition, $[a] = [b]$.

(2.7) NOTATION AND TERMINOLOGY Let A be a set and let R be an equivalence relation on A. Then the quotient set of A modulo R is defined by

$$A/R = \{[a]: a \in A\}$$

The symbol A/R will be read 'A mod R'.

A subset S of A will be called a complete residue system for R if for every $x \in A$ there is one and only one $s \in S$ such that xRs. Thus if S is a complete residue system for A, then

$$A/R = \{[s]: s \in S\}$$

If R is an equivalence relation on a set A then there is a natural function k from A to A/R which we will call the canonical map. The action of k on an element a of A is given by

$$k(a) = [a]$$

Note that if S is a complete residue system for R and if the function k is restricted to have domain S, then the restricted function is a bijection from S to A/R. Thus we can regard A/R = {[a]: a $\in$ S} as a family of sets.

(2.8) PROPOSITION There is a one-to-one correspondence between equivalence relations on a set A and partitions of A. This correspondence is described precisely as follows.

(a) If R is an equivalence relation on A and if S is a complete residue system for R, then {[a]: a $\in$ S} is a partition of A.

(b) If $\{A_i: i \in I\}$ is a partition of A, then we define a relation R on A by aRb if and only if there is an i $\in$ I such that $a,b \in A_i$.

(2.9) EXAMPLES Mathematics abounds with equivalence relations. We will now give a few examples. Other examples will be given in the exercises. The student is urged to construct several others.

(2.9.1) The equivalence relation induced by f on the domain of f. Let f be a function from A to B. Define an equivalence relation R on A by the rule a_1Ra_2 if and only if $f(a_1) = f(a_2)$. We will call this the equivalence relation induced by f on the domain of f.

(2.9.2) Let A be the set of all triangles in the euclidean plane. Define a relation R on A by, a_1Ra_2 if a_1 and a_2 are similar triangles. It is easy to see that R is an equivalence relation on A. Similarly, we could use congruence of triangles.

(2.9.3) Let m and n be integers. Define an equivalence relation R on **Z** by the rule mRn if m - n is even. It is easy to see that this is an equivalence relation on **Z**. Later we will call this relation congruence mod 2.

EXERCISES

(2.1) Let t $\in$ **R** and define a relation ~ on **R** as follows: if x, y $\in$ **R**, then x ~ y if there is a k $\in$ **Z** such that x - y = kt. Show that $\sim$ is an equivalence relation on **R**.

(2.2) If S is the set of all human beings, define a relation R on S as follows: if $x, y \in S$, then xRy if x and y have the same mother. Show that R is an equivalence relation on S.

(2.3) Show that the equivalence relation R defined in (2.9.3) is an equivalence relation on **Z**. Find a complete residue system for R and describe [m] for each $m \in \mathbf{Z}$.

(2.4) Let S be a set and let $\sim$ and $\square$ be equivalence relations on S. Define $\sim$ is finer than $\square$ if the following implication is valid

If $x, y \in S$, then $x \sim y$ implies $x \mathbin{\square} y$

Let $\sim$ and $\square$ be equivalence relations on S such that $\sim$ is finer than $\square$ and let $k_1: S \to S/\sim$ and $k_2: S \to S/\square$ be the canonical mappings. Use the Fundamental Factorization Theorem to show that there is a unique function $f: S/\sim \to S/\square$ such that $f \circ k_1 = k_2$.

(2.5) Let $\mathbf{Z}^*$ denote the set of all nonzero integers. Define a relation $\sim$ on the set $\mathbf{Z} \times \mathbf{Z}^*$ by

$$(a, b) \sim (c, d) \text{ if } ad = bc \quad (a, b, c, d \in \mathbf{Z})$$

Show that $\sim$ is an equivalence relation on $\mathbf{Z} \times \mathbf{Z}^*$.

(2.6) Define a relation $\sim$ on $\mathbf{R} \times \mathbf{R}$ by

$$(x, y) \sim (z, w) \text{ if } 2(x - z) - 3(y - w) = 0 \qquad (x, y, z, w \in \mathbf{R})$$

Show that $\sim$ is an equivalence relation on $\mathbf{R} \times \mathbf{R}$. Find a complete residue system for $\sim$ and describe $[(x, y)]$ as a subset of the euclidean plane.

(2.7) Define a relation R on **C** by

$$(a + bi) \mathrel{R} (c + di) \text{ if } a^2 + b^2 = c^2 + d^2$$

Show that R is an equivalence relation on **C**. Describe the equivalence class of $1 + 2i$. Find a complete residue system for R.

(2.8) Define a relation ~ on **C** - {0} as follows:

$$(a + bi) \sim (c + di) \text{ if } \tan^{-1}(b/a) = \tan^{-1}(d/c) \text{ or } a = c = 0$$

Show that ~ is an equivalence relation. Describe the equivalence class of 1 + 2i. Find a complete residue system for ~.

(2.9) In each of the following cases show that the relation ~ defined on **R** is an equivalence relation.

(a) $x \sim y$ if $x - y \in \mathbf{Q}$.

(b) $x \sim y$ if $x - y \in \mathbf{Z}$.

(2.10) Let P be the set of all polynomials in x with real coefficients and let $f(x) = x^2 + 1$. Define a relation ~ on P as follows: $g \sim h$ if there is a $q \in P$ such that $qf = g - h$. Show that ~ is an equivalence relation on P.

(2.11) Let P be the set of all polynomials in x with real coefficients. Define a relation R on P as follows: gRh if $\deg(g) = \deg(h)$. Show that R is an equivalence relation on P.

(2.12) Let $A \subseteq \mathbf{R}$ be such that $A \neq \emptyset$ and $x, y \in A$ implies $y - x \in A$. Define a relation ~ on **R** by $x \sim y$ if $x - y \in A$. Show that ~ is an equivalence relation on **R**. Does this result if **R** is replaced by **Q**? By **Z**? By P? See Exercise 2.10.

(2.13) Let R be an equivalence relation on X and let $k: X \to X/R$ be the canonical mapping. Let ~ be the equivalence relation induced on X by k. Show that ~ = R.

(2.14) Let R, S and T be relations on a set A and let $R < S$ denote R is finer than S. Show that

(a) $S < R$ if and only if $S \subseteq R$.

(b) if $R < S$ and $S < T$, then $R < T$.

(c) if $R < S$ and $S < R$, then $R = S$.

(2.15) Let $\{A_i: i \in I\}$ be a partition of a set A and let $B \subseteq A$. Show that $\{B \cap A_i: i \in I\}$ is a partition of B.

(2.16) Let A be a finite set and let $\{A_i: i \in I\}$ be a partition of A such that for every i, j $\in$ I we have $\text{Card}(A_i) = \text{Card}(A_j)$. Show that there is a k $\in$ **N** such that $k\text{Card}(A_i) = \text{Card}(A)$.

3. BINARY OPERATIONS

In this section we will define and discuss the abstract properties of binary operations on a set A. Again, the student is urged to construct several examples of each concept introduced.

(3.1) NOTATION AND TERMINOLOGY Let A be a nonempty set. A binary operation on A is any function $\star: A \times A \to A$. If $\star$ is a binary operation on A, then the action of $\star$ on the pair (a, b) will be denoted by $a \star b$.

Let $\star$ be a binary operation on A and let $e \in A$. Then e is

(a) a left identity with respect to $\star$ if for every $a \in A$, $e \star a = a$.

(b) a right identity with respect to $\star$ if for every $a \in A$, $a \star e = a$.

(c) an identity with respect to $\star$ if e is both a left identity with respect to $\star$ and a right identity with respect to $\star$.

Let $\star$ be a binary operation on A. Then an element $a \in A$ is said to be idempotent with respect to $\star$ if $a \star a = a$.

We will say that $\star$ is commutative, if for every $a,b \in A$, we have

$$a \star b = b \star a$$

We will say that $\star$ is associative, if for every $a,b,c \in A$, we have

$$a \star (b \star c) = (a \star b) \star c$$

If e is an identity for A with respect to $\star$ then an element $b \in A$ is said to be an inverse for a with respect to $\star$ if

$$\text{(I)} \quad a \star b = b \star a = e$$

In the event that b is uniquely determined by (I), we will sometimes write $b = a^{-1}$. The symbol a^{-1} is read 'a inverse'. There is some ambiguity in the notation a^{-1} since there is no indication of the dependence on the operation $\star$. Thus, we will only use this notation when the ambient operation is clear.

If A is a set, $\star$ is a binary operation on A and $B \subseteq A$, then we will say that B is closed with respect to $\star$ if

(Closure property) $x,y \in B$ implies $x \star y \in B$

Thus, B is closed with respect to $\star$ if and only if the function $\star$ restricts to a function from $B \times B$ to B.

(3.2) PROPOSITION Let A be a nonempty set and let $\star$ be a binary operation on A. If e is a left identity for A with respect to $\star$ and f is a right identity for A with respect to $\star$, then $e = f$.

PROOF Since e is a left identity for A with respect to $\star$, we have $e \star f = f$. Since f is a right identity for A with respect to $\star$, we have $e \star f = e$. It follows that $e = f$.

(3.3) COROLLARY Let A be a nonempty set and let $\star$ be a binary operation on A. If e and f are identities for A with respect to $\star$, then $e = f$.

It follows immediately that if A has two distinct left identities, then it can have no identity with respect to $\star$. We will see in an exercise that a set can have infinitely many left identities with respect to an operation.

(3.4) EXAMPLES We will include here only a few of many possible examples of binary operations. The student is urged to construct other examples and to refer to the exercises for still more examples.

(3.4.1) Operations on **R**. The usual operations $+$, $-$, $\cdot$ are all binary operations on **R**. However, $\div$ is not a binary operation on **R**. The reason for this is simply that division by zero is undefined. Thus, the domain of $\div$ is not $\mathbf{R} \times \mathbf{R}$. If we define $\mathbf{R}^{\star} = \mathbf{R} - \{0\}$, then

$\div$ can be regarded as an operation on $\mathbf{R}^*$. In the sequel, when we refer to the operation $\div$ we will mean the operation on $\mathbf{R}^*$.

Note that 0 is an identity for $\mathbf{R}$ with respect to + and 1 is an identity for $\mathbf{R}$ with respect to $\cdot$. Further note that 0 is a right identity for $\mathbf{R}$ with respect to $-$, but there is no $x \in \mathbf{R}$ such that x is a left identity for $\mathbf{R}$ with respect to $-$.

As we well know, both operations + and $\cdot$ are commutative and associative. The student can easily verify that both operations $-$ and $\div$ are neither commutative nor associative.

Note that addition can be regarded as a operation on $\mathbf{Z}$ as well as $\mathbf{N}$. We will refer to the operation addition on $\mathbf{Z}$ as the operation obtained by restricting to $\mathbf{Z}$, or the restricted operation. We will also denote this operation, somewhat ambiguously, by +. Similarly, if $*$ is an operation on a set A and $B \subseteq A$ is such that the function $*$ restricted to $B \times B$ is a function from $B \times B$ to B, then we will refer to this restricted function as the restricted operation $*$.

(3.4.2) Certainly, any function from $\mathbf{R} \times \mathbf{R}$ to $\mathbf{R}$ will define a binary operation on $\mathbf{R}$. As another example, for every $x,y \in \mathbf{R}$ we define

$$x * y = x + y - xy$$

We leave as an exercise to the student to show that this operation is both commutative and associative and that 0 is an identity for $\mathbf{R}$ with respect to $*$.

If $a \in \mathbf{R}$ and $a \neq 1$, then note that $a * a/(a - 1) = 0$. Thus $a/(a - 1)$ is an inverse for a with respect to $*$. It is also noteworthy that for every $a \in \mathbf{R}$ we have $a * 1 = 1 * a = 1$.

(3.4.3) Some operations on $\mathbf{R}^n$. We define

$$\mathbf{R}^n = \{(x_1, x_2, \ldots, x_n): x_1, \ldots, x_n \in \mathbf{R}\}$$

Thus, $\mathbf{R}^2$ is euclidean two-space, $\mathbf{R}^3$ is euclidean three-space, etc.

We define an operation + on $\mathbf{R}^n$ by

$$(x_1, x_2, \ldots, x_n) + (y_1, y_2, \ldots, y_n)$$
$$= (x_1 + y_1, x_2 + y_2, \ldots, x_n + y_n)$$

It is easy to show that + is commutative and associative on $\mathbf{R}^n$ and that the n-tuple (0,0, ..., 0) is an identity for $\mathbf{R}^n$ with respect to +. Note that an inverse for $(x_1, x_2, \ldots, x_n)$ with respect to + is given by

$$(-x_1, -x_2, \ldots, -x_n)$$

As another example of an operation on $\mathbf{R}^3$ we define the cross product of vectors as follows.

$$(a, b, c) \times (d, e, f) = (bf - ce, cd - af, ae - bd)$$

Cross product is an operation on $\mathbf{R}^3$ which is neither commutative nor associative. But as the student knows from studying three dimensional calculus, it is a very useful operation.

(3.4.4) Operations on **P**(S). Let S be a set, then we can regard $\cap$, $\cup$, −, and ∇ as operations on **P**(S). We have previously seen that $\cap$, $\cup$ and ∇ are both commutative and associative. We will leave as an exercise to the student to find identities for **P**(S) with respect to these operations.

(3.4.5) Operations on $M_2(\mathbf{R})$. Let $M_2(\mathbf{R})$ denote the set of all 2 × 2 matrices with real entries. We will now recall the definitions of matrix multiplication and matrix addition.

$$\begin{pmatrix} a & b \\ c & d \end{pmatrix} + \begin{pmatrix} e & f \\ g & h \end{pmatrix} = \begin{pmatrix} a + e & b + f \\ c + g & d + h \end{pmatrix}$$

$$\begin{pmatrix} a & b \\ c & d \end{pmatrix} \begin{pmatrix} e & f \\ g & h \end{pmatrix} = \begin{pmatrix} ae + bg & af + bh \\ ce + dg & cf + dh \end{pmatrix}$$

It is easy, but computational to show that both operations are associative and addition is commutative. However, as can be shown by example, multiplication is not commutative. The student should find identities for both operations.

(3.4.6) Composition of Functions. Let A be a nonempty set and denote

$$F(A) = \{f: f \text{ is a function from } A \text{ to } A\}$$
$$B(A) = \{f: f \in F(A) \text{ and } f \text{ is a bijection}\}$$

Then composition of functions can be regarded as a binary operation on F(A). The restricted operation is an operation on B(A). We have already seen that composition of functions is an associative operation. Note that composition of functions is not a commutative operation on either F(A) or B(A) unless Card(A) < 3. We should further note that 1_A is an identity for both F(A) and B(A) with respect to composition.

(3.4.7) Semigroups and Monoids. Let A be a nonempty set and let $\star$ be an associative binary operation on A. Then the pair $(A, \star)$ will be called a semigroup. If, in addition, there is an element $e \in A$ such that e is an identity for A with respect to $\star$, then the pair $(A, \star)$ will be called a monoid. Note in particular that the pair $(\mathbf{Z}, \cdot)$ is a monoid and the pair $(2\mathbf{Z}, \cdot)$ is a semigroup.

(3.5) NOTATION AND TERMINOLOGY The Cayley Table for an Operation. We will occasionally specify an operation of a finite set by listing what we will refer to as the Cayley table for that operation. An example for such a table is given below for the set {a, b, c}.

$\star$	a	b	c
a	a	c	b
b	b	a	a
c	c	b	a

In this table, we can find the value of $a \star b$ by reading the value in the row corresponding to a and the column corresponding to b. Of course this value is c. Note that a is a right identity with respect to $\star$ but a is not an identity with respect to $\star$.

The student will observe that operations which are constructed from Cayley tables are almost never commutative and almost never

associative. Inspection of the Cayley table is sufficient to
determine whether that operation is commutative. The associativity
of an operation by inspecting the Cayley table is in general a very
tedious task.

(3.6) DEFINITION Let A be a set and let $*$ be a binary operation on
A. Then an equivalence relation on A will be called a congruence
relation on A with respect to $*$ if

$\quad$ aRb implies (a $*$ c)R(b $*$ c) and (c $*$ a)R(c $*$ b)

for every a,b,c $\in$ A.

(3.7) PROPOSITION   Let A be a non empty set and let $*$ be a binary
operation on A. If R is a congruence relation on A with respect to $*$
and if a,b,c,d $\in$ A are such that aRb and cRd, then (a $*$ c)R(b $*$ d).
PROOF Let a,b,c,d $\in$ A and let aRb and cRd. Then

$\quad$ (a $*$ c)R(b $*$ c)   (definition of congruence relation)
$\quad$ (b $*$ c)R(b $*$ d)   (definition of congruence relation)
$\quad$ (a $*$ c)R(b $*$ d)   (R is transitive)

We will see that if R is a congruence relation on A with respect
to some operation that there is a natural way to introduce an
operation on the factor set A/R.

(3.8) PROPOSITION Let $*$ be a binary operation on a set A and let R
be a congruence relation on A with respect to $*$. Let k be the
canonical map, k: A $\to$ A/R. Then the function k $\circ$ $*$: A $\times$ A $\to$ A/R
factors through the function k $\times$ k. In particular, there is a
function, which we will specify as $\circ$, making the following diagram
commutative.

$$
\begin{array}{ccc}
A \times A & \xrightarrow{\ *\ } & A \\
{\scriptstyle k \times k}\downarrow & & \downarrow{\scriptstyle k} \\
A/R \times A/R & \xrightarrow{\ \circ\ } & A/R
\end{array}
$$

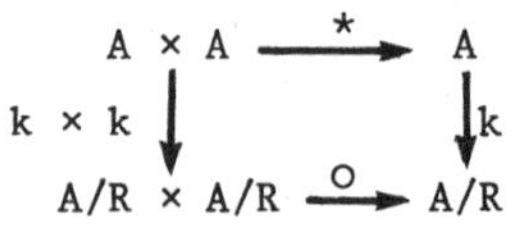

PROOF If $(k \times k)(a, b) = (k \times k)(c, d)$, then aRc and bRd. It follows from the previous proposition that $(a * b)R(c * d)$. Thus $(k \circ *)(a, b) = (k \circ *)(c, d)$. By the fundamental factorization theorem, $k \circ *$ factors through $k \times k$.

(3.9) DEFINITION The operation resulting in (3.8) will be called the operation on A/R induced by *. Note that the action of $\circ$ is given by

(1) $[a] \circ [b] = [a * b]$

In the sequel, we will often abuse the notation somewhat and let * denote both the operation on A and the operation on A/R induced by *. We will sometimes say that the operation defined by formula (1) is 'well defined'.

(3.10) PROPOSITION Let A be a non empty set and let * be a binary operation on A. Let R be a congruence relation on A with respect to * and let $\circ$ be the induced operation on A/R.

(a) If * is a commutative operation, then $\circ$ is a commutative operation.

(b) If * is an associative operation, then $\circ$ is an associative operation.

(c) If e is a left identity for A with respect to *, then [e] is a left identity for A/R with respect to $\circ$.

(d) If e is a right identity for A with respect to *, then [e] is a right identity for A/R with respect to $\circ$.

(e) If e is an identity for A with respect to *, then [e] is an identity for A/R with respect to $\circ$.

PROOF We will prove only parts (a) and (c). First, to prove part (a), assume that * is a commutative operation on A. Then for every $a,b \in A$ we have

$$\begin{aligned} [a] \circ [b] &= [a * b] && \text{(definition of } \circ \text{)} \\ &= [b * a] && \text{(* is commutative)} \\ &= [b] \circ [a] && \text{(definition of } \circ \text{)} \end{aligned}$$

Therefore, $\circ$ is a commutative operation on A/R.

To prove (c), let $a \in A$. Then

$$[e] \circ [a] = [e * a] \quad \text{(definition of } \circ \text{)}$$
$$= [a] \quad \text{(e is a left identity for A)}$$

Thus, [e] is a left identity for A/R with respect to $\circ$.

The proofs of (b), (d) and (e) will be left as an exercise to the student.

EXERCISES

(3.1) Let x and y be real numbers and define a binary operation $*$ on **R** by the rule

$$x * y = x + y + xy$$

Show that $*$ is a commutative, associative operation on **R**. Show that 0 is an identity for **R** with respect to $*$.

(3.2) Prove (3.10) parts (b), (c) and (d).

(3.3) Let x and y be real numbers and define

$$x * y = x^2y - xy$$

Show that $*$ is neither commutative nor associative.

(3.4) Let x and y be real numbers and define

$$x * y = x - 2y$$

Show that $*$ is neither commutative nor associative.

(3.5) Let P be a polynomial function of two real variables and let

$$x * y = P(x, y)$$

Find all polynomials P such that $*$ is a commutative and associative operation on **R**.

(3.6) Let S be the set of all 2 × 2 matrices of the form

$$\begin{pmatrix} a & b \\ 0 & 0 \end{pmatrix}$$

where a, b ∈ **R**.

(a) Show that S is closed with respect to matrix multiplication.

(b) Show that for each b ∈ R, the matrix

$$\begin{pmatrix} 1 & b \\ 0 & 0 \end{pmatrix}$$

is a left identity for S with respect to the restricted operation. Thus conclude that S has no right identity.

(3.7) Show that the relation congruence mod 2 as defined in example (2.9.3) is a congruence relation with respect to addition and multiplication. [cf. Exercise (2.3)]

(3.8) Let n be a positive integer. Then two integers k and l are said to be congruent mod n if there is an integer t such that k - l = tn. In this case we will write

$$k \equiv l \pmod{n}$$

Show that congruence mod n is a congruence relation with respect to addition and multiplication.

(3.9) Let $\mathbf{Z}^*$ denote the set of all nonzero integers. Define a relation ~ on $\mathbf{Z} \times \mathbf{Z}^*$ as follows

$$(a, b) \sim (c, d) \text{ if } ad = bc \quad (a,b,c,d \in \mathbf{Z})$$

Define operations addition and multiplication on $\mathbf{Z} \times \mathbf{Z}^*$ as follows

$$(a, b) + (c, d) = (ad + bc, bd) \qquad (a,b,c,d \in \mathbf{Z})$$

$$(a, b)(c, d) = (ac, bd) \qquad (a,b,c,d \in \mathbf{Z})$$

Show that ~ is a congruence relation on $\mathbf{Z} \times \mathbf{Z}^*$ with respect to addition and with respect to multiplication.

(3.10) Define a relation ~ on $\mathbf{R}^2$ as follows

$(x, y) \sim (z, w)$ if $3(x - z) + 2(y - w) = 0$

Show that ~ is a congruence relation on $\mathbf{R}^2$ with respect to addition.

(3.11) Show that the equivalence relation R defined in exercise (2.1) is a congruence relation on **R** with respect to +.

(3.12) Show that the equivalence relations ~ defined in exercise (2.9) are congruence relations on **R** with respect to +.

(3.13) Show that the equivalence relations discussed in exercise (2.12) are congrence relations on **R** with respect to +.

(3.14) Let $\emptyset \neq A \subseteq \mathbf{R}^*$ be such that $x, y \in A$ implies $x/y \in A$. Define a relation ~ on $\mathbf{R}^*$ by

$x \sim y$ if $x/y \in A$

Show that ~ is a congruence relation with respect to multiplication. Does the same result obtain if $\mathbf{R}^*$ is replaced by the set of nonzero complex numbers? By the set of nonzero rational numbers? By the set of nonzero integers?

(3.15) Construct a Cayley table of an operation on a set with four elements to show that the inverse of an element need not be unique.

(3.16) (a) Define $T: A \times A \to A \times A$ by $T(a, b) = (b, a)$. Show that T is a bijection.

(b) Define $S: A \times (A \times A) \to (A \times A) \times A$ by $S(a, (b, c)) = ((a, b), c)$. Show that S is a bijection.

(c) Let * be a binary operation on A. Show that * is commutative if and only if the following diagram commutes.

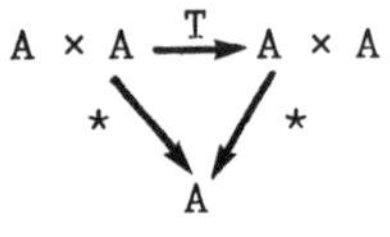

(d) Let $\star$ be a binary operation on A. Show that $\star$ is associative if and only if the following diagram is commutative.

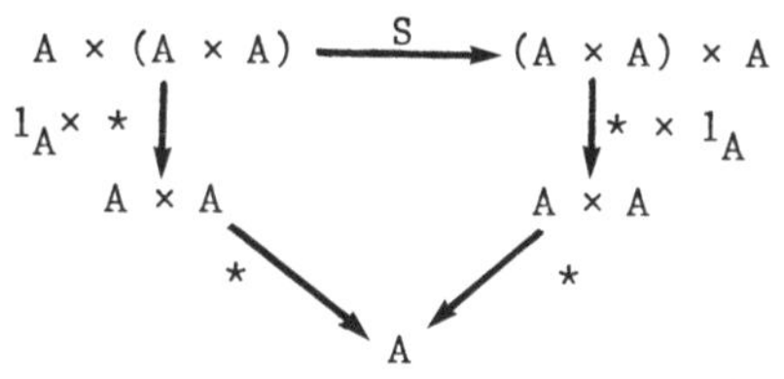

(3.17) Let A be the set $\mathbf{R}^2$ and define an operation $\star$ on A by the rule

$$(x, y) \star (z, w) = (xz, yz + w)$$

Show that the pair $(A, \star)$ is a semigroup. [See Example (3.4.7).] Show that the pair (1, 0) is an identity for A with respect to $\star$. Thus deduce that the pair $(A, \star)$ is a monoid. Is $\star$ a commutative operation?

(3.18) Let A be a nonempty set and let $x \in A$. Define an operation $\star$ on A by $a \star b = x$ for every $a,b \in A$. Show that $(A, \star)$ is a semigroup. Is $(A, \star)$ a monoid? Is $\star$ a commutative operation?

(3.19) Let A be a nonempty set and let $\le$ be a binary relation on A. Then, $\le$ is said to be a partial ordering on A if the following three conditions are satisfied for each $a,b,c \in A$:

(a) $a \le a$

(b) If $a \le b$ and $b \le a$, then $a = b$.

(c) If $a \le b$ and $b \le c$, then $a \le c$.

If a, b are elements of A, then and element c of A is said to be a least upper bound for a and b if the following conditions are satisfied:

(i) $a \le c$ and $b \le c$

(ii) If $d \in A$ and $a \le d$, $b \le d$, then $c \le d$.

Show the following

(1) If a and b are elements of A, then the least upper bound of

a and b is unique. We will denote the least upper bound of a and b by $a \vee b$ and we read 'a join b'. If any two elements of A have a least upper bound, then A is called a join semilattice. If A is a join semilattice, show that.

(2) $\vee$ is a binary operation on A.

(3) $(A, \vee)$ is a semigroup.

(4) $\vee$ is commutative.

(5) Give conditions on an element $e \in A$ so that e is an identity for A with respect to $\vee$.

(6) Show that similar definitions and statements can be made using the greatest lower bound of elements in A. The greatest lower bound of a and b is usually denoted $a \wedge b$ and read 'a meet b'. A join semilattice which is also a meet semilattice is called a lattice.

4. THE INTEGERS

In this section we will develop many of the properties of the positive integers which we have previously accepted without proof. The intent here is to develop two very important concepts. First, the technique of proof by induction and second, the idea of recursive definition. While reading this section, the student will be asked to supply many details. The student is cautioned to ignore all previous knowledge about the integers and approach this section in the abstract. We begin with the Peano axioms for **N**.

(4.1) NOTATION AND TERMINOLOGY We will assume the existence of a set, which we will call the natural numbers and denote **N**, together with an injective function $s: \mathbf{N} \to \mathbf{N}$, called the successor function, which satisfies the following axioms.

(A) There is a unique $1 \in \mathbf{N}$ such that for every $x \in \mathbf{N}$, we have $1 \neq s(x)$.

(B) Axiom of Induction. If $M \subseteq \mathbf{N}$ satisfies

(1) $1 \in M$

(2) If $x \in M$, then $s(x) \in M$.

Then $M = \mathbf{N}$.

In exercise (1.24) we defined a set S to be infinite if there is an injective function $f\colon S \to S$ which is not surjective. From axiom (A), **N** is an infinite set.

We begin by showing how exponents can be defined for each $n \in \mathbf{N}$ in an abstract set with respect to a binary operation. The technique which we employ here is called recursive definition.

(4.2) DEFINITION Let A be a set and let $*$ be a binary operation on A, then we will define, for each $x \in A$

(a) $x^1 = x$

(b) For each $n \in \mathbf{N}$, $x^{s(n)} = x * x^n$.

We will show that we have now defined x^n for every $n \in \mathbf{N}$. Let

$$M = \{n\colon x^n \text{ is defined}\}$$

Then, by (a), $1 \in M$. By (b), if $n \in M$, then we can find $x^{s(n)}$, and hence, $s(n) \in M$. Now, by axiom (B), we have $M = \mathbf{N}$.

Note that there are two parts to a recursive definition. First, a concept is defined for 1, then we define the concept for s(n) by using n. Then induction is used to show that the concept is well defined.

We will now show how recursive definition can be used to define addition of elements of **N**.

(4.3) DEFINITION Let $n \in \mathbf{N}$ and define

(a) $n + 1 = s(n)$

(b) For each $k \in \mathbf{N}$, $n + s(k) = s(n + k)$.

Thus, $n + s(1) = s(n + 1) = s(s(n))$, etc.

Recursive definition lends itself readily to applications of the axiom of induction. We will illustrate this by proving the following proposition.

(4.4) PROPOSITION Addition is a binary operation on **N** which is commutative and associative.

PROOF The proof that addition is well defined is exactly like that given in the previous proposition and will be left as an exercise to the student.

We will first show that addition is commutative. The technique of proof used here will be called proof by induction.

First, we claim that $n + 1 = 1 + n$ for every $n \in \mathbf{N}$. To show this, let

$$M = \{n: n \in \mathbf{N} \text{ and } n + 1 = 1 + n\}$$

Then, since $1 + 1 = 1 + 1$, $1 \in M$. Let $k \in M$, that is $k + 1 = 1 + k$. Then

$$\begin{aligned} 1 + s(k) &= s(1 + k) && \text{[use (b) with } n = 1] \\ &= s(k + 1) && (k \in M.) \\ &= s(s(k)) && \text{[use (a) with } n = k] \\ &= s(k) + 1 && \text{[use (a) with } n = s(k)] \end{aligned}$$

Therefore, by axiom (B), we have $M = \mathbf{N}$ and our claim is established.

Next, we claim that for each $n,m \in \mathbf{N}$, we have

$$n + s(m) = s(n) + m$$

To show this, let $n \in \mathbf{N}$ and let

$$M = \{m: m \in \mathbf{N} \text{ and } n + s(m) = s(n) + m\}$$

Observe that $1 \in M$, since

$$\begin{aligned} n + s(1) &= s(n + 1) && \text{[use (b) with } k = 1] \\ &= s(s(n)) && \text{[use (a)]} \\ &= s(n) + 1 && \text{[use (a) with } n = s(n)] \end{aligned}$$

Next assume that $k \in M$, i.e., $n + s(k) = s(n) + k$, then

$$\begin{aligned} n + s(s(k)) &= s(n + s(k)) && \text{[use (b) with } k = s(k)] \\ &= s(s(n) + k) && (k \in M) \\ &= s(n) + s(k) && \text{[use (b) with } n = s(n)] \end{aligned}$$

Hence, $s(k) \in M$ and $M = \mathbf{N}$, as required.

Finally, let $n \in \mathbf{N}$ and set

$$M = \{m: m \in \mathbf{N} \text{ and } n + m = m + n\}$$

We have seen that $1 \in M$, so assume that $k \in M$, that is $n + k = k + n$, then

$$\begin{aligned} n + s(k) &= s(n + k) && \text{[use (b)]} \\ &= s(k + n) && (k \in M) \\ &= k + s(n) && \text{[use (b)]} \\ &= s(k) + n && \text{(use the previous fact)} \end{aligned}$$

Thus $M = \mathbf{N}$, and addition is commutative on $\mathbf{N}$.

We will now sketch the proof that addition is associative. The student should complete the proof. Let $m,k \in \mathbf{N}$ and set

$$M = \{n: n \in \mathbf{N} \text{ and } n + (m + k) = (n + m) + k\}$$

Use the technique of induction to show that $M = \mathbf{N}$.

We will now show how multiplication can be defined recursively on $\mathbf{N}$.

(4.5) DEFINITION Let $n \in \mathbf{N}$. Then

(a) $1n = n$

(b) $s(k)n = kn + n$

The resulting operation, which we will continue to write by juxtaposition, is called multiplication.

(4.6) PROPOSITION Multiplication is a binary operation on $\mathbf{N}$ which is commutative and associative. Furthermore, multiplication distributes over addition, that is,

$$n(m + k) = nm + nk$$

for every $n,m,k \in \mathbf{N}$.

PROOF The proof of this theorem will be broken into steps and left to the student in the exercises.

The student who is interested in foundations can at this point see how the set of integers can be constructed from $\mathbf{N}$ and how addition and multiplication can be extended to the set of integers. We will leave this construction as an exercise.

(4.7) DEFINITION Let $m,n \in \mathbf{N}$. Then we will say that m is less than n, and write $m < n$ if there is a $k \in \mathbf{N}$ such that $n = m + k$. We will write $m \leq n$ to mean either $m < n$ or $m = n$.

(4.8) PROPOSITION Let $m,n,k \in \mathbf{N}$. Then

(a) $m \leq m$

(b) If $m \leq n$ and $n \leq m$, then $m = n$.

(c) If $m \leq n$ and $n \leq k$, then $m \leq k$.

(d) If $m \leq n$, then $m + k \leq n + k$.

(e) If $m \leq n$, then $mk \leq nk$.

(f) Exactly one of the following is true:

(1) $m < n$

(2) $n < m$

(3) $n = m$

(4.9) PROPOSITION The Division Algorithm. Let $m,n \in \mathbf{N}$. Then there are unique nonnegative integers q, r such that

(a) $n = qm + r$

(b) $0 \leq r < m$

PROOF If $n \leq m$ the result is clear. We will proceed by induction on n, that is we will fix the value of m and set

$$M = \{n: n \text{ and } m \text{ satisfy (a) and (b)}\}$$

Then $1 \in M$, since $1 \leq m$ for every $m \in \mathbf{N}$.

If $n \in M$, then we must show that $n + 1 \in M$. The desired result will be true if $n + 1 \leq m$, hence assume that $m < n + 1$. Since $n \in M$, then there are $q,r \in \mathbf{N}$ such that

$$n = qm + r \text{ and } 0 \leq r < m$$

Adding 1 to both sides, we get

$$n + 1 = qm + (r + 1)$$

If $r + 1 < m$, we are done. If $r + 1 = m$, then

$$n + 1 = (q + 1)m$$

Hence, $n + 1 \in M$ and $M = \mathbf{N}$ as required.

We will leave the proof of uniqueness of q and r as an exercise to the student.

(4.10) DEFINITION Let $m,n \in \mathbf{N}$. Then m is said to divide n, and we write $m|n$ if there is a $k \in \mathbf{N}$ such that $mk = n$.

(4.11) PROPOSITION Let $m,n,k \in \mathbf{N}$. Then

(a) If $m|n$ and $n|k$, then $m|k$.

(b) $1|n$

(c) If $m|n$ and $m|k$, then $m|(n + k)$.

(d) If $m|n$, then $m|kn$.

PROOF The proofs of these facts are elementary and will be left as exercises.

(4.12) DEFINITION Let $m,n \in \mathbf{N}$. Then an element $d \in \mathbf{N}$ is said to be the greatest common divisor of m and n if

(a) $d|m$, $d|n$ and

(b) if $k \in \mathbf{N}$ is such that $k|m$ and $k|n$, then $k|d$.

In this case, we will write $d = (m, n)$.

A fundamental fact about $\mathbf{N}$ which follows from the Peano axioms is the well ordering principle. The well ordering principle simply says that any non empty subset of $\mathbf{N}$ has a smallest element. We will state this principle here without proof and show how this together with the division algorithm can be used to show the existence of the greatest common divisors of any two elements in $\mathbf{N}$.

(4.13) PROPOSITION Let $m,n \in \mathbf{N}$. Then there is a $d \in \mathbf{N}$ such that $d = (m, n)$. Moreover, there are integers $a,b \in \mathbf{Z}$ such that $d = am + bn$.

PROOF Let

$$S = \{x: x = am + bn \text{ for some } a,b \in \mathbf{Z} \text{ and } x \in \mathbf{N}\}$$

Then S is a nonempty subset of $\mathbf{N}$ and hence, by the well ordering principle, S has a smallest element. Let $am + bn = d$ be the smallest element of S. We claim that d is the greatest common divisor of m and n. First, to show that $d|m$, let $q, r \in \mathbf{N}$ be such that

$$m = qd + r \text{ and } 0 \leq r < d$$

Then $r = m - qd = (1 - qa)m - qbn$. Hence, $r \in S$. Since $r < d$ we must have $r = 0$. Therefore, $d|m$. Similarly, $d|n$.

Finally, if $k \in \mathbf{N}$ is such that $k|m$ and $k|n$, then we must show that $k|d$. Let $e,f \in \mathbf{N}$ be such that $m = ek$ and $n = fk$. Now

$$d = am + bn = aek + bfk = (ae + bf)k$$

and hence, $k|d$ as required.

(4.14) PROPOSITION Let m, n, and k be integers such that

(a) $(m, n) = 1$

(b) $m|(nk)$

Then $m|k$.

PROOF If $(m, n) = 1$, then there are integers a and b such that

$$am + bn = 1$$

Now, $amk + bnk = k$. Since $m|(amk)$ and $m|(bnk)$, we have $m|k$.

(4.15) DEFINITION An integer p is said to be prime if

(a) $p \neq 1$, $p \neq -1$

(b) The only divisors of p are 1, -1, p, and -p.

(4.16) COROLLARY If p is a prime integer and m, n are integers such that $p|(mn)$, then $p|m$ or $p|n$.

EXERCISES

(4.1) Show that the set $\{0, 1, 2, \ldots, n - 1\}$ is a complete residue system with respect to congruence mod n.

(4.2) Let $\mathbf{Z}_n$ be the set of all equivalence classes of $\mathbf{Z}$ with respect to congruence mod n. Show that the following properties of the induced operations of addition and multiplication on $\mathbf{Z}_n$ obtain.

(a) Addition and multiplication are both commutative and associative.

(b) [0] is an identity with respect to addition for $\mathbf{Z}_n$.

(c) [1] is an identity with respect to multiplication for $\mathbf{Z}_n$.

(d) For every $k \in \mathbf{Z}$, [-k] is an inverse for [k] with respect to addition.

(e) For every $a,b,c \in \mathbf{Z}$,

$$[a]([b] + [c]) = [a][b] + [a][c]$$

(4.3) Define $\mathbf{Z}_n^\times$ to be the set of all elements [k] of $\mathbf{Z}_n$ such that [k] has an inverse with respect to multiplication. Show that $[k] \in \mathbf{Z}_n^\times$ if and only if $(k, n) = 1$. We will denote the cardinality of $\mathbf{Z}_n^\times$ by $\phi(n)$.

(4.4) Let $f: A \to A$ be a bijection. Define the orbit of an element $a \in A$ by

$$or(a) = \{x: x = a \text{ or } x = f^n(a) \text{ or } x = f^{-n}(a) \text{ for some } n \in \mathbf{N}\}$$

Show that the set of all orbits of elements of A constitute a partition of A.

(4.5) Use induction and the recursive definitions of addition and multiplication given in this section to show the following facts.

(a) $x(y + z) = xy + xz \qquad (x,y,z \in \mathbf{N})$

(b) $x1 = x \qquad (x \in \mathbf{N})$

(c) Multiplication is commutative.

(d) Multiplication is associative.

(4.6) Let $m,n \in \mathbf{N}$ be such that $m|n$. Show that congruence mod n is finer than congruence mod m.

(4.7) Let R be the relation congruence mod 5 on the set of integers. Construct a Cayley table of the operations induced by addition and multiplication on the set **Z**/R. Do the same for congruence mod 6.

(4.8) Prove the uniqueness assertion in (4.9).

(4.9) Prove (4.11).

(4.10) Prove (4.8).

(4.11) Prove the following by induction.

(a) $1 + 2 + 3 + \cdots + n = n(n - 1)/2$

(b) $1 + 3 + 5 + \cdots + (2n - 1) = n^2$

(c) $1 + 4 + 9 + \cdots + n^2 = n(n + 1)(2n + 1)/6$

(4.12) Prove that for any real number x such that $-1 \le x$ and for any $n \in \mathbf{N}$, $1 + nx \le (1 + x)^n$.

(4.13) Let $n,k \in \mathbf{N}$ be such that $k < n$. Show that either $k|n$ or there is a positive integer m such that $mk < n < (m + 1)k$.

(4.14) The Fibonacci numbers can be defined as the following sequence.

$$x_1 = 1$$
$$x_2 = 1$$
$$x_k = x_{k-1} + x_{k-2}$$

Let $r_k = x_{k-1}/x_k$ and show that r_k has limit $(\sqrt{5} - 1)/2$ as k tends to infinity.

(4.15) Given the product formula for derivatives and given that $Dx = 1$, prove the power formula for positive integers.

(4.16) Use the well ordering principle to prove the following equivalent form of the axiom of induction.

If $M \subseteq \mathbf{N}$ satisfies

(1) $1 \in M$

(2) If $k \in \mathbf{N}$ and for every $n < k$ we have $n \in M$, then $k \in M$.

Then $M = \mathbf{N}$.

(4.17) Permutations. Let $f: A \to A$ and define

$$M_f = \{x: x \in A \text{ and } f(x) \neq x\}$$
$$F_f = \{x: x \in A \text{ and } f(x) = x\}$$

(a) Show that $\{M_f, F_f\}$ is a partition of A.

(b) Let $f \in S_n$. Then, we will call f a cycle if there is a k such that the orbit of k with respect to f is M_f. Show that there is an m such that

(1) $f^m(k) = k$;

(2) if $1 \le i, j \le m$, then $f^i(k) \neq f^j(k)$;

(3) $M_f = \{k, f(k), \ldots, f^{m-1}(k)\}$.

In this case we will write

$$f = (k, f(k), \ldots, f^{m-1}(k))$$

and we will call f an m-cycle. Observe that this symbol completely determines the action of f since all elements outside the parenthesis are fixed by f and action of f on an element inside the parenthesis is either the element immediately following or k. For example, we can consider the 4-cycle in S_6 defined by

$$f = (1, 3, 5, 4)$$

Then using our previous notation,

$$f = \begin{pmatrix} 1 & 2 & 3 & 4 & 5 & 6 \\ 3 & 2 & 5 & 1 & 4 & 6 \end{pmatrix}$$

Note also that we can write

$$f = (3, 5, 4, 1)$$

(c) If $f \in S_n$ and $k \in M_f$, then let O be the orbit of k with respect to f. If $O = \{k, f(k), \ldots, f^m\}$, then define $g,h \in S_n$ by

$$g = (k, f(k), \dots, f^m(k))$$

$$h(m) = \begin{cases} f(m) & \text{if } m \notin 0 \\ m & \text{if } m \in 0 \end{cases}$$

Show that

(1) $f = g \circ h = h \circ g$;

(2) $M_h \subseteq M_f$;

(3) $M_h \neq M_f$.

(4) Use the induction principle discussed in exercise (4.14) to show that f can be written as a product of cycles whose orbits are pairwise disjoint. In this case, we will say that f has been written as a product of disjoint cycles.

(d) Represent the following as products of disjoint cycles.

$$\begin{pmatrix} 1 & 2 & 3 & 4 & 5 & 6 \\ 2 & 1 & 3 & 6 & 4 & 5 \end{pmatrix}$$
$$\begin{pmatrix} 1 & 2 & 3 & 4 & 5 & 6 \\ 2 & 3 & 1 & 4 & 6 & 5 \end{pmatrix}^{-1} \begin{pmatrix} 1 & 2 & 3 & 4 & 5 & 6 \\ 1 & 3 & 2 & 6 & 4 & 5 \end{pmatrix}$$
$$(1, 2, 3, 5)(6, 1, 4, 3)$$

(e) If $f = (1, 2, 5, 6)(4, 7, 8)$, then find f^{-1}.

(f) If $f = (a_1, a_2, \dots, a_n)$, then show that

$$f = (a_n, a_{n-1})(a_n, a_{n-2}) \cdots (a_n, a_1)$$

Thus deduce that every permutation can be written as a product of 2-cycles. In the sequel we will refer to a 2-cycle as a transposition.

For brevity, we will sometimes delete commas and spaces between numbers if $n < 10$. Thus the cycle (1, 2, 3, 4) would be denoted (1234).

(4.18) The Schroeder-Bernstein Theorem. Let A and B be sets, then we will write $A \leq B$ if there is an injective function f from A to B. Show that:

(a) For any set A, $A \leq A$.

(b) If A, B and C are sets such that $A \leq B$ and $B \leq C$, then $A \leq C$.

(c) For the rest of this exercise we will assume that A and B are sets, $A \le B$ and $B \le A$. Furthermore, we will assume that $f: A \to B$ and $g: B \to A$ are injective functions. We will be showing that there is a bijection from A to B. We will say that an element $a \in A$ is parentless if $g^{-1}(a)$ is empty. Similarly, we define an element $b \in B$ to be parentless if $f^{-1}(b)$ is empty. If $x, y \in A \cup B$, then x is said to be an ancestor of y if there is a finite composition h of functions f and g such that $h(x) = y$. Define

A_1 = {x: $x \in A$ and x has infinitely many ancestors}
B_1 = {x: $x \in B$ and x has infinitely many ancestors}
A_2 = {x: $x \in A$ and x has its last ancestor in A}
B_2 = {x: $x \in B$ and x has its last ancestor in A}
A_3 = {x: $x \in A$ and x has its last ancestor in B}
B_3 = {x: $x \in B$ and x has its last ancestor in B}

Show that:

(a) $\{A_1, A_2, A_3\}$ is a partition of A.
(b) $\{B_1, B_2, B_3\}$ is a partition of B.
(c) The restriction of f to A_1 is a bijection from A_1 to B_1.
(d) The restriction of f to A_2 is a bijection from A_2 to B_2.
(e) The restriction of g to B_3 is a bijection from B_3 to A_3.
(f) Define a function h from A to B as follows

$$h(x) = \begin{cases} f(x) & \text{if } x \in A_1 \\ f(x) & \text{if } x \in A_2 \\ g^{-1}(x) & \text{if } x \in A_3 \end{cases}$$

Show that h is a bijection from A to B. Thus we have shown that if A and B are sets such that $A \le B$ and $B \le A$, then there is a bijection from A to B.

2
Groups

In this chapter the fundamental definitions and propositions of group theory are developed and illustrated. The main exposition is by design concise. To fully understand the theory the student must carefully study the examples and work through all the exercises.

1. GROUPS AND SUBGROUPS

(1.1) DEFINITION A group $<G, *>$ is a nonempty set G together with a binary operation $*$ on G satisfying the following properties.

(i) Associativity: For all $x, y, z \in G$, $(x * y) * z = x * (y * z)$.

(ii) Existence of identity: There is an element $e \in G$ such that $e * x = x = x * e$ for all $x \in G$.

(iii) Existence of inverses: For all $x \in G$ there exists a $y \in G$ such that $x * y = e = y * x$.

Recalling I, (3.3) we note that the identity e of G guaranteed by (ii) is necessarily unique. We sometimes write e_G for the identity of G to emphasize the ambient group.

Frequently we speak simply of a group G rather than $<G, *>$ when the operation is implicitly clear from the context. Similarly we write xy for $x * y$ when there is no danger of confusion.

(1.2) PROPOSITION Let G be a group and $x \in G$. Define two maps ι_x, r_x: $G \to G$ (left and right multiplication by x, respectively) by

$$\iota_x(y) = xy \quad \text{and} \quad r_x(y) = yx \quad (y \in G)$$

Then the following assertions hold.

(i) For all $x,y \in G$, $\iota_x \circ \iota_y = \iota_{xy}$.

(ii) $\iota_e = 1_G$, the identity map on G.

(iii) For all $x \in G$, the map ι_x is bijective.

(iv) For all $x,y \in G$, $r_x \circ r_y = r_{yx}$.

(v) $r_e = 1_G$.

(vi) For all $x \in G$, the map r_x is bijective.

PROOF (i) Let $y \in G$. Then

$$\begin{aligned} \iota_x \circ \iota_y(z) &= \iota_x(\iota_y(z)) && \text{(definition of composition)} \\ &= \iota_x(yz) && \text{(definition of } \iota_y) \\ &= x(yz) && \text{(definition of } \iota_x) \\ &= (xy)z && \text{(associativity)} \\ &= \iota_{xy}(z) && \text{(definition of } \iota_{xy}) \end{aligned}$$

and so $\iota_x \circ \iota_y = \iota_{xy}$.

(ii) Let $z \in G$. Then

$$\begin{aligned} \iota_e(z) &= ez && \text{(definition of } \iota_e) \\ &= z && \text{(group identity)} \\ &= 1_G(z) && \text{(definition of } 1_G) \end{aligned}$$

Hence $\iota_e = 1_G$.

(iii) According to I, (1.8), to show that ι_x is bijective, it suffices to exhibit an inverse map. By (1.1), (iii), there exists an element $y \in G$ such that $xy = e = yx$. Then by parts (i) and (ii) above, we have

$$\iota_x \circ \iota_y = \iota_{xy} = \iota_e = 1_G$$
$$\iota_y \circ \iota_x = \iota_{yx} = 1_e = 1_G$$

Therefore ι_y is the required inverse map for ι_x.

The arguments for (iv), (v), and (vi) are similar and should be carried out in detail by the reader.

(1.3) COROLLARY Let G be a group. Then G satisfies the following properties.

(i) Left cancellation: If $x,y,z \in G$ with $xy = xz$, then $y = z$.

(ii) Right cancellation: If $x,y,z \in G$ with $yx = zx$, then $y = z$.

PROOF If $x,y,z \in G$ with $xy = xz$, then $\iota_x(y) = \iota_x(z)$ and hence, since ι_x is injective, $y = z$. The second statement follows from the injectivity of the map r_x.

(1.4) PROPOSITION Let G be a group. Then the following assertions hold.

(i) Inverses are unique; that is, for every $x \in G$ there exists a unique $y \in G$ such that $xy = e = yx$. Henceforth we denote the unique inverse of x in G by x^{-1}.

(ii) If $x,y \in G$ with $xy = e$, then $x = y^{-1}$ and $y = x^{-1}$. Hence to verify inverses, one need only check on one side.

(iii) $(x^{-1})^{-1} = x$ for all $x \in G$.

(iv) $(xy)^{-1} = y^{-1}x^{-1}$ for all $x,y \in G$.

(v) If $x \in G$, then $xx = x$ if and only if $x = e$.

PROOF (i) Suppose that y and y' both act as inverses for a given group element x. Then, in particular, $xy = e$ and $xy' = e$. Thus, by left cancellation, $y = y'$.

(ii) If $xy = e$ then, since $xx^{-1} = e$ also, by left cancellation $y = x^{-1}$. Similarly $y^{-1}y = e$, so by right cancellation, $x = y^{-1}$.

(iii) Let $x \in G$. Then by definition $xx^{-1} = e$. Now apply (ii) (with x^{-1} in place of y) to obtain $x = (x^{-1})^{-1}$.

(iv) Consider the product $(xy)(y^{-1}x^{-1})$:

$$\begin{aligned}
(xy)(y^{-1}x^{-1}) &= x[y(y^{-1}x^{-1})] && \text{(associativity)} \\
&= x[(yy^{-1})x^{-1}] && \text{(associativity)} \\
&= x(ex^{-1}) && \text{(definition of inverse)} \\
&= xx^{-1} && \text{(definition of identity)} \\
&= e && \text{(definition of inverse)}
\end{aligned}$$

Thus $(xy)(y^{-1}x^{-1}) = e$ and by (ii) it follows that $y^{-1}x^{-1} = (xy)^{-1}$.

(v) If $xx = x$, then since $xe = x$, by left cancellation $x = e$. Obviously $ee = e$, whence the assertion follows.

In general, if $*$ is a binary operation on a set S, we say that an element $s \in S$ is <u>idempotent</u> if $s * s = s$. Accordingly, part (v) of this last proposition states that e is the unique idempotent element of a group G.

(1.5) NOTATION AND TERMINOLOGY Parentheses and Integral Exponents. Henceforth we omit parentheses from extended group products in the manner permissible to all associative operations, except possibly when they retain strategic value. For instance, we shall simply write xyz for the common value of $(xy)z = x(yz)$.

If $x \in G$ and $n \in \mathbf{N} \cup \{0\}$, x^n is defined inductively as follows:

$$x^0 = e$$
$$x^n = xx^{n-1}$$

Thus informally speaking, $x^n = xx \cdots x$ (n-times); that is, x^n is the n-fold product of x with itself. Further define $x^{-n} = (x^{-1})^n$. Then x^n is defined for all $n \in \mathbf{Z}$, and the fundamental property of exponentiation holds:

$$x^m x^n = x^{m+n} \text{ for all } m,n \in \mathbf{Z}$$

WARNING. The familiar rule $(xy)^n = x^n y^n$ for all $x,y \in G$, $n \in \mathbf{Z}$, fails in general since it depends upon commutativity. For instance, $(xy)^2 = xyxy$ while $x^2y^2 = xxyy$. The inner factors yx and xy may not be equal. [Note that this failure is already suggested in (1.4), (iv): we do not claim $(xy)^{-1} = x^{-1}y^{-1}$.]

Let G be a group. If the cardinality of the set G is finite, we call G a finite group or a group of finite order. In this case Card (G), the number of elements in G, is called the order of G and denoted o(G). If Card (G) is infinite, we say G is an infinite group or a group of infinite order.

A group <G, $*$> such that $*$ is commutative is called a commutative or <u>abelian</u> group. More explicitly, G is abelian if

$$x * y = y * x \text{ for all } x,y \in G$$

We often employ a special notation - the so-called additive

notation - when working with the case of an abelian group G. If $x,y \in G$ we write $x + y$ for the product of x and y. In this case, we will call G an <u>additive</u> group. In the sequel, all additive groups are abelian. Accordingly, the associative law is then expressed

$$(x + y) + z = x + (y + z)$$

for all $x,y,z \in G$. The identity of G in this modified notation becomes 0 (zero). Hence the second group axiom now reads

$$x + 0 = x = 0 + x$$

for all $x \in G$. Furthermore, the inverse of x is rendered $-x$, and we abbreviate $x + (-y)$ to $x - y$. Note finally that exponential expressions assume coefficients; thus

$$nx = x + x + \cdots + x$$

It is still desirable at times to retain the ordinary (multiplicative) notation for abelian groups; this is a matter of context.

(1.6) EXAMPLES The following examples illustrate the theory so far developed. Most of these will recur throughout this and subsequent chapters, and therefore each must be carefully considered. Observe that the earlier examples in the sequence tend to be familiar objects; later examples are more exotic.

(1.6.1) The integers **Z** form an infinite abelian group under the operation of addition. The associativity and commutativity of addition are well known, 0 is the identity, and the additive inverse of an element $n \in \mathbf{Z}$ is of course $-n$. (Note that our usual arithmetic conventions are precisely concurrent with the additive notation for an abelian group.) Technically we should denote this group $\langle \mathbf{Z}, + \rangle$, but generally we speak simply of the group **Z**. Since **Z** does not form a group under multiplication (Why?), no confusion can result.

Similarly, **Q**, **R** and **C** are infinite abelian groups under addition.

(1.6.2) The set $\{+1, -1\}$ forms an abelian group of order 2 with respect to ordinary multiplication. (The reader should verify closure and identify the identity and inverses.) This group is sometimes denoted $\mathbf{Z}^{\times}$ and called the group of units of **Z**. We shall explore this type of group further in later chapters. For the present we simply introduce the following related abelian groups. All are infinite.

$\mathbf{Q}^{*}$, the group of nonzero rational numbers under multiplication
$\mathbf{R}^{*}$, the group of nonzero real numbers under multiplication
$\mathbf{C}^{*}$, the group of nonzero complex numbers under multiplication

(1.6.3) Let $G = \mathbf{R} - \{1\} = \{x \in \mathbf{R}: x \neq 1\}$ and define $*$ on G by

$$x * y = x + y - xy \quad (x,y \in G)$$

The reader should verify that $*$ is an associative, commutative binary operation on G (note that <u>closure</u> must be verified) with identity 0. If $x \in G$ then $x^{-1} = x/(x - 1) \in G$ [cf. I, (3.4.2)]. Hence $\langle G, *\rangle$ is an abelian group.

(1.6.4) Let n be a positive integer. Then from I, (3.10), and exercise I, (4.2), $\langle \mathbf{Z}_n, +\rangle$ is a finite abelian group of order n. From exercise I, (4.3), $\mathbf{Z}_n^{\times}$ is a finite abelian group under multiplication.

(1.6.5) Let n be a positive integer. Then S_n, the symmetric group on n elements, is a finite group of order n! under composition of permutations. According to I, (3.4.6), the group S_n is <u>not</u> abelian for $n > 2$. More generally, for any set T, B(T) the set of all bijections from T to T, is a group which is finite if and only if T is finite, and nonabelian whenever Card (T) > 2.

(1.6.6) Recall from calculus that if f and g are real-valued functions defined on a common domain, then their sum f + g is defined by

$$(f + g)(x) = f(x) + g(x)$$

for all x in the given domain. [This is generalized in exercise (1.9).] Let $C^0(\mathbf{R})$ denote the set of all real-valued continuous functions with domain **R**. One knows from the elementary theory of continuity that the sum of two continuous functions is again continuous. Hence $C^0(\mathbf{R})$ is closed under addition. The reader should verify the following assertions.

(i) Addition is both associative and commutative on $C^0(\mathbf{R})$.

(ii) The constant function $z(x) = 0$ for all x is the identity element for $\langle C^0(\mathbf{R}), +\rangle$.

(iii) Let $f \in C^0(\mathbf{R})$. Define $-f$ by $(-f)(x) = -[f(x)]$. Then $-f \in C^0(\mathbf{R})$, and $-f$ is the additive inverse of f in $C^0(\mathbf{R})$. (Each statement follows at once from the corresponding property of **R**.) Thus $\langle C^0(\mathbf{R}), +\rangle$ is yet another infinite abelian group.

(1.6.7) Consider the set S consisting of the symbols $e = s^0$, $s = s^1, s^2, \ldots, s^{n-1}$ where n is a positive integer. Define an operation $*$ on S by the formula

$$s^j * s^k = \begin{cases} s^{j+k} & \text{if } j + k < n \\ s^{j+k-n} & \text{if } j + k \geq n \end{cases}$$

Then $\langle S, *\rangle$ is a finite abelian group of order n. We leave the verification of associativity and commutativity to the reader; $e = s^0$ is the identity, and the inverse of s^j is s^{n-j}. This group is generally denoted C_n and called the cyclic group of order n.

Note that the expression s^j, $1 < j < n$ has two interpretations in this context. On the one hand, s^j is simply one of the symbols contained in S. On the other hand, we might interpret s^j as $s * s * \cdots * s$ (j-times). However, one sees easily that s^j in the exponential sense does indeed resolve to s^j in the symbolic sense, so the ambiguity is harmless.

We point out that our use of the symbol s in the construction of C_n is completely inessential. The reader should attempt at this point to identify formally the resemblence of C_n and $\mathbf{Z}_n$ in anticipation of matters raised in Section 2.

A Cayley Table for C_6

*	e	s	s^2	s^3	s^4	s^5
e	e	s	s^2	s^3	s^4	s^5
s	s	s^2	s^3	s^4	s^5	e
s^2	s^2	s^3	s^4	s^5	e	s
s^3	s^3	s^4	s^5	e	s	s^2
s^4	s^4	s^5	e	s	s^2	s^3
s^5	s^5	e	s	s^2	s^3	s^4

(1.6.8) We construct a nonabelian group of order 8 called the dihedral group D_4. The elements of D_4 are the symbols e, s, s^2, s^3, t, st, s^2t, s^3t. Products in D_4 are governed by the following rules.

(i) Products involving only powers of s resolve as if in C_4, the cyclic group of order 4. In particular, $s^4 = e$.

(ii) Products involving only powers of t resolve as if in C_2, the cyclic group of order 2. In particular, $t^2 = e$.

(iii) $ts = s^{-1}t = s^3t$. Consequently $ts^j = s^{-j}t$ for all j. (Why?)

A Cayley Table for D_4

*	e	s	s^2	s^3	t	st	s^2t	s^3t
e	e	s	s^2	s^3	t	st	s^2t	s^3t
s	s	s^2	s^3	e	st	s^2t	s^3t	t
s^2	s^2	s^3	e	s	s^2t	s^3t	t	st
s^3	s^3	e	s	s^2	s^3t	t	st	s^2t
t	t	s^3t	s^2t	st	e	s^3	s^2	s
st	st	t	s^3t	s^2t	s	e	s^3	s^2
s^2t	s^2t	st	t	s^3t	s^2	s	e	s^3
s^3t	s^3t	s^2t	st	t	s^3	s^2	s	e

The identity of D_4 is e. The existence of inverses is apparent in the Cayley table; associativity is not so apparent, but this follows from general principles established in exercise (4.32). One sees easily that D_4 is nonabelian: $ts = s^3t \neq st$.

It emerges in the exercises that the product defined on D_4 is neither so exotic nor so complicated as it may now appear. D_4 is but one early member of the family of so-called dihedral groups which arise naturally in geometry. Moreover, these dihedral groups are in turn especially naive examples of a general group theoretic construction called the semi-direct product. See exercise (4.32).

(1.6.9) A 2 × 2 matrix with real entries is a square array of the form

$$\begin{pmatrix} a & b \\ c & d \end{pmatrix}$$

where a,b,c,d, $\in \mathbf{R}$. The set of all such matrices is denoted $M_2(\mathbf{R})$. In I, (3.4.5), we defined addition and multiplication of 2 × 2 matrices and we noted that addition is commutative and associative and multiplication is associative but not commutative. Exercise (1.12) shows that $\langle M_2(\mathbf{R}), +\rangle$ is an infinite abelian group. We will now consider a subset of $M_2(\mathbf{R})$ which forms an infinite nonabelian group under multiplication.

Define a map $\det: M_2(\mathbf{R}) \to \mathbf{R}$ by

$$\det \begin{pmatrix} a & b \\ c & d \end{pmatrix} = ad - bc$$

This map is called the <u>determinant</u>. One may verify with a straightforward calculation that for all $A,B \in M_2(\mathbf{R})$,

$$(*) \quad \det(AB) = \det(A)\det(B)$$

Define a subset $GL_2(\mathbf{R})$ of $M_2(\mathbf{R})$ by

$$GL_2(\mathbf{R}) = \{A \in M_2(\mathbf{R}): \det(A) \neq 0\}$$

We claim $GL_2(\mathbf{R})$ is a group under matrix multiplication. We will call this group the general linear group (of rank 2 matrices over $\mathbf{R}$).

There are four points to establish.

(1) Closure: Matrix multiplication does indeed restrict to an operation on $GL_2(\mathbf{R})$. Suppose $A,B \in GL_2(\mathbf{R})$. Then $\det(A) \neq 0$, $\det(B) \neq 0$, and so by (*) $\det(AB) = \det(A)\det(B) \neq 0$. Hence, by definition, $AB \in GL_2(\mathbf{R})$.

(2) Associativity: This is inherited from $M_2(\mathbf{R})$.

(3) Existence of Identity: One verifies easily that the matrix

$$I_2 = \begin{pmatrix} 1 & 0 \\ 0 & 1 \end{pmatrix} \in GL_2(\mathbf{R})$$

is an identity for the group $GL_2(\mathbf{R})$.

(4) Existence of Inverses: Let

$$A = \begin{pmatrix} a & b \\ c & d \end{pmatrix} \in GL_2(\mathbf{R})$$

Define $B \in Gl_2(\mathbf{R})$ by

$$B = \begin{pmatrix} d/\det(A) & -b/\det(A) \\ -c/\det(A) & a/\det(A) \end{pmatrix}$$

Note how the construction of B presumes $\det(A) \neq 0$. One calculates that $AB = I_2 = BA$. Moreover, it follows from (*) that $\det(A)\det(B) = \det(I_2) = 1$, whence $\det(B) \neq 0$. Hence B is the required inverse of A in $GL_2(\mathbf{R})$.

We shall see later that in some sense $GL_2(\mathbf{R})$ plays the same role in $M_2(\mathbf{R})$ that $\mathbf{Z}^\times$ plays in $\mathbf{Z}$. [See IV, (1.9.3).]

(1.7) Direct Products. Let there be given two groups $\langle G, * \rangle$, $\langle H, \square \rangle$. Then we may form the direct product of the sets G and H to form the set

$$G \times H = \{(g, h): g \in G \text{ and } h \in H\}$$

Moreover, we may define a binary operation on $G \times H$ by

$$(g, h)(g', h') = (g * g', h \square h')$$

for $g,g' \in G$, $h,h' \in H$. We leave it to the reader to verify that in fact $G \times H$ is a group under this operation. The group $G \times H$ is called the direct product of $\langle G, *\rangle$ and $\langle H, \square\rangle$. Omitting the operation symbols $*$ and $\square$, (as we usually do), we speak simply of $G \times H$, the (group-theoretic) direct product of G and H.

The direct product construction extends in an obvious way to more than two factors. As with sets, the direct product of groups $G_1, \ldots, G_n$ is denoted

$$G_1 \times G_2 \times \cdots \times G_n \quad \text{or} \quad \prod_{i=1}^{n} G_i$$

The group operation is again defined componentwise. In the special case $G_1 = G_2 = \cdots = G_n = G$, we write

$$G^n = G \times G \times \cdots \times G \quad \text{(n-copies)}$$

One can also construct direct products for arbitrary infinite families of groups. See exercise (4.33).

(1.8) DEFINITION A subset H of a group G is a subgroup of G if H itself is a group with respect to the operation of G.

Observe that if G is a group, then it contains the subgroups $\{e\}$ and G itself. A subgroup $H \neq \{e\}$ is called <u>nontrivial</u>. A subgroup $H \neq G$ is called <u>proper</u>. Observe further that if e_H is the identity of a subgroup H, then $e_H = e_G$ since e_G is the unique idempotent of G by (1.4), (v).

(1.9) PROPOSITION A nonempty subset H of a group G is a subgroup of G if and only if the following two conditions hold.

(i) If $x, y \in H$, then $xy \in H$.

(ii) If $x \in H$, then $x^{-1} \in H$.

Moreover, if H is finite, (i) suffices.

PROOF Let H be a subgroup of G. Then (i) holds by definition, and (ii) follows from (1.4), (i) and our remark that $e_H = e_G$.

Now let H be a nonempty subset of G satisfying (i) and (ii). Then by (i), H is closed under the operation defined on G, which of course retains its associativity under restriction. Since H is nonempty, let $x \in H$. Then by (ii), $x^{-1} \in H$, and again by (i), $xx^{-1} = e \in H$. Thus H has the required identity and inverses.

To secure the final assertion, suppose H is a finite, nonempty subset of G which satisfies (i). We shall show that H also satisfies (ii), and by our previous arguments, this suffices. Let $x \in H$. Then by (1.2) $\iota_x: G \to G$ [defined by $1_x(y) = xy$] is bijective. By (i), ι_x restricts to a map $H \to H$; under restriction, injectivity persists and since H is finite, it follows that $\iota_x|_H: H \to H$ is likewise bijective. Therefore, there exists a $y \in H$ such that $\iota_x(y) = x$; that is, $xy = x$. But then $y = e$ and $e \in H$. The same reasoning guarantees the existence of $z \in H$ such that $\iota_x(z) = e$; that is, $xz = e$. But then $z = x^{-1}$ and $x^{-1} \in H$. Hence (ii) holds.

From the last proposition we can deduce another criterion for subgroups which is often more efficient to apply.

(1.10) PROPOSITION Let H be a nonempty subset of a group G. Then H is a subgroup of G if and only if

(*) $x, y \in H$ implies $xy^{-1} \in H$

PROOF Subgroups clearly satisfy (*). Conversely, suppose a nonempty subset H satisfies the stated condition. Let $x \in H$. Then applying (*) in the special case $x = y$, we find that $xx^{-1} = e \in H$. Next applying (*) to e and x, we conclude that $ex^{-1} = x^{-1} \in H$. Therefore H satisfies provision (ii) of (1.9). There remains only to establish provision (i). Suppose that $y \in H$. Then, as we have seen, $y^{-1} \in H$, and again by (*), $x(y^{-1})^{-1} = xy \in H$ as (1.9), (i) requires.

(1.11) EXAMPLES We present some elementary examples at this point. More examples will be given in (1.20).

(1.11.1) We first consider some subsets of **Z**.

(i) **N** is not a subgroup of **Z** since it does not contain the identity 0.

(ii) The set of nonnegative integers contains the identity 0 and is closed under the operation of addition but does not contain inverses; it therefore is not a subgroup of **Z**.

(iii) The set of even integers is a subgroup of **Z**.

(1.11.2) **C** admits the following chain of subgroups.

$$\mathbf{Z} \subseteq \mathbf{Q} \subseteq \mathbf{R} \subseteq \mathbf{C}$$

The multiplicative group $\mathbf{C}^*$ admits the following chain of subgroups.

$$\mathbf{Z}^\times \subseteq \mathbf{Q}^* \subseteq \mathbf{R}^* \subseteq \mathbf{C}^*$$

(1.11.3) Let us consider some subsets of the (additive) group $\mathbf{R} \times \mathbf{R}$ [cf. I, (3.4.3)].

(i) The set $H = \{(x, y): (x, y) \in \mathbf{R} \times \mathbf{R} \text{ and } y = x + 1\}$ is not a subgroup of $\mathbf{R} \times \mathbf{R}$ since it does not contain the identity (0, 0).

(ii) The set $H = \{(x, y): (x, y) \in \mathbf{R} \times \mathbf{R} \text{ and } y = 3x\}$ is a subgroup of $\mathbf{R} \times \mathbf{R}$.

(iii) The set $H = \{(x, y): (x, y) \in \mathbf{R} \times \mathbf{R} \text{ and } y = x^2\}$ is not a subgroup of $\mathbf{R} \times \mathbf{R}$. (Why?)

(1.12) PROPOSITION Let C be a collection of subgroups of G. Then $\bigcap_{H \in C} H$ is also a subgroup of G.

PROOF Let $H_0 = \bigcap_{H \in C} H$. We will use (1.10). Suppose that $x, y \in H_0$. Then $x, y \in H$ for all $H \in C$. Since each H is a subgroup of G, $xy^{-1} \in H$ for all $H \in C$. Hence $xy^{-1} \in H_0$ as required.

(1.13) PROPOSITION Let X be a subset of G, and let

$$C = \{H: X \subseteq H \text{ and } H \text{ is a subgroup of } G\}$$

Define $\langle X \rangle$ by

$$\langle X \rangle = \bigcap_{H \in C} H$$

Then the following assertions hold.

(i) $\langle X \rangle$ is a subgroup of G which contains X.

(ii) If K is any subgroup of G which contains X, then $\langle X \rangle \subseteq K$.

PROOF (i) By (1.12) $\langle X \rangle$ is a subgroup. By set theory $X \subseteq \langle X \rangle$.

(ii) If K is any subgroup of G which contains H, then by

definition $K \in C$. It follows by set theory that

$$\langle X\rangle = \bigcap_{H\in C} H \subseteq K$$

Thus $\langle X\rangle$ is the minimal subgroup of G containing X, and we call $\langle X\rangle$ the subgroup <u>generated</u> by X. If $G = \langle X\rangle$, then G is generated by X, and if this occurs for some finite set X, we say G is finitely generated. The next proposition gives a concrete description of $\langle X\rangle$.

(1.14) PROPOSITION Let X be a subset of a group G. Then the following assertions hold.

(i) $\langle X\rangle$ consists of all products of the form

$$(*)\quad x_1 x_2 \cdots x_n$$

where $n \in \mathbf{N}$, and either $x_j \in X$ or $x_j^{-1} \in X$ for $1 \le j \le n$. (Here we interpret a product of length 0 as e).

(ii) If G is finite, then we need only consider $x_j \in X$.

PROOF Let K denote the collection of all elements of the form (*). We leave it to the reader to verify that K is a subgroup of G containing X. Hence by (1.13), (ii), $\langle X\rangle \subseteq K$. Conversely, every element of K is a product of elements of X and their inverses. Since the subgroup $\langle X\rangle$ is closed under inverses and products, it follows that $K \subseteq \langle X\rangle$. Therefore, $K = \langle X\rangle$ as claimed.

To establish (ii) we recall from (1.9) that if G is finite then to verify that a subset of G is a subgroup of G we need only verify that it is closed under products.

The special case when X consists of a single element $x \in G$ is of utmost importance. We write $\langle x\rangle$ for $\langle\{x\}\rangle$ and call $\langle x\rangle$ the <u>cyclic subgroup</u> generated by $x \in G$. Thus

$$\langle x\rangle = \{x^n : n \in \mathbf{Z}\}$$

and, if G is finite,

$$\langle x\rangle = \{x^n : n \in \mathbf{N}\}$$

We observe that if G is an additive abelian group and $x \in G$, then

$$\langle x \rangle = \{nx: n \in \mathbf{Z}\}$$

and, if G is finite,

$$\langle x \rangle = \{nx: n \in \mathbf{N}\}$$

If, in fact, $G = \langle x \rangle$ for some $x \in G$, we say that G is a <u>cyclic group</u>. It is easily shown that every cyclic group is abelian.

WARNING In general, if H and K are subgroups of G, $H \cup K$ is <u>not</u> a subgroup of G [cf. exercise (1.36)]. We write $H \vee K$ (read 'H join K') for $\langle H \cup K \rangle$. Thus $H \vee K$ is the smallest subgroup of G containing H and K. We then have the following Hasse diagram of set inclusions (here $H \wedge K = H \cap K$).

It is easily verified that if G is an additive abelian group and H and K are subgroups of G, then

$$H \vee K = H + K = \{x \in G: x = h + k \text{ for some } h \in H, k \in K\}$$

(1.15) NOTATION AND TERMINOLOGY The Order of an Element. Let x be a member of a group G. Define a subset X of the positive integers by

$$X = \{m > 0: x^m = e\}$$

If $X = \emptyset$, that is, if no positive power of x is the identity, we say x has infinite order and write $o(x) = \infty$. If X is nonempty, it has a smallest element n, the smallest positive power of x which yields the identity (or, if G is an additive group, the smallest positive multiple of x which yields 0). In this case, we write $o(x) = n$ and say x has order n. The connection between the notions of order for elements and subgroups is made presently.

(1.16) PROPOSITION Let $x \in G$ and suppose n is a positive integer.

Then the following four statements are equivalent.

(i) $o(x) = n$.

(ii) For all $m \in \mathbf{Z}$, $x^m = e$ if and only if $n|m$.

(iii) The elements $e, x, x^2, \ldots, x^{n-1}$ are distinct and in fact

$$\langle x\rangle = \{e, x, x^2, \ldots, x^{n-1}\}$$

(iv) $o(\langle x\rangle) = n$; that is, the order of the cyclic subgroup generated by x is n.

PROOF (i)$\Longrightarrow$(ii) Suppose that $x \in G$ and $o(x) = n$. Let $m \in \mathbf{N}$. By the division algorithm, $m = qn + r$ for some $q \in \mathbf{Z}$ and $0 \le r < n$. Then

$$x^m = x^{qn+r} = (x^n)^q x^r = x^r$$

Since n is the smallest positive power of x which yields the identity and $0 \le r < n$, $x^m = e$ if and only if $r = 0$, that is, if and only if $n|m$.

(ii)$\Longrightarrow$(iii) Clearly $\{e, x, x^2, \ldots, x^{n-1}\} \subseteq \langle x\rangle$. If $x^j = x^k$ with $1 \le j < k \le n$, then $x^{k-j} = e$ with $0 < k-j < n$, contradicting the minimality of n. Hence the indicated elements are distinct. Finally, if $m \in \mathbf{Z}$, $m = qn + r$ for some $q \in \mathbf{Z}$, and $0 \le r < n$; hence, since $x^n = e$, $x^m = x^r$.

The remaining implications are left to the reader.

(1.17) PROPOSITION Let $x \in G$. Then the following statements are equivalent.

(i) $o(x) = \infty$.

(ii) For all $m \in \mathbf{Z}$, $x^m = e$ if and only if $m = 0$.

(iii) For all $m,n \in \mathbf{Z}$, $x^m = x^n$ if and only if $m = n$.

(iv) $\langle x\rangle$ is infinite.

PROOF Exercise.

(1.18) PROPOSITION Let G be a cyclic group with generator x and H a subgroup of G. Then H is a cyclic group with generator x^n where n is the smallest positive element of the set

$$X = \{m \in \mathbf{Z}: x^m \in H\}$$

PROOF We first observe that since H is nonempty and closed under inverses the set X does contain a positive integer. Then, since $x^n \in H$, $\langle x^n \rangle \subseteq H$. We must show that $H \subseteq \langle x^n \rangle$. Suppose that $y \in H$. Then, since $G = \langle x \rangle$, $y = x^m$ for some $m \in \mathbf{Z}$. By the division algorithm, $m = qn + r$ with $q,n \in \mathbf{Z}$ and $0 \leq r < n$. Then $y = x^m = (x^n)^q x^r$. But then $x^r = (x^n)^{-q} y \in H$ (since $x^n, y \in H$) and hence, by the minimality of n, $r = 0$. It follows that $y \in \langle x^n \rangle$ (for all $y \in H$) and hence $H = \langle x^n \rangle$ as claimed.

The preceding proposition enables us to completely determine the subgroup structure of **Z**.

(1.19) PROPOSITION (i) $\langle \mathbf{Z}, + \rangle$ is cyclic with generator 1.

(ii) Let H be a nontrivial subgroup of **Z**. Then

$$H = \langle n \rangle = n\mathbf{Z} = \{m \in \mathbf{Z}: m = nq \text{ for some } q \in \mathbf{Z}\}$$

where n is the smallest positive element of H.

(iii) If k and m are positive integers, then

$$k\mathbf{Z} \cap m\mathbf{Z} = n\mathbf{Z} \text{ where } n = \text{the least common multiple of } k \text{ and } m$$
$$k\mathbf{Z} \vee m\mathbf{Z} = d\mathbf{Z} \text{ where } d = (k, m)$$

PROOF Exercise.

For example, $8\mathbf{Z} \cap 10\mathbf{Z} = 40\mathbf{Z}$ and $8\mathbf{Z} \vee 10\mathbf{Z} = 2\mathbf{Z}$. Hence we have the following Hasse diagram of subgroups of **Z**.

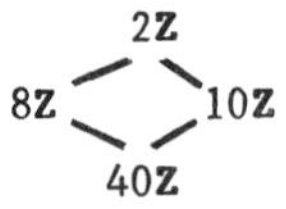

(1.20) EXAMPLES

(1.20.1) We first consider subgroups of the additive group $\mathbf{Z}_{10}$. Since $\mathbf{Z}_{10}$ is finite, every element has finite order and for any $x \in \mathbf{Z}_{10}$ the cyclic subgroup generated by x consists of all distinct positive multiples of x. Thus, for example,

$$\langle [2] \rangle = \{[2], [4], [6], [8], [0]\}$$

and $o([2]) = 5$. The table below lists all cyclic subgroups of $\mathbf{Z}_{10}$ and their generators; the reader is encouraged to verify each entry.

n	$\langle [n] \rangle$
0	$\{[0]\}$
1, 3, 7, 9	$\mathbf{Z}_{10}$
2, 4, 6, 8	$\{[0], [2], [4], [6], [8]\}$
5	$\{[0], [5]\}$

Observe that different group elements may give rise to the same subgroup. Furthermore, $\mathbf{Z}_{10}$ is itself a cyclic group with four different generators. We now pose some questions: Are these all the subgroups of $\mathbf{Z}_{10}$? Is there any common property shared by the group elements which generate the full group? Is there any significance to the fact that the order of every group element and of every subgroup divides the order of $\mathbf{Z}_{10}$? The answers to these questions will emerge shortly in the main exposition and the exercises.

(1.20.2) We list the cyclic subgroups of the symmetric group S_3. Since S_3 is a finite group, for each $\alpha \in S_3$ we need only take the successive powers $\alpha^1, \alpha^2, \ldots$, stopping when we first reach the identity.

$$\langle (1) \rangle = \{(1)\}$$
$$\langle (1, 2) \rangle = \{(1), (1, 2)\}$$
$$\langle (1, 3) \rangle = \{(1), (1, 3)\}$$
$$\langle (2, 3) \rangle = \{(1), (2, 3)\}$$
$$\langle (1, 2, 3) \rangle = \{(1), (1, 2, 3), (1, 3, 2)\} = \langle (1, 3, 2) \rangle$$

Notice that the sizes of the subgroups (1,2, and 3) are all divisors of 6, the order of S_3. Observe further that S_3 itself is not cyclic. This is hardly surprising since, as we know, S_3 is not abelian, while every cyclic group is abelian.

S_3 is not generated by one element, but it is generated by two. In fact, we claim that $\langle(1, 2), (1, 3)\rangle = S_3$. It suffices to show that every element of S_3 can be written as a product of (1, 2)'s and (1, 3)'s [cf. (1.14)]. Clearly (1), (1, 2) and (1, 3) are in $\langle(1, 2), (1, 3)\rangle$. The reader may also verify that $(1, 2)(1, 3)(1, 2) = (2, 3)$, $(1, 3)(1, 2) = (1, 2, 3)$, and $(1, 2)(1, 3) = (1, 3, 2)$. This completes the argument.

(1.20.3) Let C_n be the cyclic group of order n, introduced in (1.6.7). Then $C_n = \langle s\rangle$, so that cyclic groups are cyclic.

(1.20.4) Let D_4 be the dihedral group of order 8 introduced in (1.6.8). We leave it to the reader to determine all cyclic subgroups of D_4. From the definition of D_4 it follows at once that $D_4 = \langle\{s, t\}\rangle$.

(1.20.5) Let $C^1(\mathbf{R})$ denote the set of all differentiable functions from **R** to **R** with continuous derivative. Since differentiability implies continuity, we have $C^1(\mathbf{R}) \subseteq C^0(\mathbf{R})$. Recall (1.6.6). If $f,g \in C^1(\mathbf{R})$, then Df and Dg exist and are continuous by assumption. Therefore,

$$D(f - g) = Df - Dg$$

exists and is continuous, whence $f - g \in C^1(\mathbf{R})$. Hence, by (1.10), $C^1(\mathbf{R})$ is a subgroup of $C^0(\mathbf{R})$.

More generally, let $C^n(\mathbf{R})$ denote the set of all n-times differentiable functions with continuous n-th derivative and let $\mathbf{R}[x]$ denote the set of all polynomial functions with real coefficients. Then we have a chain of subgroups

$$\mathbf{R}[x] \subseteq C^n(\mathbf{R}) \subseteq C^{n-1}(\mathbf{R}) \subseteq \dots \subseteq C^0(\mathbf{R})$$

In fact these inclusions are strict.

Finally we observe that none of the indicated subgroups is cyclic. In fact, if $f \in C^0(\mathbf{R})$, then $\langle f\rangle = \{cf: c \in \mathbf{Z}\}$, the set of integer multiples of f.

(1.20.6) We introduce two important subgroups of $GL_2(\mathbf{R})$. [Recall (1.6.9).]

Let $T_2(\mathbf{R})$ denote the subset of $M_2(\mathbf{R})$ consisting of elements of the form:

$$(*) \quad \begin{pmatrix} a & b \\ 0 & d \end{pmatrix} \quad (a,b,d \in \mathbf{R})$$

Elements of $T_2(\mathbf{R})$ are called upper triangular matrices. Let $B_2(\mathbf{R}) = T_2(\mathbf{R}) \cap GL_2(\mathbf{R})$. Then $B_2(\mathbf{R})$ consists of upper triangular matrices with nonzero determinant. Since the determinant of a matrix of the form (*) is just the product ad, we see that $B_2(\mathbf{R})$ consists of precisely those upper triangular matrices with nonzero diagonal entries. Finally, let $\mathbf{R}I_2$ denote the set of all matrices of the form

$$(**) \quad \begin{pmatrix} a & 0 \\ 0 & a \end{pmatrix} \quad (a \in \mathbf{R})$$

(such matrices are called scalar matrices) and let $\mathbf{R}^*I_2$ denote the set of nonzero elements of $\mathbf{R}I_2$ (so that $a \neq 0$). The reader may then verify the following facts.

(i) $T_2(\mathbf{R})$ is a subgroup of the additive group $M_2(\mathbf{R})$.

(ii) We have the following chain of subgroups of the multiplicative group $GL_2(\mathbf{R})$

$$\mathbf{R}^*I_2 \subseteq B_2(\mathbf{R}) \subseteq GL_2(\mathbf{R})$$

(1.20.7) The Center of a Group. Given a group G, we define a subset Z(G) of G, called the <u>center of G</u>, by

$$Z(G) = \{x \in G: xy = yx \text{ for all } y \in G\}$$

Claim: Z(G) is a subgroup of G.

Proof Since Z(G) is nonempty it suffices to establish (1.9), (i), (ii). Let $x,y \in Z(G)$. Then by definition $xz = zx$, $yz = zy$ for all $z \in G$. Thus, for all $z \in G$,

$$xyz = xzy = zxy$$

and therefore $xy \in Z(G)$. Thus (1.9), (i) holds. For (1.9), (ii), suppose that $x \in Z(G)$. Then for all $y \in G$, $xy = yx$. Then (multiplying by x^{-1} on the left and on the right) $yx^{-1} = x^{-1}y$. Thus $x^{-1} \in Z(G)$.

The reader may verify the following examples.

(i) $Z(S_3) = \{(1)\}$ [cf. Example (1.6.5)].

(ii) $Z(D_4) = \{e, s^2\} = \langle s^2 \rangle$ [cf. Example (1.6.8)].

(iii) $Z(GL_2(\mathbf{R})) = \mathbf{R}^* I_2$ [cf. Example (1.6.9)].

(1.20.8) Subgroups of Direct Products.

(i) Let H and H' be subgroups of groups G and G', respectively. Then $H \times H'$ is a subgroup of the direct product $G \times G'$. As special cases $G \times \{e'\}$ and $\{e\} \times G'$ are subgroups of $G \times G'$, where e and e' are the identities of G and G' respectively. These subgroups are called factor subgroups of the direct product.

(ii) The Diagonal Subgroup. Let G be a group. Define a subset D of the direct product $G \times G$ by

$$D = \{(x, x): x \in G\}$$

Then D is a subgroup of $G \times G$ called the diagonal subgroup of $G \times G$.

EXERCISES

(1.1) Find all subgroups of S_3.

(1.2) Find all subgroups of D_4.

(1.3) Show that there is only one way to complete the following Cayley table so that {s, t, u} forms a group.

*	s	t	u
s	s		
t			
u			

(1.4) Show that there is only one way to complete the following Cayley table so that the set {s, t, u, v} forms a group.

*	s	t	u	v
s			u	
t				
u			v	
v				

(1.5) Interpret Proposition (1.2) in the special case $G = \mathbf{Z}$, and directly check the validity of the resulting statements.

(1.6) Write out a Cayley table for C_8 the cyclic group of order eight.

(1.7) Write out a Cayley table for the direct product of groups $\mathbf{Z}_2 \times \mathbf{Z}_3$.

(1.8) Verify that $\{1, i, -1, -i\}$ is a subgroup of $\mathbf{C}^*$.

(1.9) Let X be a nonempty set and let $\langle A, +\rangle$ be an additive group; that is an abelian group written additively. Let $M = A^X$, the set of all functions from X to A. Define addition + on M by

$$(f + g)(x) = f(x) + g(x)$$

for $f,g \in M$, $x \in X$. Show that $\langle M, +\rangle$ is an additive group.

(1.10) (a) Let $I = \{f: f \in C^0(\mathbf{R}), f(0) = 0\}$. Show that I is a subgroup of $C^0(\mathbf{R})$.

(b) Let X be any subset of R and let

$$I(X) = \{f: f \in C^0(\mathbf{R}), f(x) = 0 \text{ for all } x \in X\}$$

Show that I(X) is a subgroup of $C^0(\mathbf{R})$.

(c) Suppose that $X \subseteq Y \subseteq \mathbf{R}$. Show that I(Y) is a subgroup of I(X). Does $X \neq Y$ imply $I(X) \neq I(Y)$?

(1.11) Show that $C^{n+1}(\mathbf{R}) \neq C^n(\mathbf{R})$.

(1.12) Verify that $\langle M_2(\mathbf{R}), +\rangle$ is indeed an infinite additive group.

(1.13) Using (1.6.8) as a guide, construct a nonabelian group of order 10.

(1.14) Give an example of a group G and a subset H of G which is closed under the operation of G, but H is not a subgroup of G.

(1.15) Show that $\mathbf{Z}_n^{\times}$ is a group under multiplication [cf. exercise I, (4.3)].

(1.16) Let G be a group and define a function f: G $\to$ G by the rule $f(x) = x^{-1}$. Show that $f \circ f = 1_G$. Conclude that f is bijective.

(1.17) Explicitly calculate the inverses of the following elements of $GL_2(\mathbf{R})$.

(a) $\begin{pmatrix} 2 & 1 \\ 1 & 2 \end{pmatrix}$ (b) $\begin{pmatrix} e & 3 \\ 2 & e \end{pmatrix}$

(1.18) Show that a matrix $A \in M_2(\mathbf{R})$ has a multiplicative inverse if and only if $A \in GL_2(\mathbf{R})$.

(1.19) Let $A \in M_2(\mathbf{R})$ and suppose there exists a $B \in M_2(\mathbf{R})$ such that $AB = I_2$. Show that $A, B \in GL_2(\mathbf{R})$ and $B = A^{-1}$. Thus $AB = I_2$ implies $BA = I_2$. Prove that this property obtains in any group.

(1.20) Show that $Z(GL_2(\mathbf{R})) = \mathbf{R}^{*}I_2$.

(1.21) Show that the group S_n has a trivial center for all $n > 2$.

(1.22) Evaluate f^4 for each of the permutations $f \in S_4$ listed below.

(a) (1, 2, 3) (b) (1, 4, 3, 2) (c) (1, 2, 3)(1, 4, 3, 2)

(1.23) Find the order of each element in

(a) C_{10}, the cyclic group of order 10;

(b) C_{11}, the cyclic group of order 11;

(c) D_4, the dihedral group of order 8 constructed in (1.6.8).

(1.24) Find the orders of the following elements of the given group.

(a) $1 \in \mathbf{Z}$.

(b) $1 \in \mathbf{Z}^{\times}$.

(c) $e^{2\pi i/3} \in \mathbf{C}$.

(d) $e^{2\pi i/3} \in \mathbf{C}^{*}$.

(1.25) Show that the group $\mathbf{Q}$ is not cyclic.

(1.26) Let $z = e^{\Pi i/3}$. Describe the cyclic subgroup of $\mathbf{C}^{*}$ generated by z.

(1.27) Consider the elements

$$A = \begin{pmatrix} 0 & -1 \\ 1 & 0 \end{pmatrix} \qquad B = \begin{pmatrix} 0 & 1 \\ -1 & -1 \end{pmatrix}$$

of $GL_2(\mathbf{R})$. Show that $o(A) = 4$, $o(B) = 3$, and $o(AB) = \infty$.

(1.28) Let $f \in S_n$ be a cycle of length k. Show that $o(f) = k$.

(1.29) Let x and y be elements of a group G, with $xy = yx$. Suppose that x and y have finite order, and $\langle x \rangle \cap \langle y \rangle = \{e\}$. Show that $o(xy) = k$ where k is the least common multiple of $o(x)$ and $o(y)$.

(1.30) Use the previous two exercises to deduce a formula for the order of a permutation in S_n in terms of the lengths of the disjoint cycles appearing in its cycle decomposition.

(1.31) (a) Show that for all n, S_n is generated by the transpositions (1, 2), (1, 3), ... , (1, n).

(b) Use (a) to show that for all n, S_n is generated by the (n-1)-cycle (1, 2, ..., n-1) and the transposition (n-1, n).

(1.32) Let H be a subgroup of G and let $x \in G$. Show that xHx^{-1} is also a subgroup of G.

(1.33) Let G be an abelian group. Define a subset T of G by

$$T = \{x: x \in G, o(x) \text{ is finite}\}$$

Show that T is a subgroup of G. T is called the <u>torsion subgroup</u> of G. Use exercise (1.27) to show that this conclusion fails in general if G is nonabelian.

(1.34) Show that a nontrivial group which has no proper subgroups is necessarily finite of prime order.

(1.35) Let H and K be subgroups of an abelian group G. Show that

$H \vee K = \{hk: h \in H, k \in K\}$

(1.36) (a) Let H and K be subgroups of G. Show that $H \cup K$ is a subgroup of G if and only if either $H \subseteq K$ or $K \subseteq H$.

(b) Conclude that a group is never expressible as the union of two proper subgroups.

(c) Show by example that a group may be expressible as a union of three proper subgroups.

(1.37) Show that every subgroup of a cyclic group is cyclic.

(1.38) Let $C_n = \langle s \rangle$ be the cyclic group of order n. Show that $\langle s^k \rangle = C_n$ if and only if k and n are relatively prime.

(1.39) Show that a group G is abelian if and only if $(xy)^2 = x^2y^2$ for every $x, y \in G$.

(1.40) Suppose that $x^2 = e$ for every x in a group G. Prove that G is abelian.

(1.41) Let G be a finite group of even order. Show that there exists an $x \in G$ such that $o(x) = 2$.

(1.42) Let G be a finite abelian group and suppose that for all positive integers n the equation $x^n = e$ has at most n solutions in G. Show that G is cyclic.

(1.43) Let G be a group such that the equation $(xy)^n = x^n y^n$ holds for every $x, y \in G$ for three consecutive integers n. Show that G is abelian.

(1.44) Let $n \in \mathbf{N}$ and H a subgroup of the symmetric group S_n which contains the cycles $\alpha = (1, 2, \ldots, n)$ and $\tau = (1, 2)$.

(a) Show that H contains all transpositions of the form $(m, m + 1)$ [hint: consider $\alpha^{-1}\tau\alpha$].

(b) Use (a) to show that H contains all transpositions (1, m).

(c) Use (b) to show that H = G.

(1.45) Show that if H is a subgroup of the symmetric group S_n and H contains an n-cycle and a transposition then $H = S_n$.

2. HOMOMORPHISMS

(2.1) DEFINITION Let $\langle G, *\rangle$ and $\langle G', \square\rangle$ be groups. A function $f: G \to G'$ is called a homomorphism of groups if

(*) $f(x * y) = f(x) \square f(y)$

for all $x,y \in G$. If f is moreover bijective, we say f is an isomorphism of groups. If such an isomorphism exists, we call G and G' isomorphic and write $G \cong G'$.

Generally, when there is no danger of confusion, (*) is expressed simply as $f(xy) = f(x)f(y)$.

(2.2) PROPOSITION Let G and G' be groups with identities e and e' respectively. Suppose that $f: G \to G'$ is a homomorphism. Then the following assertions hold.

(i) $f(e) = e'$.

(ii) $f(x^{-1}) = [f(x)]^{-1}$ for all $x \in G$.

PROOF (i) Since f is a homomorphism, $f(e) = f(ee) = f(e)f(e)$. Then, since e' is the unique idempotent in the group G' [cf. (1.4), (v)], $f(e) = e'$.

(ii) $f(x)f(x^{-1}) = f(xx^{-1}) = f(e) = e'$ by (i) and hence, by the uniqueness of inverses [Proposition (1.4), (i)], $f(x^{-1}) = [f(x)]^{-1}$.

(2.3) EXAMPLES We now give some elementary examples which illustrate the definition of homomorphism and Proposition (2.2). More examples will be given after we have developed further theory.

(i) The map $f: \mathbf{Z} \to \mathbf{Z}$ defined by $f(m) = 5m$ for $m \in \mathbf{Z}$ is a homomorphism of groups since

$$f(m + n) = 5(m + n) = 5m + 5n = f(m) + f(n)$$

for all $m,n \in \mathbf{Z}$.

(ii) The map $f: \mathbf{Z} \to \mathbf{Z}$ defined by $f(m) = m + 1$ is not a homomorphism of groups [observe that $f(0) \neq 0$].

(iii) The map $f: \mathbf{R} \to \mathbf{R}^*$ defined by $f(x) = 2^x$ is a homomorphism of groups since, for all $x,y \in \mathbf{R}$,

$$f(x + y) = 2^{(x+y)} = 2^x 2^y$$

(recall that the operation used in $\mathbf{R}$ is addition while the operation of $\mathbf{R}^*$ is multiplication). Verifying Proposition (2.2) we see that $f(0) = 2^0 = 1$ [and hence (2.2), (i) holds], and $f(-x) = 2^{-x} = (2^x)^{-1} = [f(x)]^{-1}$ [so that (2.2), (ii) holds].

(2.4) PROPOSITION (i) $1_G: G \to G$ is an isomorphism of groups for all groups G.

(ii) Let $f: G \to G'$ and $g: G' \to G''$ be group homomorphisms. Then $g \circ f: G \to G''$ is likewise a group homomorphism.

(iii) If $f: G \to G'$ is an isomorphism of groups then so is $f^{-1}: G' \to G$.

PROOF (i) follows from the definition of 1_G.

(ii) Suppose that f,g are homomorphisms as indicated. Let $x,y \in G$.

$$\begin{aligned} g \circ f(xy) &= g(f(xy)) && \text{(definition of composition)} \\ &= g(f(x)f(y)) && \text{(since f is a homomorphism)} \\ &= g(f(x))g(f(y)) && \text{(since g is a homomorphism)} \\ &= (g \circ f)(x)(g \circ f)(y) && \text{(definition of composition)} \end{aligned}$$

Thus $g \circ f$ is a homomorphism also.

(iii) Let $x',y' \in G'$, and suppose that $f^{-1}(x') = x$ and $f^{-1}(y') = y$. Then by definition of the inverse map, $f(x) = x'$ and $f(y) = y'$. Since f is a homomorphism,

$$f(xy) = f(x)f(y) = x'y'$$

Hence $f^{-1}(x'y') = xy$ and f^{-1} is a homomorphism. Since f^{-1} is a priori invertible and therefore bijective, it is an isomorphism.

Note that (ii) implies that the composition of isomorphisms is again an isomorphism.

(2.5) COROLLARY (i) $G \cong G$ for all groups G.

(ii) If $G \cong G'$, then $G' \cong G$.

(iii) If $G \cong G'$ and $G' \cong G''$, then $G \cong G''$.

Thus isomorphism is an equivalence relation on the class of all groups.

(2.6) NOTATION AND TERMINOLOGY Let G and G' be groups. Then Hom(G, G') denotes the set of all homomorphisms from G to G'. Since the trivial map, which sends every element $x \in G$ to e', is a homomorphism of groups, Hom(G, G') is not empty. Proposition (2.4), (ii) shows that for three groups G, G', G'', composition defines a mapping

$$\text{Hom}(G', G'') \times \text{Hom}(G, G') \to \text{Hom}(G, G'') \quad \text{by} \quad (g, f) \mapsto g \circ f$$

In the special case G = G', Hom(G, G) is abbreviated to End(G); elements of End(G) (homomorphisms of G into itself) are called <u>endomorphisms</u>. Composition of endomorphisms defines a map

$$\text{End}(G) \times \text{End}(G) \to \text{End}(G) \quad \text{by} \quad (f, f') \mapsto f \circ f'$$

and thus composition is a binary operation on End(G). This operation is moreover associative and has an identity, namely 1_G. However, since endomorphisms need not be invertible, End(G) will not in general form a group with respect to composition. Bijective endomorphisms; that is, isomorphisms from G onto G; are called <u>automorphisms</u> of G. The set of all such automorphisms is denoted Aut(G). Since $1_G \in \text{Aut}(G)$, Aut(G) is not empty. Using (2.4) we may in fact show that Aut(G) is a subgroup of the group B(G), the group of all bijective mappings (not necessarily homomorphisms) from G to G.

(2.7) DEFINITION Let $f \in \mathrm{Hom}(G, G')$. Then the kernel of f, ker f, is the inverse image of the identity of G' under f. Explicitly, if e' is the identity of G', then

$$\ker f = \{x \in G: f(x) = e'\}$$

(2.8) PROPOSITION Let $f \in \mathrm{Hom}(G, G')$. Then f is injective if and only if ker f = {e}.
PROOF By (2.2), f(e) = e'; hence $e \in \ker f$. If f is injective, then e' can have only one preimage in G and hence ker f = {e}. Conversely, suppose that ker f = {e} and $x,y \in G$ are such that f(x) = f(y). Then

$$f(xy^{-1}) = f(x)f(y)^{-1} = e'$$

and hence $xy^{-1} \in \ker f$. By hypothesis, $xy^{-1} = e$; hence x = y (multiply both sides by y on the right).

(2.9) PROPOSITION Let $f \in \mathrm{Hom}(G, G')$. Then the following assertions hold.

(i) If H is a subgroup of G then f(H) is a subgroup of G'. Hence, in particular, im f is a subgroup of G'.

(ii) If H' is a subgroup of G', then $f^{-1}(H')$ is a subgroup of G.

(iii) Ker f is a subgroup of G.

PROOF We will use Proposition (1.10).

(i) Since e' = f(e) by (2.2), $H' \neq \emptyset$. Let H' = f(H) and $x',y' \in H'$. Then by definition, x' = f(x) and y' = f(y) for some $x,y \in H$. Since H is a subgroup of G, $xy^{-1} \in H$. Hence

$$x'y'^{-1} = f(x)f(y)^{-1} = f(xy^{-1}) \in f(H) = H'$$

(ii) Let $H = f^{-1}(H')$ and $x,y \in H$ ($H \neq \emptyset$ since $e \in H$). Then if x' = f(x) and y' = f(y), $x',y' \in H'$. Since H' is a subgroup of G', $x'y'^{-1} \in H'$. Then, once again using the fact that $f(xy^{-1}) = f(x)f(y)^{-1} = x'y'^{-1}$, we see that $xy^{-1} \in H$ as required.

(iii) Ker f = $f^{-1}(\{e'\})$ and the result now follows from (ii).

(2.10) THEOREM (The Correspondence Theorem) Let $f \in \mathrm{Hom}(G, G')$ be surjective. Then the mappings, whose actions are specified below,

$$H \mapsto f(H) \quad \text{and} \quad H' \mapsto f^{-1}(H')$$

are a pair of inverse maps between the set of all subgroups H of G containing ker f and the set of all subgroups H' of G'.

PROOF By (2.9) if H is a subgroup of G then $H' = f(H)$ is a subgroup of G'. Suppose that H' is a subgroup of G'. Then, again by (2.9), $H = f^{-1}(H')$ is a subgroup of G. Since $e' \in H'$,

$$\ker f = f^{-1}(\{e'\}) \subseteq f^{-1}(H') = H$$

It remains only to show that the maps given above are inverse maps. We must establish two facts.

(1) Let H be a subgroup of G containing ker f. We must show that $(f^{-1} \circ f)(H) = H$. By set theory $H \subseteq f^{-1}(f(H))$. We must then verify the opposite inclusion. Let $x \in f^{-1}(f(H))$. Then by definition, $f(x) \in f(H)$; hence there exists a $y \in H$ such that $f(x) = f(y)$. But then $f(xy^{-1}) = f(x)f(y)^{-1} = e'$ so that $xy^{-1} \in \ker f$. By assumption, $\ker f \subseteq H$. Now, since y and $xy^{-1} \in H$ and H is a subgroup of G, $x = xy^{-1}y \in H$. Hence equality is established.

(2) Let H' be a subgroup of G'. We must show that $(f \circ f^{-1})(H') = H'$. We leave it to the reader to verify that this equality follows merely by set theory, the inclusion $H' \subseteq (f \circ f^{-1})(H')$ depending on the surjectivity of f.

Remark. The student should verify that if $H = \langle x \rangle$ is a cyclic subgroup of G, then $f(H) = \langle f(x) \rangle$ and hence f(H) is also cyclic.

If $f: G \to G'$ is an injective homomorphism of groups then we say that f is an <u>embedding</u> of the group G into the group G'. In some cases we will identify each $x \in G$ with its image f(x) in G', thereby identifying G with its image f(G) in G'. For example, we identify **R** with the subgroup of **R**[x] consisting of all constant polynomial functions.

(2.11) EXAMPLES

(2.11.1) If $n \in \mathbf{N}$ then the canonical map $f: \mathbf{Z} \to \mathbf{Z}_n$ is a surjective homomorphism of groups with ker $f = n\mathbf{Z}$. According to the Correspondence Theorem, f induces a bijective correspondence between the subgroups of $\mathbf{Z}$ containing $n\mathbf{Z}$ and the subgroups of $\mathbf{Z}_n$. Note that if $n, m \in \mathbf{Z}$, then $n\mathbf{Z} \subseteq m\mathbf{Z}$ if and only if $m|n$. Thus, since $\mathbf{Z}$ is cyclic, the indicated correspondence is given by

$$m\mathbf{Z} = \langle m \rangle \mapsto \langle f(m) \rangle = \langle [m] \rangle \quad \text{where } m|n$$

In particular, if n is prime, $\mathbf{Z}_n$ has no nontrivial proper subgroups.

The correspondence between the subgroups of $\mathbf{Z}$ containing $10\mathbf{Z}$ and the subgroups of $\mathbf{Z}_{10}$ is illustrated below.

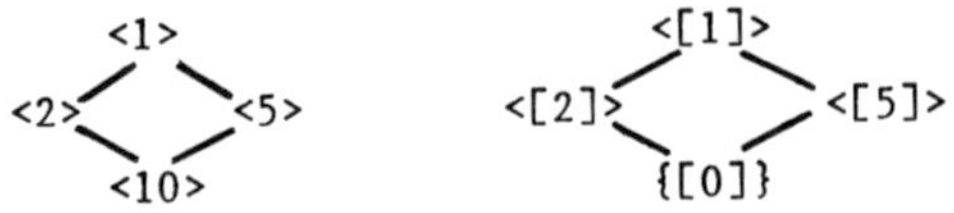

This discussion shows, by the way, that our list of subgroups of $\mathbf{Z}_{10}$ in (1.20.1) is complete.

(2.11.2) Let $f: \mathbf{R}^* \to \mathbf{R}^*$ be the absolute value map, $f(x) = |x|$, restricted to $\mathbf{R}^*$. Then, since $|xy| = |x||y|$ for all $x, y \in \mathbf{R}^*$, $f \in \text{End}(\mathbf{R}^*)$. Clearly ker $f = \{-1, 1\}$ and im $f = \mathbf{R}^+$ where $\mathbf{R}^+$ denotes the set of positive reals. Note that ker f and im f are indeed subgroups of $\mathbf{R}^*$.

The absolute value map extends to the complex numbers by

$$|x + iy| = \sqrt{x^2 + y^2} \quad (x, y \in \mathbf{R})$$

One thus obtains a map $T: \mathbf{C}^* \to \mathbf{R}^*$ with $T(x + iy) = |x + iy|$. It is easily verified that $T \in \text{Hom}(\mathbf{C}^*, \mathbf{R}^*)$. By definition,

$$\ker T = \{x + iy: x, y \in \mathbf{R} \text{ and } x^2 + y^2 = 1\}$$

a subgroup of $\mathbf{C}^*$ usually denoted S^1. S^1 is called the <u>circle group</u> since it consists of all points in the complex plane of distance 1 from the origin.

(2.11.3) The (natural) exponential map,

$$\mathbf{R} \overset{\exp}{\to} \mathbf{R}^+ \quad \text{by} \quad x \mapsto e^x$$

is an isomorphism of groups from $\langle \mathbf{R}, + \rangle$ to $\langle \mathbf{R}^+, \bullet \rangle$. The inverse of exp is the natural logarithm, $\log \in \mathrm{Hom}(\mathbf{R}^+, \mathbf{R})$.

(2.11.4) For those familiar with the complex exponential, we may now consider the map $f: \mathbf{R} \to \mathbf{C}^*$, defined by

$$f(x) = e^{2\pi i x}$$

which is again a homomorphism of groups. Recalling the identity

$$e^{2\pi i x} = \cos(2\pi x) + i \sin(2\pi x)$$

we see that im $f = S^1$, the circle group, and ker $f = \mathbf{Z}$.

(2.11.5) C_n, the cyclic group of order n, is isomorphic to $\mathbf{Z}_n$ via the map $f: C_n \to \mathbf{Z}_n$ defined by

$$f(s^m) = [m]$$

Let $0 \le k, m < n$. We consider two cases:

(i) If $k + m < n$, then

$$\begin{aligned} f(s^k s^m) &= f(s^{k+m}) \\ &= [k + m] \\ &= [k] + [m] \\ &= f(s^k) + f(s^m) \end{aligned}$$

(ii) If $k + m \ge n$, then

$$\begin{aligned} f(s^k s^m) &= f(s^{k+m-n}) \\ &= [k] + [m] + [-n] \\ &= [k] + [m] \\ &= f(s^k) + f(s^m) \end{aligned}$$

(2.11.6) Differentiation, $D: C^1(\mathbf{R}) \to C^0(\mathbf{R})$, is a homomorphism of groups. This assertion amounts to the well-known formula

$$D(f + g) = Df + Dg$$

According to the Fundamental Theorem of Calculus, D is surjective; for given $f \in \mathbf{C}^0(\mathbf{R})$,

$$f(x) = D\left(\int_0^x f(t)dt\right)$$

Furthermore, it is a well-known consequence of the Mean Value Theorem that

$$\ker(D) = \mathbf{R}$$

where we identify $\mathbf{R}$ with the constant functions from $\mathbf{R}$ to $\mathbf{R}$ as above.

(2.11.7) The map det: $GL_2(\mathbf{R}) \to \mathbf{R}^*$, introduced in (1.6.9), is a surjective homomorphism of groups. By definition,

$$\ker(\det) = \{A \in GL_2(\mathbf{R}) : \det(A) = 1\}$$

This important subgroup of $GL_2(\mathbf{R})$ is denoted $SL_2(\mathbf{R})$, the special linear group of rank 2 matrices over R.

(2.11.8) The Sign Homomorphism. We introduce the <u>sign homomorphism</u>, sgn: $S_n \to \{\pm 1\}$, on S_n. Define the sign of a k-cycle in S_n to be $(-1)^{k+1}$. Recall that every element of S_n may be factored uniquely into the product of disjoint cycles. Now define the sign, sgn(f), of an arbitrary permutation f in S_n to be the product of the signs of the cycles appearing in this decomposition. The reader may calculate directly that sgn(ft) = sgn(f)sgn(t) for any transposition t; hence sgn is a homomorphism since every permutation is the product of transpositions.

The existence of the sign homomorphism implies that if there is a factorization of f in S_n into an even (odd) number of transpositions, then every factorization of f into transpositions likewise involves an even (odd) number of factors. The kernel of sgn: $S_n \to \{\pm 1\}$ is thus a subgroup of S_n consisting of the so-called even permutations. This subgroup is denoted A_n and is called the <u>alternating group</u>.

(2.12) SOME GENERAL EXAMPLES

(2.12.1) <u>Inclusion of Subgroups</u> Let H be a subgroup of G. Then the inclusion map $H \to G$ which sends every element of H to itself is a homomorphism of groups.

(2.12.2) The Universal Property of Z Let x be an arbitrary element of an arbitrary group G. Define a map $f: \mathbf{Z} \to G$ by $f(n) = x^n$. Then f is the unique homomorphism from $\mathbf{Z}$ to G which sends 1 onto x.

(2.12.3) Projections and Injections Let $G \times G'$ be a direct product of groups. Recall that the projection maps, $p_1: G \times G' \to G$ and $p_2: G \times G' \to G'$, are defined by $p_1(x, x') = x$ and $p_2(x, x') = x'$. Then p_1 and p_2 are surjective homomorphisms of groups (called projection onto the first and second factor, respectively).

We can also define injection mappings $i_1: G \to G \times G'$ and $i_2: G' \to G \times G'$ by $i_1(x) = (x, e')$ and $i_2(x') = (e, x')$. Then i_1 and i_2 are injective homomorphisms of groups (verify that they have trivial kernels.)

(2.12.4) The Diagonal Map Let G be a group. Then the map $d: G \to G \times G$ defined by $d(x) = (x, x)$ is an injective homomorphism of groups with $\text{im}(d) = D$, the diagonal subgroup of $G \times G$. Hence $G \cong D$. The map d is called the diagonal map.

(2.12.5) Cayley's Theorem Let G be a group and $x \in G$. By (1.2) the map $\iota_x: G \to G$ defined by $\iota_x(y) = xy$ is an element of B(G). We may therefore define a map $L: G \to B(G)$ by $L(x) = \iota_x$. By (1.2), (ii), L is a homomorphism of groups, for if $x,y \in G$,

$$L(xy) = \iota_{xy} = \iota_x \circ \iota_y = L(x) \circ L(y)$$

If $x \in \ker L$, then $\iota_x = 1_G$; hence, in particular, $x = xe = \iota_x(e) = e$. It follows by (2.8) that L is injective and thus $G \cong \text{im}(L)$. Consequently, every group G is isomorphic to a subgroup of the symmetric group B(G). This simple but striking and important result is known as Cayley's Theorem.

(2.12.6) Conjugation; Inner Automorphisms Let G be a group and $x \in G$. Define a map $\alpha_x: G \to G$ by $\alpha_x(y) = xyx^{-1}$ ($y \in G$) (the map α_x

is called conjugation by x). The following facts may be verified.

(i) $\alpha_{xy} = \alpha_x \circ \alpha_y$ for all $x, y \in G$.

(ii) $\alpha_x \in \mathrm{Aut}(G)$ for all $x \in G$.

(iii) The map α: $G \to \mathrm{Aut}(G)$ defined by $\alpha(x) = \alpha_x$ is a homomorphism of groups with kernel $Z(G)$, the center of G.

The image of α is a subgroup of Aut(G) denoted Inn(G). Elements of Inn(G) - that is, automorphisms arising from conjugation - are called inner automorphisms.

EXERCISES

(2.1) Find a nontrivial homomorphism (or prove that there are none)

(a) From C_2 to C_4.

(b) From C_4 to C_2.

(c) From C_2 to C_3.

(d) From C_4 to S_3.

(e) From S_3 to C_4.

(f) From C_4 to D_4.

(g) From D_4 to C_4.

(2.2) Show that $C_6 \cong C_2 \times C_3$.

(2.3) (a) Let k be an even positive integer. Show that there is a nontrivial homomorphism from C_2 to C_k.

(b) Show that $C_2 \cong \{1, -1\}$.

(c) Let $n > 1$. Show that there exists a nontrivial homomorphism from S_n to C_k (k even).

(2.4) (a) Show that the second derivative map

$$D^2: C^2(\mathbf{R}) \to C^0(\mathbf{R})$$

is a homomorphism of groups. What is ker (D^2)?

(b) Deduce that more generally the n-th derivative map

$$D^n: C^n(\mathbf{R}) \to C^0(\mathbf{R})$$

is a homomorphism of groups. What is $\ker(D^n)$?

(2.5) Let $a \in \mathbf{R}$. Show that the map

$$\iota_a: C^0(\mathbf{R}) \to C^0(\mathbf{R}) \quad \text{by} \quad f \mapsto af$$

is an automorphism of $C^0(\mathbf{R})$.

(2.6) Let $n \in \mathbf{N}$, $a \in \mathbf{R}$. Show that the map $aD^n: C^n(\mathbf{R}) \to C^0(\mathbf{R})$ defined by $(aD^n)(f) = a(D^n f)$, is a homomorphism of groups. What is $\ker(aD^n)$ if $a \neq 0$?

(2.7) Define a map $L: C^n(\mathbf{R}) \to C^0(\mathbf{R})$ by

$$L(f) = a_n D^n f + a_{n-1} D^{n-1} f + \cdots + a_1 D^1 f + a_0 f$$

where $a_0, a_1, \ldots, a_n \in \mathbf{R}$. Show that L is a homomorphism of groups. This homomorphism is called the n-th degree differential operator. Describe $\ker(L)$ as a solution to a linear differential equation.

(2.8) (a) Let $n \in \mathbf{Z}$. Show that the map $f_n: \mathbf{Z} \to \mathbf{Z}$ defined by $f_n(m) = nm$ is an endomorphism. For which n is f_n injective? Surjective? An isomorphism?

(b) Conclude that a group can be isomorphic to a proper subgroup. Can this happen for a finite group?

(c) Show that every endomorphism of $\mathbf{Z}$ arises as a map f_n for some $n \in \mathbf{Z}$.

(2.9) Generalize the preceding exercise [part (c)] to $\mathbf{Q}$.

(2.10) Find an injective homomorphism from the group $\mathbf{Z}$ into the circle group S^1.

(2.11) Define a subset G of $M_2(\mathbf{R})$ by

$$G = \left\{ \begin{pmatrix} a & b \\ -b & a \end{pmatrix} : a, b \in \mathbf{R} \right\}$$

Let G^* denote the nonzero elements of G.

(a) Show that $\langle G, + \rangle$ is a group and $G \cong \mathbf{C}$.

(b) Show that $\langle G^*, \cdot \rangle$ is a group and $G^* \cong \mathbf{C}^*$.

(2.12) Show that $\mathbf{C} \cong \mathbf{R}^+ \times S^1$. Here, $\mathbf{R}^+$ is the group of positive real numbers under multiplication.

(2.13) Let G be a group of order 6. Show that either $G \cong C_6$ or $G \cong S_3$.

(2.14) Let $n > 1$ be an integer. Show that there exist at least 2 nonisomorphic abelian groups of order n^2.

(2.15) Interpret Proposition (2.2) when $f: G \to H$ is in fact

$$\det: GL_2(\mathbf{R}) \to \mathbf{R}^*$$

Verify assertions (i) and (ii) directly.

(2.16) Apply the Correspondence Theorem to the surjective homomorphism det to exhibit a proper subgroup H of $GL_2(\mathbf{R})$ which properly contains $SL_2(\mathbf{R})$.

(2.17) Show that if G is a nontrivial group then End(G) is not a group with respect to composition.

(2.18) Let $f \in \mathrm{Hom}(G, H)$, where G and H are groups.

(a) Show that for all $x \in G$, $o(f(x))$ divides $o(x)$.

(b) Show that $o(f(x)) = o(x)$ if and only if f is injective.

(2.19) (a) Let C_n be the cyclic group of order n. Show that the order of every element of C_n divides n.

(b) Conclude that if $(m, n) = 1$, then $\mathrm{Hom}(C_m, C_n)$ consists of only the trivial map.

(c) Prove that conversely, if $(m, n) > 1$, then there is a nontrivial homomorphism from C_m to C_n.

(2.20) Let n be a nonnegative integer and let G be an abelian group. Define a map $E_n: G \to G$ by $E_n(x) = x^n$.

(a) Show that $E_n \in \mathrm{End}(G)$.

(b) Let $T_n(G) = \{x: x \in G, x^n = e\}$. Show that $T_n(G) = \ker E_n$. Conclude that $T_n(G)$ is a subgroup of G.

(2.21) Let G and H be groups, and suppose that G is generated by some subset S. Show that two homomorphisms f, g: G → H are equal if and only if they agree on S; that is, f(s) = g(s) for every s ∈ S.

(2.22) Let f: G → H be a group homomorphism.

(a) Show by example that the inclusion $f(Z(G)) \subseteq Z(H)$ fails in general.

(b) Show that this inclusion obtains if f is surjective.

(c) Show that $f(Z(G)) = Z(H)$ if f is an isomorphism.

(2.23) Let p_1, p_2 be the projection maps from G × H to G, H respectively. Find ker p_1 and ker p_2.

(2.24) Let G, H, K be groups and let f: K → G, g: K → H be group homomorphisms. Let h: K → G × H be the unique map which makes the following diagram commutative [cf. I, exercise (1.19)].

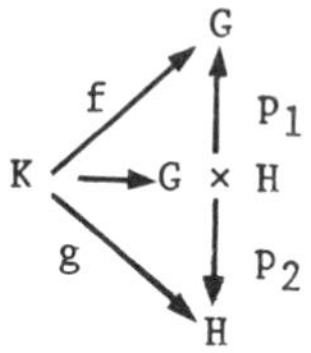

Show that h is a group homomorphism.

(2.25) Let T be a set and let S be a subset of T. Define a subset H of B(T) by

$$H = \{f: f \in B(T),\ f(s) = s \text{ for all } s \in S\}$$

Show that H is a subgroup of B(T) which is isomorphic to B(T - S).

(2.26) Let f: G → G be an automorphism of the finite group G. Suppose that

(i) f(x) = x if and only if x = e;

(ii) $f \circ f = 1_G$.

Prove that G is abelian. Hint: show that the map which sends x to $xf(x^{-1})$ is injective.

(2.27) Let G be a group, and let C denote the collection of all nontrivial subgroups of G. Suppose that the intersection of C is not {e}. Show that for every x in G, o(x) is finite.

(2.28) Show that every finite group G or order greater than two admits an automorphism different than the identity.

3. LAGRANGE'S THEOREM AND THE QUOTIENT SET

Let G be a group with subgroup H. In this section we define an equivalence relation on G relative to H. The resulting quotient set of equivalence classes is then denoted G/H. In Section 4 we will determine conditions on H which guarantee that the given equivalence relation is in fact a congruence relation with respect to the group operation on G; in this case H will be called a normal subgroup of G and the operation induced on the factor set G/H gives a group structure to G/H. We then speak of the quotient group G/H. In many cases (especially in inductive arguments), the quotient G/H is far more tractable than the original group G.

(3.1) DEFINITION Let H be a subgroup of G. Given $x, y \in G$, we write

$$x \equiv y \pmod{H}$$

if $xy^{-1} \in H$. In this case we say x is congruent to y modulo H.

(3.2) PROPOSITION Let H be a subgroup of G. Then congruence modulo H is an equivalence relation on G.

PROOF We check the three defining properties of an equivalence relation.

(i) Reflexive: Let $x \in G$. Then $xx^{-1} = e \in H$; hence $x \equiv x \pmod{H}$.

(ii) Symmetric: Suppose $x, y \in G$ and $x \equiv y \pmod{H}$. Then $xy^{-1} \in H$, and since subgroups are closed with respect to inverses, $(xy^{-1})^{-1} \in H$; hence $yx^{-1} \in H$ and $y \equiv x \pmod{H}$.

(iii) Transitive: Suppose $x, y, z \in G$ with $x \equiv y \pmod{H}$ and $y \equiv z \pmod{H}$. Then $xy^{-1} \in H$ and $yz^{-1} \in H$. Since H is closed under products, $xy^{-1}yz^{-1} \in H$; that is, $xz^{-1} \in H$. Hence $x \equiv z \pmod H$.

As in Chapter I we write [x] for the equivalence class of x modulo H. We recall that the distinct equivalence classes partition the set G.

(3.3) NOTATION AND TERMINOLOGY Let S and T be subsets of a group G. We define the subset ST of G by

$$ST = \{x \in G: x = st \text{ for some } s \in S,\ t \in T\}$$

In case $S = \{s\}$, we abbreviate $\{s\}T$ to sT, and similarly we write St for $S\{t\}$.

Now let H be a subgroup of G. The reader should verify the following facts:

(i) HH = H;

(ii) Hx = H if and only if $x \in H$;

(iii) He = H.

(3.4) PROPOSITION Let H be a subgroup of G and congruence modulo H be defined as above. Then, if $x \in G$,

$$[x] = Hx = \{z \in G: z = yx \text{ for some } y \in H\}$$

[The set Hx is called the right coset of H (in G) represented by x.]

PROOF Let $z \in G$. Then $z \equiv x \pmod H$ if and only if $zx^{-1} = y$ for some $y \in H$; that is, if and only if $z = yx$ for some $y \in H$. Hence $[x] = Hx$.

REMARK If G is an additive abelian group with subgroup H, then congruence modulo H is defined by

$$x \equiv y \pmod H \text{ if } x - y \in H$$

and, in this case,

$$[x] = H + x = \{z \in G: z = y + x \text{ for some } y \in H\}$$

Observe that if $G = \mathbf{Z}$ and $H = n\mathbf{Z}$, then congruence modulo H is precisely congruence modulo n as defined in the first chapter.

We recall the following facts concerning equivalence relations from Chapter I.

(i) $[x] = [y]$ if and only if $x \equiv y \pmod{H}$.

(ii) $[x] \cap [y] = \emptyset$ if $x \not\equiv y \pmod{H}$.

(iii) G is the disjoint union of the distinct equivalence classes modulo H.

(3.5) PROPOSITION Let H be a subgroup of G and $x \in H$. Then card (Hx) = card (H); hence all cosets of H in G have the same cardinality.

PROOF We recall from (1.2) that the map $r_x: G \to G$ defined by $r_x(y) = yx$ is bijective. Hence r_x restricted to H is injective. Since the image of this map is the right coset Hx the result follows.

(3.6) The Quotient Set. Henceforth we denote the quotient set of G by the equivalence relation congruence modulo H as G/H. Hence

$$G/H = \{[x]: x \in G\} = \{Hx: x \in G\}$$

A complete residue system is in this case called a complete system of right coset representatives of H in G. Thus such a system of representatives consists of precisely one element from each distinct right coset.

Let us now define (G: H), the index of H in G, by

$$(G: H) = \text{Card } (G/H)$$

Hence (G: H) is the number of distinct right cosets of H in G. In exercise (3.19) we show that (G: H) is also the number of distinct left cosets of H in G. Hence index is independent of parity. If (G: H) is finite, we say H is of finite index in G.

(3.7) THEOREM (Lagrange) Let H be a subgroup of the finite group G. Then

$$o(G) = (G: H)\, o(H)$$

In particular, $o(H) | o(G)$.

PROOF By I, (2.8), G is the disjoint union of the distinct equivalence classes modulo H; that is, the disjoint union of the distinct right cosets modulo H. By (3.5) each of these cosets has the same number of elements, namely Card (H). Since (G: H) is the number of distinct right cosets, the result now follows.

(3.8) COROLLARY Let $K \subseteq H \subseteq G$ be a chain of subgroups of the finite group G. Then

$$(G: K) = (G: H)(H: K)$$

PROOF By Lagrange's Theorem we have

(1) $o(G) = (G: H)\, o(H)$,

(2) $o(H) = (H: K)\, o(K)$,

(3) $o(G) = (G: K)\, o(K)$.

Substituting (2) into (1) yields

(4) $o(G) = (G: H)\,(H: K)\, o(K)$;

then comparing (3) and (4) and cancelling $o(K)$ yields the desired result.

(3.9) COROLLARY If G is a finite group and $x \in G$, then $o(x) | o(G)$.

PROOF By (1.16), $o(x) = o(\langle x \rangle)$ and the result now follows by Lagrange's Theorem.

(3.10) EXAMPLE Let us prove that the multiplicative group $\mathbf{Z}_{11}^{\times}$ is cyclic. Since $o(\mathbf{Z}_{11}^{\times}) = 10$, by Lagrange's Theorem every element has order 1, 2, 5 or 10. Computing the first five successive powers of [2] we get [2], [4], [8], [5], [10]. Hence, since we have not yet reached [1], [2] does not have order 1, 2 or 5. We conclude therefore that [2] has order 10 and hence [2] is a generator for the group $\mathbf{Z}_{11}^{\times}$.

(3.11) COROLLARY Let G be a finite group of order n. Then $x^n = e$ for all $x \in G$.

PROOF Let $o(x) = m$. By (3.9), $m|n$; hence $n = mk$ for some $k \in \mathbf{N}$. Then

$$x^n = x^{mk} = (x^m)^k = e^k = e$$

The next two corollaries should be familiar to the student of number theory.

(3.12) COROLLARY (Euler) Let n be a positive integer and let

$$\phi(n) = \text{Card } \{m: 0 < m < n \text{ and } (m, n) = 1\}$$

Then for all m such that $(m, n) = 1$, we have

$$m^{\phi(n)} \equiv 1 \pmod{n}$$

PROOF By Example (1.6.4), $\mathbf{Z}_n^\times$ forms a group under multiplication of order $\phi(n)$. Hence if m is relatively prime to n, $[m]^{\phi(n)} = [1]$ by (3.9). The result now follows.

(3.13) COROLLARY Let $p > 0$ be prime. Then for all $m \in \mathbf{Z}$, $m^p \equiv m \pmod{p}$.

PROOF If $p|m$, then $m \equiv 0 \pmod{p}$ whence $m^p \equiv 0 \equiv m \pmod{p}$. If $p \nmid m$, then $(m, p) = 1$ and (3.12) applies. Since $\phi(p) = p - 1$, this gives $m^{p-1} \equiv 1 \pmod{p}$; multiplying both sides of this congruence by m, we obtain the desired result.

The corollary above is sometimes called Fermat's Little Theorem.

(3.14) COROLLARY Let G be a finite group of prime order. Then G is cyclic, hence abelian. In fact, G is generated by any element which is not the identity.

PROOF Let $o(G) = p$ where p is a prime. Let $x \in G$ with $x \neq e$. Then $o(x)$ divides p and is not equal to 1; hence $o(x) = p$. Thus $\langle x \rangle = G$ as claimed.

If $f \in \text{Hom}(G, G')$ and $K = \ker f$ then congruence modulo K takes on a special significance.

(3.15) PROPOSITION Let G and G' be groups, $f \in \mathrm{Hom}(G, G')$, and $K = \ker f$. Then the following assertions hold.

(i) $Kx = Ky$ (or, equivalently, $[x] = [y]$) if and only if $f(x) = f(y)$ for $x,y \in G$.

(ii) If $x \in G$ and $f(x) = x'$, then $f^{-1}(x') = Kx$.

PROOF (i) By (3.4), $Kx = Ky$ if and only if $xy^{-1} \in K$; that is, if and only if $f(xy^{-1}) = e'$. But, since f is a homomorphism, $f(xy^{-1}) = f(x)f(y)^{-1}$. Thus $Kx = Ky$ if and only if $f(x)f(y)^{-1} = e'$, or, equivalently, if and only if $f(x) = f(y)$.

(ii) By definition, $y \in f^{-1}(x')$ if and only if $f(y) = f(x)$. By (i), this occurs if and only if $Ky = Kx$; that is, if and only if $y \in Kx$.

REMARK Part (ii) of the preceding proposition may be reformulated as follows. The general solution to the equation $f(x) = x'$ is given by Kx where x is any particular solution and K is the general solution to $f(x) = e$. Cast thus, the statement should have a familiar ring to the student of differential equations or linear algebra [see (3.16.2) below].

(3.16) EXAMPLES

(3.16.1) Let n be a positive integer. As mentioned previously, congruence modulo the subgroup $n\mathbf{Z}$ is exactly the usual equivalence relation of congruence modulo n. It follows that the equivalence classes mod n are precisely the (left or right) cosets of $n\mathbf{Z}$ in $\mathbf{Z}$. For example, if $n = 5$,

$$[4] = 5\mathbf{Z} + 4 = \{\ldots, -1, 4, 9, 13, \ldots\}$$

We observe that

$$(\mathbf{Z}: n\mathbf{Z}) = o(\mathbf{Z}_n) = n \text{ for } n > 0$$

(3.16.2) The mapping $L: C^2(\mathbf{R}) \to C^0(\mathbf{R})$ defined by

$$L(f) = D^2f - 5Df + 4f$$

is an example of a so-called linear differential operator. The reader may check that L is a homomorphism of groups. By definition, ker L is exactly the solution set to the ordinary differential equation

$$D^2f - 5Df + 4f = 0$$

One knows from the elementary theory of such equations that in fact

$$\ker L = \{c_1e^x + c_2e^{4x}: c_1, c_2 \in \mathbf{R}\}$$

Now one can show directly that $L(x^2) = 4x^2 - 10x + 2$; that is, x^2 is a particular solution to the equation

$$(*) \quad D^2f - 5Df + 4f = 4x^2 - 10x + 2$$

Hence, by (3.15), the complete inverse image of $4x^2 - 10x + 2$ under L [that is, the general solution to (*)] is given by

$$x^2 + \ker L = x^2 + c_1e^x + c_2e^{4x}$$

where c_1, c_2 are arbitrary constants.

(3.15.3) Let $G = D_4$, the dihedral group of order 8 constructed in (1.6.8). By Lagrange's Theorem, every proper subgroup of G has order 2 or 4. Let

$$K = \langle s \rangle = \{e, s, s^2, s^3\}$$
$$H = \langle t \rangle = \{e, t\}$$

Since $(G: H) = o(G)/o(H) = 4$, H has exactly four distinct right cosets in G. We calculate:

$$He = \{e, t\}$$
$$Hs = \{s, ts\} = \{s, s^3t\}$$
$$Hs^2 = \{s^2, ts^2\} = \{s^2, s^2t\}$$
$$Hst = \{st, tst\} = \{st, s^3\}$$

Observe the following facts:

(i) Each coset has two elements (the number of elements in H).

(ii) The distinct cosets partition G.

(iii) Since, for example, $s^3t \in Hs$, $Hs^3t = Hs$.

We now calculate the distinct left cosets of H in G.

$$eH = H = \{e, t\}$$
$$sH = \{s, st\}$$
$$s^2H = \{s^2, s^2t\}$$
$$stH = \{st, s\}$$

Observe that the left coset sH does not equal the right coset Hs.

The reader should calculate the left and right cosets of K in G. Since $o(K) = 4$, $(G: K) = 2$. Hence there should be exactly two right (or left) cosets of K in G, each containing four elements. Observe that for all $u \in G$, $Ku = uK$. As we shall see shortly, K is a special type of subgroup of G called a normal subgroup of G.

(3.15.4) This example is classical. Recall from (2.11.6) that $D: C^1(\mathbf{R}) \to C^0(\mathbf{R})$ is a surjective homomorphism with kernel $\mathbf{R}$. We illustrate (3.15).

(i) Let F(x) and $G(x) \in C^1(\mathbf{R})$. Then the two cosets $\mathbf{R} + F(x)$ and $\mathbf{R} + G(x)$ are equal if and only if DF = DG. By definition,

$$\mathbf{R} + F(x) = \mathbf{R} + G(x) \text{ if and only if } F(x) - G(x) \in \mathbf{R}$$

Thus, phrased in language more familiar to the calculus student, DF = DG if and only if the functions F and G differ by a constant.

(ii) Let $f(x) \in C^0(\mathbf{R})$, and let $F(x) \in C^1(\mathbf{R})$ be a particular preimage of f(x) under D; that is, $DF = f(x)$. Then by (3.15), the entire inverse image of f(x) under D is just the coset $\mathbf{R} + F(x)$. This fact is usually expressed as follows:

$$\int f(x)\,dx = F(x) + c$$

where $DF = f(x)$ and c is an arbitrary, unassigned constant - the so-called constant of integration. In other words, an indefinite integral is precisely a coset of R in the group $C^1(\mathbf{R})$.

(3.15.5) Recall that the (multiplicative) group $\mathbf{Z}^\times = \{-1, 1\}$ is a subgroup of the group $\mathbf{R}^*$. If $x,y \in \mathbf{R}^*$, then $x \equiv y \pmod{\mathbf{Z}^\times}$ if and only if $xy^{-1} \in \mathbf{Z}^\times$; that is, if and only if $x = y$ or $x = -y$. Thus for each $x \in \mathbf{R}^*$, $[x] = \{x, -x\}$.

(3.15.6) We recall from (2.11.2) that the absolute value map from $\mathbf{C}^*$ to $\mathbf{R}^*$ defined by

$$|x + iy| = \sqrt{x^2 + y^2}$$

is a surjective homomorphism of groups with kernel S^1 (the circle group). Hence, if $z = x + iy \in \mathbf{C}^*$, the right coset is the set

$$S^1 z = [z] = \{w \in \mathbf{C}^*: |w| = |z|\}$$

Pictured in the complex plane, $S^1 z$ is then the circle centered at the origin with radius $r = |z|$. Observe that the distinct cosets partition the complex plane.

EXERCISES

(3.1) Let G, H be as specified below. List all right cosets of H in G.

(a) $G = C_{10}$, $H = \{e, s^5\}$.

(b) $G = S_3$, $H = \{e, (1, 2, 3), (1, 3, 2)\}$.

(c) $G = C_{12}$, $H = \{e, s^3, e^6, e^9\}$.

(d) $G = A_4$, $H = \{e, (1, 2)(3, 4), (1, 3)(2, 4), (1, 4)(2, 3)\}$.

(e) $G = D_4$, $H = \{e, s, s^2, s^3\}$.

(3.2) Let $H = \{1, -1, i, -i\}$ be regarded as a subgroup of S^1 and let $z = e^{is}$. Describe the coset Hz as a subset of the complex plane.

(3.3) Let H be a subgroup of S_1 and let $z = e^{is}$. Describe Hz as a subset of the complex plane.

(3.4) Let $H = \{(t, 2t): t \in \mathbf{R}\}$ be regarded as a subgroup of $\mathbf{R}^2$. Describe the right coset of H in $\mathbf{R}^2$ represented by (a, b).

(3.5) Describe the cosets of S^1 in the group $\mathbf{C}^*$ as subsets of the complex plane.

(3.6) Let $H = \{z: z = 2^t e^{2t\pi i}, t \in \mathbf{R}\}$.

(a) Show that H is a subgroup of $\mathbf{C}^*$.

(b) If $z = a + bi$, describe the coset Hz as a subset of the complex plane.

(c) Show that $\{t: 1 \le t < 2\}$ is a complete system of coset representatives for H in $\mathbf{C}^*$.

(3.7) Consider the sets $S = \{s, t\}$ and $T = \{st, s^2\}$ as subsets of D_4. Find TS and ST. Conclude that $\text{Card}(TS) \ne \text{Card}(ST)$.

(3.8) Consider the subgroups $H = \{e, st\}$ and $K = \{e, st, s^2, s^3t\}$ of the dihedral group D_4.

(a) List all left and right cosets of H in D_4.

(b) Find all x in D_4 such that $xH \ne Hx$.

(c) List all left and right cosets of K in D_4.

(d) Conclude that, if $x \in D_4$, then $xK = Kx$.

(e) List all left and right cosets of H in K.

(f) Conclude that, if $x \in K$, then $xH = Hx$.

(3.9) Let H be a subgroup of a group G such that $(G: H) = 2$. Show that for every $x \in G$

(i) either $xH = H$ or $xH = G - H$;

(ii) either $Hx = H$ or $Hx = G - H$.

Find all $x \in G$ such that $xH = H = Hx$. Show that for every $x \in G$, $xH = Hx$.

(3.10) If H, K are subgroups of a group G such that $HK = KH$, then show that HK is a subgroup of G.

(3.11) Let m,n be elements of the group $\mathbf{Z}$. Show that

(a) $\langle m, n\rangle = m\mathbf{Z} + n\mathbf{Z}$;

(b) $m\mathbf{Z} + n\mathbf{Z} = k\mathbf{Z}$ where $k = (m, n)$.

(3.12) Let H be a subgroup of G and $\equiv$ (mod H) the equivalence relation defined by

$$x \equiv y \pmod{H} \quad \text{if } xy^{-1} \in H$$

(a) Let $G = \mathbf{R}^*$, $H = \{1, -1\}$. Show that $[s] = [t]$ if and only if $|s| = |t|$.

(b) Let $G = SL_2(\mathbf{R})$, $H = GL_2(\mathbf{R})$. Show that $[s] = [t]$ if and only if $\det s = \det t$.

(3.13) Find all subgroups of A_4. Conclude that while the order of A_4 is 12, A_4 has no subgroups of order 6. In particular, the converse of Lagrange's theorem is false.

(3.14) Find the order of all the elements of A_4. Conclude that A_4 has no element of order 4.

(3.15) Let $G = \{f: f \in B(\mathbf{R}), f(x) = ax + b \text{ with } a,b \in \mathbf{R}, a \neq 0\}$.

(a) Show that G is a subgroup of $B(\mathbf{R})$.

(b) Note that $H = \{f: f \in G, f(x) = x + b, b \in \mathbf{R}\}$ is a subgroup of G. Describe geometrically the cosets of H in G.

(c) Note that $K = \{f: f \in G, f(x) = ax, a \neq 0\}$ is a subgroup of G. Describe geometrically the cosets of K in G.

(d) Is there an $f \in G$ such that $fH \neq Hf$?

(e) Is there an $f \in G$ such that $fK \neq Kf$?

(3.16) (a) Show that $L: C^2(\mathbf{R}) \to C^0(\mathbf{R})$ defined by

$$L(f) = D^2f + Df - 2f$$

defines a homomorphism.

(b) Show that

$$\ker L = \{f: f(x) = ae^{-2x} + be^x, a,b \in \mathbf{R}\}$$

(c) Given that $f(x) = x^3 + x^2$ is a solution to

$$D^2f + Df - 2f = -2x^3 + x^2 + 8x + 2$$

find all solutions.

(3.17) Define $L: \mathbf{R}^3 \to \mathbf{R}^3$ by

$$L(x, y, z) = (x + 2y - z, 2x + y + z, x - y + 2z)$$

(a) Show that L is a homomorphism.

(b) Show that $\ker L = \{(t, -t, -t): t \in \mathbf{R}\}$.

(c) Given that (1, 2, 1) is a solution to the linear system of equations

$$x + 2y - z = 4$$
$$2x + y + z = 5$$
$$x - y + 2z = 1$$

find all solutions.

(3.18) Let H be a subgroup of G. Define a relation on G as follows

$$x \sim y \pmod{H} \text{ if } x^{-1}y \in H$$

Show that $\sim \pmod{H}$ is an equivalence relation on G with equivalence classes given by $[s] = sH$.

(3.19) Let H be a subgroup of a group G and let $H\backslash G$ denote the set of left cosets of H in G. Show that the bijection $I: G \to G$, defined by $I(x) = x^{-1}$, induces a bijection from G/H to $H\backslash G$.

(3.20) Let ϕ be defined as in (3.12). Find the following.

(a) $\phi(7)$.

(b) $\phi(16)$.

(c) $\phi(22)$.

(d) $\phi(60)$.

(3.21) Let p be a prime and k a positive integer. Show that

$$\phi(p^k) = p^k(1 - 1/p)$$

(3.22) Let n be a positive integer with distinct prime factors $p_1, p_2, \ldots, p_m$. Show that

$$\phi(n) = n(1 - 1/p_1)(1 - 1/p_2) \cdots (1 - 1/p_m)$$

Deduce that $\phi(mn) = \phi(m)\phi(n)$ for every $n, m \in \mathbf{N}$ with $(m, n) = 1$.

(3.23) Let G be a finite abelian group with order m and let n be a positive integer such that $(m, n) = 1$. Show that the mapping $f_n: G \to G$, defined by $f_n(x) = nx$, (additive notation), is a bijection.

(3.24) Suppose that n is a positive integer such that $p = 2^n + 1$ is a prime. Consider the group $Z_p^\times$ under multiplication.

(a) Show that $o([2]) = 2n$.

(b) Conclude that 2n divides $p - 1$.

(c) Show that $n = 2^k$ for some integer k.

4. NORMAL SUBGROUPS AND THE QUOTIENT GROUP

Let H be a subgroup of G. We now seek conditions on H which guarantee that the equivalence relation congruence modulo H is in fact a congruence relation on G. In this case the group operation on G induces a group operation on the quotient set G/H. The following two conditions are required:

(i) If $x,y \in G$ are such that $x \equiv y \pmod H$, then $xz \equiv yz \pmod H$ for all $z \in G$.

(ii) If $x,y \in G$ are such that $x \equiv y \pmod H$, then $zx \equiv zy \pmod H$ for all $z \in G$.

Condition (i) requires that whenever $x,y \in G$ are such that $xy^{-1} \in H$, then $xz(yz)^{-1} \in H$ for all $z \in G$. But $xz(yz)^{-1} = xzz^{-1}y^{-1} = xy^{-1}$; hence this condition is satisfied for all subgroups H of G.

Condition (ii) requires that whenever $x,y \in G$ are such that $xy^{-1} \in H$, then $zx(zy)^{-1} \in H$ for all $z \in G$. Since $zx(zy)^{-1} = zxy^{-1}z^{-1}$ and $xy^{-1} = h$ for some $h \in H$ by hypothesis, condition (ii) will be satisfied provided for all $h \in H$ and all $z \in G$, $zhz^{-1} \in H$. We are thus motivated to make the following definition.

(4.1) DEFINITION Let N be a subgroup of G. N is a normal subgroup of G if $xNx^{-1} \subseteq N$ for all $x \in G$; that is, whenever $x \in G$ and $n \in N$, $xnx^{-1} \in N$. If N is normal in G we write $N \lhd G$.

We note some general examples. Both {e} and G itself are always normal in G. If N is any subgroup contained in Z(G), the center of G, then N is normal in G. In particular, every subgroup of an abelian group is normal.

(4.2) THEOREM Let N be a normal subgroup of G.

(i) The binary operation of G induces a binary operation on the quotient set G/N defined by

(*) $[x][y] = [xy]$ for all $x,y \in G$

Under this operation G/N is a group with identity [e] and, for all $x \in G$, $[x]^{-1} = [x^{-1}]$.

(ii) The canonical map k_N: $G \to G/N$ defined by $k_N(x) = [x]$ is a surjective homomorphism of groups with ker $K_N = N$.

PROOF (i) We have seen that if N is a normal subgroup of G then the equivalence relation congruence mod N is in fact a congruence relation on the set G. Hence, by I, (3.8), the binary operation on G induces a binary operation on the quotient set G/H defined by $[x][y] = [xy]$. By I, (3.10), the induced operation is associative with identity [e]. If $x \in G$, then

$$[x][x^{-1}] = [xx^{-1}] = [e]$$

and, similarly $[x^{-1}][x] = [e]$. Hence [x] has inverse $[x^{-1}]$ in G/N.

(ii) Since N is a normal subgroup of G, $[x][y] = [xy]$ for all $x,y \in G$; hence k_N: $G \to G/N$ is a homomorphism. Clearly it is surjective. Finally, if $x \in G$, then $x \in \ker k_N$ if and only if $[x] = [e]$; that is, if and only if $xe^{-1} \in N$, or equivalently, $x \in N$.

Since $[x] = Nx$, the above theorem, written in coset notation states that if N is a normal subgroup of G, then the quotient set G/N is a group under the binary operation

(**) $NxNy = Nxy \quad (x,y \in G)$

The identity is the coset $Ne = N$; if $x \in G$, $(Nx)^{-1} = Nx^{-1}$, and, finally, the canonical map k_N: $G \to G/N$ is defined by $k_N(x) = Nx$.

We now list some equivalent conditions for a subgroup N to be normal in G. Observe, for example, that condition (iii) shows that in D_4 the subgroup $H = \langle t \rangle$ is not normal while the subgroup $K = \langle s \rangle$ is normal [cf. Example (1.6.8)].

(4.3) PROPOSITION Let N be a subgroup of G. Then the following conditions are equivalent.

(i) N is a normal subgroup of G.

(ii) $xNx^{-1} = N$ for all $x \in G$.

(iii) $Nx = xN$ for all $x \in G$.

(iv) The product of two right (respectively, left) cosets of N in G is again a right (respectively, left) coset.

(v) $(Nx)(Ny) = Nxy$ for all $x,y \in G$.

PROOF (i)$\Longrightarrow$(ii) Assume N is a normal subgroup of G and $x \in G$. By Definition (4.1) applied to the element $x^{-1} \in G$, $x^{-1}Nx \subseteq N$. We then have

$$N = x(x^{-1}Nx)x^{-1} \subseteq xNx^{-1} \subseteq N$$

the last inclusion holding since N is normal in G. Hence $N = xNx^{-1}$.

(ii)$\Longrightarrow$(iii) Suppose for each $x \in G$, $xNx^{-1} = N$. Then multiplying both sides on the right by x, this equality becomes $xN = Nx$ as claimed (this is really equality of sets - the reader should check that this works!).

(iii)$\Longrightarrow$(iv) Assume (iii) holds. Then, given $x,y \in G$, we compute $(Nx)(Ny) = N(xN)y = NNxy = Nxy$ (recall that $NN = N$ since N is a subgroup of G).

(iv)$\Longrightarrow$(v) Suppose that $NxNy = Nz$ for some $z \in G$. Then, since $x \in Nx$ and $y \in Ny$, $xy \in Nz$. Hence, since equivalence classes are either disjoint or equal, $[xy] = [z]$ or, using coset notation, $Nxy = Nz$.

(v)$\Longrightarrow$(i) Suppose that (v) holds. Then for each $x \in G$,

$$xNx^{-1} = e(xNx^{-1}) \subseteq NxNx^{-1} = Nxx^{-1} = N$$

thus N is a normal subgroup of G.

(4.4) PROPOSITION Let $f: G \to G'$ be a homomorphism of groups. Then the following assertions hold.

(i) If N' is a normal subgroup of G', then $N = f^{-1}(N')$ is a normal subgroup of G.

(ii) Ker f is a normal subgroup of G.

(iii) If f is surjective and N is a normal subgroup of G, then $N' = f(N)$ is a normal subgroup of G'.

PROOF (i) Suppose that N' is a normal subgroup of G' and $N = f^{-1}(N')$. Let $x \in G$ and $n \in N$. Then

$$f(xnx^{-1}) = f(x)f(n)[f(x)]^{-1} \in N'$$

since $f(n) \in N'$ and N' is a normal subgroup of G'. It therefore follows that $xnx^{-1} \in N$ and hence N is normal as claimed.

(ii) If e' is the identity of G', then {e'} is a normal subgroup of G' and therefore, by (i), ker $f = f^{-1}(\{e'\})$ is a normal subgroup of G.

(iii) Now suppose that f is surjective, N is a normal subgroup of G, and $N' = f(N)$. Let $x' \in G'$ and $n' \in N'$. Then $n' = f(n)$ for some $n \in N$ and, since f is surjective, $x' = f(x)$ for some $x \in G$. Now, since N is a normal subgroup of G, $xnx^{-1} \in N$; hence $x'n'(x')^{-1} = f(xnx^{-1}) \in N'$.

(4.5) THEOREM (Fundamental Theorem of Group Homomorphisms) Let $f: G \to G'$ be a surjective homomorphism of groups and let K = ker f. Then the following assertions hold.

(i) If we define $f': G/K \to G'$ by $f'([x]) = f(x)$, then f' is an isomorphism.

(ii) There is a one-to-one correspondence between subgroups H of G containing K and all subgroups H' of G', given by

$$H \to f(H) \quad \text{and} \quad H' \to f^{-1}(H')$$

(iii) Under the correspondence given in (ii), normal subgroups correspond to normal subgroups.

PROOF (i) By (4.4), (ii), K is a normal subgroup of G. We will use the Fundamental Factorization Theorem to show that f factors through the surjective map $k_K: G \to G/K$ [cf. I, (1.12) and (4.2), (ii)].

Consider the diagram below.

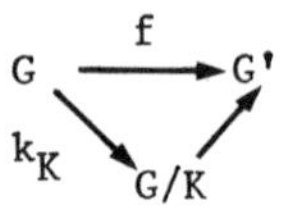

If $x,y \in G$ with $k_K(x) = k_K(y)$, then $[x] = [y]$ and hence, by (3.15), $f(x) = f(y)$. The Fundamental Factorization Theorem now asserts that there is a function $f'\colon G/K \to G'$ which makes (*) commutative. Note that the action of f' is exactly as specified. To complete the argument we must establish several facts.

We first show that f' is a group homomorphism. Let $[x], [y] \in G/K$ (with $x,y \in G$). Then

$$
\begin{aligned}
f'([x][y]) &= f'([xy]) && \text{(multiplication in } G/K) \\
&= f(xy) && \text{(definition of } f') \\
&= f(x)f(y) && \text{(since } f \text{ is a homomorphism)} \\
&= f'([x])f'([y]) && \text{(definition of } f')
\end{aligned}
$$

We next show that f' is injective. Let $[x] \in \ker f'$ $(x \in G)$. Then if e' is the identity of G', $e' = f'([x]) = f(x)$. Hence $x \in \ker f = K$ and it follows that $[x] = [e] = e_{G/K}$.

The fact that $\operatorname{im} f' = \operatorname{im} f$ now follows from the definition of f'.

Statements (ii) and (iii) follow from the Correspondence Theorem (2.10) and (4.4).

Part (i) of Proposition (4.4) is generally known as the <u>first isomorphism theorem</u> and the map $f'\colon G/K \to \operatorname{im} f$ is called the isomorphism induced by f.

Before stating the second isomorphism theorem we require a technical proposition.

(4.6) PROPOSITION Let H and K be subgroups of G with K normal in G. Then $HK = KH$ is a subgroup of G.

PROOF Since K is a normal subgroup of G, $hK = Kh$ for all $h \in H$ by (4.3). Hence $HK = KH$. Now let $x,y \in HK$. We must show that $xy^{-1} \in HK$. Since $x,y \in HK$, $x = h'k'$ and $y = hk$ for some $h,h' \in H$ and $k,k' \in$

K. Then $y^{-1} = k^{-1}h^{-1}$ and hence

$$xy^{-1} = h'k'k^{-1}h^{-1} \in h'Kh^{-1}$$

Now, since K is normal, $Kh^{-1} = h^{-1}K$ so that $xy^{-1} \in h'h^{-1}K \subseteq HK$ and the result therefore follows.

(4.7) THEOREM (The Second Isomorphism Theorem) Let H and K be subgroups of G with K normal in G. Then K is a normal subgroup of HK, $H \cap K$ is a normal subgroup of H, and

$$H/(H \cap K) \cong KH/K = HK/K$$

PROOF Since K is a normal subgroup of G, HK = KH is a subgroup of G. Then, since $K \subseteq HK$, and K is normal in G, K is also a normal subgroup of HK. We will use the first isomorphism theorem. Let $f = k_K|_H$; that is, $f: H \to G/K$ by $f(h) = Kh$. Then f is a homomorphism of groups, $\ker f = H \cap K$, and $\text{Im } f = KH/K$. Hence by the first isomorphism theorem, $H/(H \cap K) \cong KH/H = HK/H$.

(4.8) THEOREM Let $f: G \to G'$ be a surjective homomorphism of groups, and suppose that N is a normal subgroup of G which contains ker f. Let $N' = f(N)$. Then N' is a normal subgroup of G' and $G/N \cong G'/N'$.

PROOF By (4.5), N' is a normal subgroup of G'. Let $k: G' \to G'/N'$ be the canonical map. Then both k and $f: G \to G'$ are surjective homomorphisms, whence so is $k \circ f: G \to G'/N'$. Then

$$\ker (k \circ f) = f^{-1}(\ker k) = f^{-1}(N') = N$$

Hence $\ker k \circ f = N$ and the result now follows from the first isomorphism theorem.

(4.9) COROLLARY (The Fundamental Theorem of Quotient Groups)

(i) Let N be a normal subgroup of G. Then there is a one-to-one correspondence between subgroups H of G containing N and all subgroups H' of the quotient group G/N given by

$$H \to H/N \quad \text{and} \quad H' \to k_N^{-1}(H')$$

(ii) Under the correspondence of (i) if H is a subgroup of G containing N, then H is normal in G if and only if H/N is normal in G/N and in this case $(G/N)/(H/N) \cong G/H$.

PROOF We will apply (4.8) using the surjective homomorphism $k_N: G \to G/N$. If H is a subgroup of G containing N, then

$$k_N(H) = \{k_N(x): x \in H\} = \{Nx: x \in H\} = H/N$$

The result therefore follows by (4.8).

(4.10) EXAMPLES

(4.10.1) Let n be a positive integer. Then the map $f: \mathbf{Z} \to \mathbf{Z}_n$ defined by $f(m) = [m]$ is a surjective homomorphism of groups with $\ker f = n\mathbf{Z}$. Hence $\mathbf{Z}/n\mathbf{Z} \cong \mathbf{Z}_n$.

(4.10.2) We will show that $\mathbf{R}^*/\mathbf{Z}^\times \cong \mathbf{R}^+$. Define $f: \mathbf{R} \to \mathbf{R}^+$ by $f(x) = |x|$. Then f is a surjective homomorphism of groups with $\ker f = \mathbf{Z}^\times$ and the result therefore follows by the first isomorphism theorem.

(4.10.3) By Example (2.11.2) the absolute value map from $\mathbf{C}^*$ to $\mathbf{R}^*$ is a surjective homomorphism with kernel the circle group S^1. Hence by the first isomorphism theorem, S^1 is a normal subgroup of $\mathbf{C}^*$ and

(*) $\mathbf{C}^*/S^1 \cong \mathbf{R}^*$

We saw in Example (3.16.6) that if $z \in \mathbf{C}$, then the coset $S^1z = [z]$ (in the complex plane) is the circle centered at the origin of radius $r = |z|$. The isomorphism given in (*) then sends the coset S^1z to the positive number r.

(4.10.4) We recall from Example (2.11.4) that the map

$$f: \mathbf{R} \to S^1 \quad \text{by} \quad x \mapsto e^{2\pi ix}$$

is a surjective homomorphism of groups with kernel $\mathbf{Z}$. Hence by the first isomorphism theorem, f induces an isomorphism from $\mathbf{R}/\mathbf{Z}$ to S^1. This isomorphism admits a reasonable geometric interpretation. If $x,y \in \mathbf{R}$, then $\mathbf{Z} + x = \mathbf{Z} + y$ if and only if $y - x \in \mathbf{Z}$. Thus every coset $\mathbf{Z} + x$ of $\mathbf{R}/\mathbf{Z}$ is represented by a unique real number r with $0 \le r < 1$.

(4.10.5) Recall from Example (1.20.6) that

$$B_2(\mathbf{R}) = \left\{ \begin{pmatrix} a & b \\ 0 & d \end{pmatrix} : a \text{ and } d \text{ are nonzero} \right\}$$

the set of all upper triangular elements of $GL_2(\mathbf{R})$ is a group under matrix multiplication. Let

$$K = \left\{ \begin{pmatrix} 1 & b \\ 0 & d \end{pmatrix} : d \text{ is nonzero} \right\}$$

Verify that the map $f: B_2(\mathbf{R}) \to \mathbf{R}^*$ defined by

$$\begin{pmatrix} a & b \\ 0 & d \end{pmatrix} \mapsto a$$

is a surjective homomorphism of groups with ker f = K. Hence K is a normal subgroup of $B_2(\mathbf{R})$ and $B_2(\mathbf{R})/K \cong \mathbf{R}^*$.

We will now illustrate the power of the theorems that we have developed concerning normal groups and quotient groups. Let G be a finite group and N a proper, nontrivial normal subgroup of G. We then have the following facts.

(i) $o(N) < o(G)$ [in fact $o(N) | o(G)$].

(ii) G/N is a group of order less than o(G) [and, in fact, $o(G/N) | o(G)$].

Suppose that a property, which we will refer to as P, is such that if a group G satisfies P, then the groups N and G/N both satisfy P. Then statements dealing with G and property P often yield readily to inductive arguments. This is illustrated in the following theorem.

(4.11) THEOREM (Cauchy's Theorem for Abelian Groups) Let G be a finite abelian group and p a prime, If $p | o(G)$ then G has an element of order p.

PROOF We will induct on the order of G; thus we assume that the result is true for all groups of order smaller than the order of G.

Since G is finite, the set of distinct proper subgroups of G is finite. Hence there is a proper subgroup M of G which is strictly contained in no other proper subgroup of G (such a subgroup is called a <u>maximal</u> subgroup of G). If $p | o(M)$, then by our induction hypothesis M, and hence G, has an element of order p.

We may therefore suppose that $p \nmid o(M)$. Since $M \neq G$, there is an element $x \in G$ such that $x \notin M$. We will use the maximality of M (and the fact that G is abelian) to show that x has order pr for some r and hence the element $y = x^r \in G$ has order p.

Let $H = \langle x \rangle$, the cyclic subgroup generated by x. Since G is abelian, every subgroup of G is normal. Hence, by (4.6), MH is a subgroup of G. Then M is a proper subgroup of MH (note that $x \in MH$ and $x \notin M$) and hence by the maximality of M, $MH = G$.

Now, by the second isomorphism theorem and the preceding statement,

$$M/(M \cap H) \cong MH/H = G/H$$

Hence by Lagrange's Theorem (3.7)

$$o(M)/o(M \cap H) = o(G)/o(H)$$

so that $o(H)o(M) = o(G)o(M \cap H)$.

Since $p | o(G)$, $p | o(H)o(M)$. Since p is prime and $p \nmid o(M)$ (by hypothesis), it follows that $p | o(H)$. Hence, since $H = \langle x \rangle$, x has order pr for some $r \in \mathbf{N}$.

Now let $y = x^r \in G$. Then $y \neq e$ [since $r < o(x)$] and $y^p = x^{rp} = e$. Hence, by (1.16), $o(y) | p$. Since p is prime and $o(y) \neq 1$, y must be an element of G of order p and the proof is now complete.

In Chapter III we will show that in fact Cauchy's Theorem is also true for nonabelian groups.

(4.12) <u>Short Exact Sequences</u> Consider a sequence of group homomorphisms

$$N \xrightarrow{f} E \xrightarrow{g} G$$

Such a sequence is said to be exact if Im(f) = ker(g). A longer sequence of homomorphisms is called exact if it is exact at every juncture.

We examine some special cases. Let us agree to let 1 denote the trivial group in multiplicative notation (and 0 in the additive notation). Then the exactness of the sequence

$$E \xrightarrow{f} G \longrightarrow 1$$

amounts to the surjectivity of f. Likewise, the exactness of the sequence

$$1 \longrightarrow N \xrightarrow{g} E$$

amounts to the injectivity of g.

An exact sequence of groups of the type

$$(*) \quad 1 \longrightarrow N \xrightarrow{f} E \xrightarrow{g} G \longrightarrow 1$$

or, in additive notation,

$$0 \longrightarrow N \xrightarrow{f} E \xrightarrow{g} G \longrightarrow 0$$

is called a short exact sequence. If a short exact sequence (*) exists, we say E is an extension of N by G. We now consider the interpretation of (*) in light of Theorem (4.5).

Let (*) be exact. Then f is injective and hence $N \cong f(N) =$ ker g which is a normal subgroup of E. We often regard this embedding as an identification and simply write $N \subseteq E$. Then, since g is surjective,

$$(**) \quad E/N \cong G$$

The notion of exactness is fundamental to the subject of homological algebra, a branch of algebra expecially valuable to geometers and topologists. For now, for us, short exact sequences are simply an extremely efficient means to represent information about homomorphisms.

(4.13) EXAMPLES

(4.13.1) Let $f: E \to G$ be any surjective homomorphism of groups. Then there is a short exact sequence

$$1 \longrightarrow \ker f \xrightarrow{i} E \xrightarrow{f} G \longrightarrow 1$$

where i is just inclusion of subgroups. In particular, if N is a normal subgroup of G, we always have a short exact sequence

$$1 \longrightarrow N \xrightarrow{i} G \xrightarrow{k} G/N \longrightarrow 1$$

Again i denotes the inclusion mapping.

(4.13.2) Let G and G' be groups. Then there is a short exact sequence

$$1 \longrightarrow G \xrightarrow{i_1} G \times G' \xrightarrow{p_2} G' \longrightarrow 1$$

Hence, [identifying G with $i_1(G) = G \times \{e'\} = \ker(p_2)$],

$$(G \times G')/G \cong G'$$

(4.13.3) From (2.11.6) we deduce a short exact sequence

$$0 \longrightarrow \mathbf{R} \longrightarrow C^1(\mathbf{R}) \xrightarrow{D} C^0(\mathbf{R}) \longrightarrow 0$$

Thus D induces an isomorphism $D^*: C^1(\mathbf{R})/\mathbf{R} \cong C^0(\mathbf{R})$. The inverse map for D^* is the indefinite integral.

(4.13.4) By (2.12.6) if G is a group the map $f: G \to \mathrm{Inn}(G)$ defined by $f(x) = f_x$ [where $f_x(y) = xyx^{-1}$ for all $y \in G$)] is a surjective homomorphism of groups with $\ker f = Z(G)$ (the center of G). Hence f induces the following exact sequence of groups

$$1 \longrightarrow Z(G) \longrightarrow G \xrightarrow{f} \mathrm{Inn}(G) \longrightarrow 1$$

and $G/Z(G) \cong \mathrm{Inn}(G)$.

(4.13.5) By (2.11.7), $\det: GL_2(\mathbf{R}) \to \mathbf{R}^*$ is a surjective homomorphism of groups with kernel $SL_2(\mathbf{R})$. Hence det induces the following exact sequence of groups

$$1 \longrightarrow SL_2(\mathbf{R}) \longrightarrow GL_2(\mathbf{R}) \xrightarrow{\det} \mathbf{R}^* \longrightarrow 1$$

and $GL_2(\mathbf{R})/SL_2(\mathbf{R}) \cong \mathbf{R}^*$.

EXERCISES

(4.1) Conclude from exercise (3.8) that there are subgroups H and K of D_4 such that

(a) K is a normal subgroup of D_4;

(b) H is a normal subgroup of K;

(c) H is not a normal subgroup of D_4.

(4.2) Let G be a group and H a subgroup of index 2 in G. Show that H is a normal subgroup of G [cf. exercise (3.9)].

(4.3) Give an example of a group G such that

(a) $G/Z(G)$ is abelian;

(b) $G/Z(G)$ is not abelian.

(4.4) (a) Let $s \in \mathbf{Q}^*$. Show that there is a unique $n \in \mathbf{Z}$ such that

$$s = (a/b)2^n$$

where a and b are odd integers.

(b) Show that the mapping $f\colon \mathbf{Q}^* \to \mathbf{Z}$, defined by $f(x) = n$, is a surjective homomorphism.

(c) Describe ker f and show that

$$1 \longrightarrow \ker f \longrightarrow \mathbf{Q}^* \xrightarrow{f} \mathbf{Z} \longrightarrow 1$$

is exact.

(d) Conclude that $\mathbf{Z} \cong \mathbf{Q}/\ker f$.

(4.5) (a) Let $G = \mathbf{Q}/\mathbf{Z}$. Show that G is an infinite subgroup of $\mathbf{R}/\mathbf{Z}$ such that every element of G has finite order.

(b) Show that G is the set of all elements of $\mathbf{R}/\mathbf{Z}$ whose order is finite.

(4.6) Let $G = C_2 \times C_4$.

(a) Find subgroups H, K of G such that $H \cong K$ but G/H is not isomorphic to G/K.

(b) Find subgroups H, K of G such that $G/H \cong G/K$ but H is not isomorphic to K.

(4.7) Let $G = D_4$ and let $N = \{e, st, s^2, s^3t\}$.

(a) Show that the following sequences are exact.

$$1 \longrightarrow N \longrightarrow G \xrightarrow{k} G/N \longrightarrow 1$$

$$1 \longrightarrow N \xrightarrow{i_1} N \times C_2 \xrightarrow{p_2} C_2 \longrightarrow 1$$

[Recall that $i_1(x) = (x, 0)$, $p_2(x, y) = y$.]

(b) Show that $G/N \cong C_2$, but G is not isomorphic to $N \times C_2$.

(4.8) Let

$$1 \longrightarrow K \xrightarrow{f} G \xrightarrow{g} H$$

be an exact sequence of groups. Let T be a group and let $u: T \to G$ be a homomorphism such that $(g \circ u)(t) = e$ for every $t \in T$. Show that there is a unique homomorphism $u': T \to K$ making the following diagram commutative.

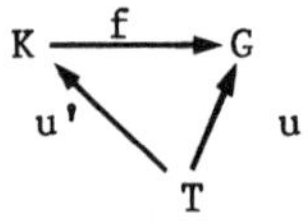

(4.9) Let

$$K \xrightarrow{f} G \xrightarrow{g} H \longrightarrow 1$$

be an exact sequence of groups. Let T be a group with $u: G \to T$ a homomorphism such that $(u \circ f)(x) = e$ for every $x \in K$. Show that there is a unique homomorphism $u': H \to T$ making the following diagram commutative.

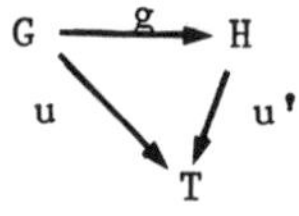

(4.10) Let $\{N_i: i \in I\}$ be a family of normal subgroups of a group G. Show that $\bigcap_{i \in I} N_i$ is a normal subgroup of G.

(4.11) Let H be a subgroup of G. Show that $\cap \{sHs^{-1}: s \in G\}$ is a normal subgroup of G.

(4.12) (a) Show that if $m|n$, then C_n has a unique subgroup of order m.

(b) Let N be a cyclic normal subgroup of a group G and let H be a subgroup of N. Show that H is a normal subgroup of G.

(4.13) Let N, M be normal sugroups of a group G. Show that NM is a normal subgroup of G.

(4.14) Let H be a subgroup of a group G such that $H \subseteq Z(G)$. Show that H is a normal subgroup of G.

(4.15) Let H be a subgroup of a group G. Show that H is a normal subgroup of G if and only if every left coset of H in G is also a right coset of H in G.

(4.16) Let H be a finite subgroup of a group G such that for any subgroup K of G, $o(K) \neq o(H)$. Show that H is a normal subgroup of G.

(4.17) Let G be a group. Show that Inn(G) is a normal subgroup of Aut(G).

(4.18) Let G be a group with $S \subseteq G$ such that $s \in S$ and $x \in G$ implies $xsx^{-1} \in S$. Show that $\langle S \rangle$ is a normal subgroup of G.

(4.19) Let G be a group and set $S = \{xyx^{-1}y^{-1}: x, y \in G\}$. Define $G' = \langle S \rangle$ and show that

(a) G' is a normal subgroup of G;

(b) G/G' is abelian;

(c) If N is a normal subgroup of G, then G/N is abelian if and only if $G' \subseteq N$.

(d) UNIVERSAL PROPERTY FOR COMMUTATOR SUBGROUPS. Let H be an abelian group and let $f: G \to H$ be a homomorphism. Show that there is a unique homomorphism k' making the following triangle commutative.

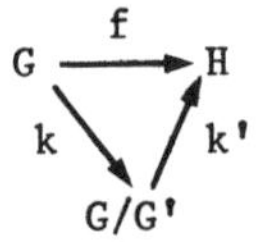

We will call G' the commutator subgroup of G or the first derived group of G.

(4.20) Let

$$G = \left\{ \begin{pmatrix} a & b \\ 0 & d \end{pmatrix} : ad \neq 0,\ a,b,d \in \mathbf{R} \right\}$$

$$H = \left\{ \begin{pmatrix} 1 & b \\ 0 & 1 \end{pmatrix} : b \in \mathbf{R} \right\}$$

Show that

(a) G is a subgroup of $GL_2(\mathbf{R})$;

(b) H is a normal subgroup of G;

(c) $G/H \cong \mathbf{R}^* \times \mathbf{R}^*$ (and is therefore abelian);

(d) H = G'.

(4.21) Let H be a subgroup of a group G. Define

$$N(H) = \{x: x \in G,\ xHx^{-1} = H\}$$

Show that

(a) N(H) is a subgroup of G;

(b) H is a normal subgroup of N(H);

(c) If K is a subgroup of G such that H is a normal subgroup of K, then $K \subseteq N(H)$;

(d) H is a normal subgroup of G if and only if G = N(H).

N(H) is called the normalizer of H in G.

(4.22) Let S be a subset of a group G and define

$$C(S) = \{x: x \in G,\ xs = sx \text{ for every } s \in S\}$$

Show that

(a) $C(S) = \bigcap_{s \in S} C(s)$;

(b) C(S) is a subgroup of G.

C(S) is called the centralizer of S.

(4.23) Let N, M be normal subgroups of a group G such that $N \cap M = \{e\}$. Show that if $x \in N$, $y \in M$, then $xy = yx$.

(4.24) Let H be a subgroup of G such that $H \subseteq Z(G)$ and G/H is cyclic. Show that G is abelian.

(4.25) Let G be a group, $x \in G$ such that $o(x) = 2$ and $\langle x \rangle$ is a normal subgroup of G. Show that $\langle x \rangle \subseteq Z(G)$. Conclude that if $G/\langle x \rangle$ is cyclic, then G is abelian.

(4.26) Let N be a normal subgroup of a finite group G such that $((G\colon N), o(N)) = 1$. Show that if $x \in G$, then $x \in N$ if and only if $x^{o(N)} = e$.

(4.27) Let H, K be distinct subgroups of a finite group G such that $(G\colon H) = (G\colon K) = 2$. Show that $H \cap K$ is a normal subgroup of G and $G/(H \cap K)$ is not cyclic.

(4.28) <u>Chinese Remainder Theorem</u>. Let n,m be positive integers and let $f\colon \mathbf{Z} \to \mathbf{Z}_n$, $g\colon \mathbf{Z} \to \mathbf{Z}_m$ be the canonical mappings. Define

$$h\colon \mathbf{Z} \to \mathbf{Z}_n \times \mathbf{Z}_m \text{ by } h(x) = (f(x), g(x))$$

(a) Let $d = \operatorname{lcm}(n, m)$. Show that $\ker h = d\mathbf{Z}$.

(b) Show that the induced injection

$$N\colon \mathbf{Z}_d \to \mathbf{Z}_n \times \mathbf{Z}_m$$

is an isomorphism if and only if $(n, m) = 1$.

(c) Conclude that if $(n, m) = 1$, then for all $a,b \in \mathbf{Z}$ the simultaneous congruences

$$x \equiv a \pmod{n}$$
$$x \equiv b \pmod{m}$$

admit a solution x. Conversely, if all such systems admit a solution, then $(n, m) = 1$.

(d) Generalize the results given here to an arbitrary number of moduli. The resulting generalization of (c) is called the Chinese Remainder Theorem.

(4.29) Let $N = \{(t, 2t): t \in \mathbf{R}\}$. Show that N is a normal subgroup of $\mathbf{R}^2$ and that $\mathbf{R}^2/N \cong \mathbf{R}$.

(4.30) Let $G = \{f: f \in B(\mathbf{R}), f(x) = ax + b \text{ with } a,b \in \mathbf{R}, a \neq 0\}$ [cf. exercise (3.15)] and $H = \{f: f \in G, f(x) = x + b, b \in \mathbf{R}\}$. Show that H is a normal subgroup of G and $G/H \cong \mathbf{R}^*$.

EXTENDED EXERCISES

(4.31) Sylow Decomposition for Finite Abelian Groups Let G be a group, which for the moment need be neither finite nor abelian.

(a) Suppose there exist subgroups H_1, H_2 of G such that

(i) $H_1 \cap H_1 = \{e\}$;

(ii) H_1 and H_2 are normal subgroups of G;

(iii) $G = H_1H_2$.

Show that $G \cong H_1 \times H_2$. [Hint: Recall exercise (4.23).] Conversely, show that if G is isomorphic to a direct product of two groups, then G contains a pair of subgroups satisfying (i), (ii), (iii).

(b) Suppose there exist subgroups $H_1, H_2, \ldots, H_n$ of G such that

(i) $H_j \cap (H_{j+1}H_{j+2} \cdots H_n) = \{e\}$, $1 \le j < n$;

(ii) H_j is normal in G, $1 \le j \le n$;

(iii) $G = H_1H_2 \cdots H_n$.

Show that $G \cong H_1 \times H_2 \times \cdots \times H_n$. Conversely, show that if G is isomorphic to a direct product of n groups, then G contains a family of n subgroups satisfying (i), (ii), (iii).

(c) Assume now that G is abelian. If p is a positive prime, define a subset A_p of G by

$$A_p = \{s \in G: o(s) \text{ is a positive power of } p\}$$

Show that A_p is a subgroup of G for all primes p and that $A_p \cap A_q = \{e\}$ if p and q are distinct primes.

(d) Assume further that G is finite and abelian, and let o(G) have the following prime decomposition.

$$o(G) = p_1^{\alpha_1} p_2^{\alpha_2} \cdots p_n^{\alpha_n}$$

Write A_j for A_{p_j}. Show that

$$G \cong A_1 \times A_2 \times \cdots \times A_n$$

Conclude that $o(A_j) = p_j^{\alpha_j}$. [Hint: Reduce the problem of showing that $G = A_1A_2 \cdots A_n$ to the case of a cyclic group.]

(4.32) Semidirect Products Let N and H be groups, and let there be given a homomorphism $\theta(h)\colon H \to \mathrm{Aut}(N)$. For $h \in H$, $n \in N$, we agree to abbreviate $\theta(h)(n)$ (the value at n of the automorphism associated to h) by ${}^hn$. Define $N \rtimes_\theta H$, the semidirect product of N and H via θ, to be the set $N \times H$ together with the operation

$$(n, h)\,(m, g) = (n\,{}^hm, hg)$$

for $n,m \in N$, $h,g \in H$. (When θ is fixed, we write simply $N \rtimes H$ for $N \rtimes_\theta H$.)

(a) Show that ${}^h(nm) = {}^hn\,{}^hm$ for all $h \in H$, $n,m \in N$, and that ${}^h({}^gn) = {}^{(hg)}n$ for all $h,g \in H$, $n \in N$.

(b) Let $G = N \rtimes H$. Show that G is a group and that in fact G is just the direct product of N and H when θ is trivial.

(c) Define maps i and j as follows

$$i\colon N \to G \text{ by } n \mapsto (n, e)$$
$$j\colon H \to G \text{ by } h \mapsto (e, h)$$

Show that both i and j are embeddings of groups. Regarding these as identifications, show that ${}^hn = hnh^{-1}$ for all $h \in H$, $n \in N$.

(d) Let $\rho\colon G \to H$ be defined by $\rho((n, h)) = h$. Show that $\rho \in \mathrm{Hom}(G, H)$ and that the sequence

$$1 \longrightarrow N \xrightarrow{i} G \xrightarrow{\rho} H \longrightarrow 1$$

is exact. What is $\rho \circ j$?

(e) Universal Property. Let E be a group, and suppose there exist $\alpha \in \mathrm{Hom}(N, E)$ and $T \in \mathrm{Hom}(H, E)$ such that

$$T(h)\alpha(n) = \alpha({}^hn)T(h)$$

for all $h \in H$, $n \in N$. Show that there exists a unique homomorphism

$\tau: G \to E$ such that $\alpha = \tau \circ i$ and $T = \tau \circ j$; that is, such that the following diagram commutes.

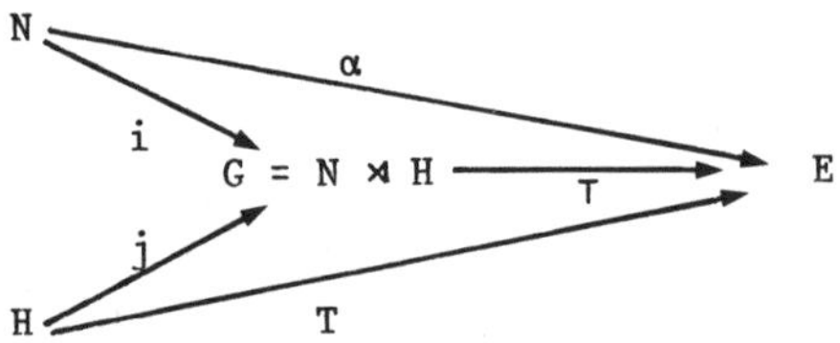

(f) Now let G be an arbitrary group and suppose there exist subgroups N and H of G such that the following conditions hold.

(i) $N \cap H = \{e\}$.

(ii) N is normal in G.

(iii) $G = NH$.

Define a map $\theta: H \to \text{Aut}(N)$ by conjugation. That is, ${}^{h}n = hnh^{-1}$ for all $h \in H$, $n \in N$. Show that $G \cong N \rtimes_\theta H$. Conversely, show that if G is the semi-direct product of two groups, then G contains a pair of subgroups satisfying (i), (ii), (iii).

(g) The Dihedral Groups. Let $C_n = \langle s \rangle$, $n \le \infty$, denote the cyclic group of order n. Show that there exists a homomorphism $\theta: C_2 \to \text{Aut}(C_n)$ such that $\theta(t)(s) = s^{-1}$. Now define the dihedral group D_n by

$$D_n = C_n \rtimes_\theta C_2$$

Show that D_n is nonabelian for all $n > 2$ and has order 2n for $n < \infty$. Exhibit an isomorphism between D_4 and the group G of Example (1.6.8).

Note that for all n, $D_n = \langle s, t \rangle$ with s and t subject to the relations $t^2 = e$, $tst = s^{-1}$. If $n < \infty$, then also $s^n = e$. (Here we have identified C_n and C_2 with subgroups of D_n.)

(h) Find $Z(D_n)$ for all n.

(i) Exhibit a nonabelian group of order 55. [Hint: $3^5 \equiv 1 (\text{mod } 11)$.]

(j) Generalize (i): State conditions under which one can form a nonabelian semidirect product $C_n \rtimes C_m$.

(4.33) <u>Direct Products and Direct Sums</u> Let I be a nonempty set and let $\{G_i\}_{i \in I}$ be a family of groups indexed by I. Recall that $\prod_{i \in I} G_i$ as a product of sets consists of indexed families $(s_i)_{i \in I}$ where $s_i \in G_i$.

(a) Let $G = \prod_{i \in I} G_i$. Define an operation on G by

$$(s_i)(t_i) = (u_i) \text{ where } u_i = s_i t_i \in G_i$$

for all $i \in I$. Show that G is a group with respect to this operation. We thus speak of the direct product of the family $\{G_i\}$.

(b) Universal Property. For $i \in I$, define a map $p_i\colon G \to G_i$ by $p_i((s_j)_{j \in I}) = s_i$. Show that $p_i \in \mathrm{Hom}(G, G_i)$ for all i. Let $\{\tau_i\colon H \to G_i\}_{i \in I}$ be a family of homomorphisms from a fixed arbitrary group H into the G_i. Show that there exists a unique homomorphism $\tau\colon H \to G$ such that $\tau_i = p_i \circ \tau$ for all i; that is, such that the following diagram commutes for all i.

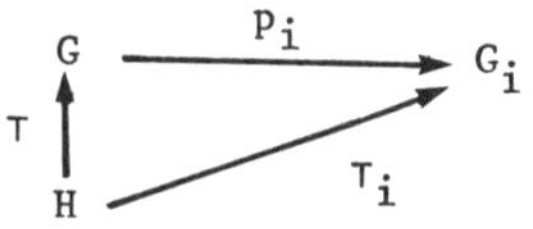

Now consider the special case of a family of <u>additive groups</u> $\{A_i\}_{i \in I}$. Define a subset $\bigoplus_{i \in I} A_i$ of $\prod_{i \in I} A_i$ by

$$\bigoplus_{i \in I} A_i = \{(s_i) \in \sum_{i \in I} A_i\colon s_i = 0 \text{ for all but finitely many } i \in I\}$$

(Note that $\bigoplus A_i = \prod A_i$ if I is a finite set.)

(c) Let $A = \bigoplus A_i$. Show that A is a subgroup of the direct product $\prod A_i$. We call A the <u>direct sum</u> of the family $\{A_i\}$.

(d) Universal Property. For $i \in I$, define a map $u_i\colon A_i \to A$ such that $u_i(t) = (s_j)_{j \in I}$ where

$$s_j = \begin{cases} t \text{ if } i = j \\ 0 \text{ if } i \neq j \end{cases}$$

(The map u_i is called inclusion onto the i-th factor.) Show that u_i is an injective homomorphism for all i. Let $\{\alpha_i\colon A_i \to B\}_{i \in I}$ be a family of homomorphisms from the A_i into a fixed additive group B. Show that there exists a unique homomorphism $\alpha: A \to B$ such that $\alpha_i = \alpha \circ u_i$ for all i; that is, such that the following diagram commutes for all i.

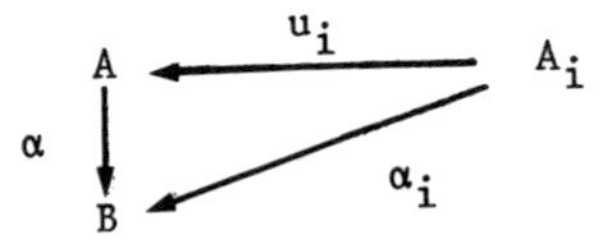

(4.34) <u>Homology</u> Let $\{A_n\}_{n \in \mathbf{Z}}$ be a family of additive groups indexed by **Z**. We call $\{A_n\}$ a **Z**-graded family. The direct sum $A = \oplus A_n$ is called the **Z**-graded additive group associated with the family $\{A_n\}$. Let $\{B_n\}$ be another **Z**-graded family of additive groups, and set $B = \oplus B_n$. Then a **Z**-graded morphism $\alpha: A \to B$ of degree r is a homomorphism of groups such that

$$\alpha(A_n) \subseteq B_{n+r}$$

for all n. (Here we have identified A_n and B_n with subgroups of A and B, respectively.) Clearly such a morphism is equivalent to a family of homomorphisms $\{\alpha_n: A_n \to B_{n+r}\}$.

(a) Let $\{C_n\}$ be a third **Z**-graded family with associated **Z**-graded additive group C. Let $\alpha: A \to B$ and $\tau: B \to C$ be **Z**-graded morphisms of degrees r and r', respectively. Show that $\tau \circ \alpha: A \to C$ is a **Z**-graded morphism of degree r + r'.

A <u>complex</u> $\{(C_n, d_n)\}$ of additive groups is a **Z**-graded family $\{C_n\}$ together with a family of homomorphisms $\{d_n: C_n \to C_{n+1}\}$ such that $d_n \circ d_{n-1} = 0$ (i.e., the trivial map). Thus if $C = \oplus C_n$ is the associated **Z**-graded additive group, we can view the family $\{d_n\}$ collectively as a morphism $d: C \to C$ of degree +1 such that $d \circ d = 0$. We frequently abbreviate $\{(C_n, d_n)\}$ to (C, d) or just C.

(b) Let C be a complex. For each integer n, define subgroups $\mathbf{Z}_n(C)$ and $B_n(C)$ of C_n as follows.

$$Z_n(C) = \ker (d_n: C_n \to C_{n+1})$$
$$B_n(C) = \text{Im} (d_{n-1}: C_{n-1} \to C_n)$$

Show that $B_n(C) \subseteq Z_n(C)$ for all n.

We now define the <u>n-th homology group</u>, $H_n(C)$, of the complex C by

$$H_n(C) = Z_n(C)/B_n(C)$$

Hence $\{H_n(C)\}$ is a **Z**-graded family of additive groups, and we may form the associated **Z**-graded additive group $H(C) = \oplus H_n(C)$. H(C) is called the homology of C.

Let $\{(C_n, d_n)\}$ and $\{(C_n', d_n')\}$ be complexes. A <u>morphism of complexes</u> of degree r is a **Z**-graded morphism $\alpha: C \to C'$ of degree r on the associated **Z**-graded additive groups such that $d_n' \circ \alpha = \alpha \circ d_n$ for all n; that is, the following diagram commutes.

$$\begin{array}{ccc} C_n & \xrightarrow{\alpha} & C'_{n+r} \\ {\scriptstyle d_n}\downarrow & & \downarrow{\scriptstyle d_n'} \\ C_{n+1} & \xrightarrow{\alpha} & C'_{n+r+1} \end{array}$$

(c) Show that if $\alpha: C \to C'$ is a morphism of complexes of degree 0, then

$$\alpha(Z_n(C)) \subseteq Z_n(C')$$
$$\alpha(B_n(C)) \subseteq B_n(C')$$

Deduce that α induces homomorphisms $\alpha_n: H_n(C) \to H_n(C')$ and hence a **Z**-graded morphism $\alpha_*: H(C) \to H(C')$ of degree 0. (In general, the lower * will denote the induced map on the homology.)

A <u>short exact sequence of complexes</u>,

$$0 \longrightarrow (C', d') \xrightarrow{\alpha} (C, d) \xrightarrow{\tau} (C'', d'') \longrightarrow 0$$

is a sequence of morphisms of complexes which is exact as a sequence of group homomorphisms. If α and τ have degree 0, such a sequence gives rise to a commutative diagram of the following type with exact rows.

$$\begin{array}{ccccccccc}
0 & \longrightarrow & C'_n & \xrightarrow{\alpha} & C_n & \xrightarrow{\tau} & C''_n & \longrightarrow & 0 \\
 & & {\scriptstyle d'}\downarrow & & {\scriptstyle d}\downarrow & & \downarrow{\scriptstyle d''} & & \\
0 & \longrightarrow & C'_{n+1} & \xrightarrow{\alpha} & C_{n+1} & \xrightarrow{\tau} & C''_{n+1} & \longrightarrow & 0
\end{array}$$

(d) The Connecting Map. Let there be given an exact sequence as above, and let $z'' \in Z_n(C'')$. Prove each of the following statements.

(i) There exists an element $c \in C_n$ such that $\tau(c) = z''$.

(ii) Suppose also $\tau(c_1) = z''$ for some $c_1 \in C_n$. Then $c_1 = c + \alpha(c')$ for some unique $c' \in C'_n$.

(iii) There exist unique elements z' and z_1' in $Z_{n+1}(C')$ such that $\alpha(z') = d(c)$ and $\alpha(z_1') = d(c_1)$.

(iv) In fact, $z_1' = z' + d(c')$.

Conclude that both z' and z_1' are congruent modulo $B_{n+1}(C')$, and hence that the relation

$$z' = \alpha^{-1} \circ d \circ \tau^{-1}(z'')$$

in fact induces a well-defined homomorphism $\rho\colon Z_n(C'') \to H_{n+1}(C')$. Next show that $B_n(C'') \subseteq \ker \rho$, so that ρ in turn induces a homomorphism $H_n(C'') \to H_{n+1}(C')$ which we also call ρ. [Thus ρ is a $\mathbf{Z}$-graded morphism $H(C'') \to H(C')$ of degree +1.]

(e) <u>The Long Exact Sequence</u> Show that for all n the sequence

$$\cdots \longrightarrow H_n(C') \xrightarrow{\alpha_*} H_n(C) \xrightarrow{\tau_*} H_n(C'') \xrightarrow{\rho} H_{n+1}(C') \longrightarrow \cdots$$

is exact. This is called the long exact sequence of homology. One can also express this by asserting the exactness of the following sequence of $\mathbf{Z}$-graded morphisms

$$\begin{array}{ccc}
H(C') & \xrightarrow{\alpha_*} & H(C) \\
{\scriptstyle \rho}\nwarrow & & \swarrow{\scriptstyle \tau_*} \\
 & H(C'') &
\end{array}$$

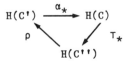

3
Group Actions and Solvable Groups

In this chapter we will study more advanced topics in the theory of groups. In Section 1, we will use group actions to develop the Sylow theorems and Cauchy's theorem. In Section 2, we will develop the theory of solvable groups which will be used later in Galois theory. Much of the group theory developed in this chapter will be for finite groups only. Since many of the arguments are counting arguments and involve cardinalities of sets, we will abbreviate the previous notation for cardinality of a set. Throughout this chapter, we will denote Card(A) by $|A|$. Also, if H is a subgroup of a group G, then we will often write $H < G$.

1. GROUP ACTIONS

In this section we will study a very useful tool, group actions. Actions of various algebraic structures on sets appear in several other contexts later in this book. For now, group actions will provide us with a very efficient, sufficiently general, tool for counting. We begin with the definition of a group action

(1.1) DEFINITION A group G is said to act on a nonempty set X if there is a function from $G \times X$ to X whose action is specified by

$$(g, x) \mapsto g * x$$

and satisfies the following axioms.

(i) For every $x \in X$, $e * x = x$.

(ii) For every $g, g' \in G$ and $x \in X$, $(gg') * x = g * (g' * x)$.

The resulting function is called a group action.

If a group G acts on a set X and $x \in X$, then the orbit of x is defined by

$$or(x) = \{y: y \in X \text{ and } g * x = y \text{ for some } g \in G\}$$

Also the stabilizer of x is defined by

$$St(x) = \{g: g \in G \text{ and } g * x = x\}$$

(1.2) EXAMPLES We will now present several examples of group actions. In the sequel, the reader is urged to interpret each theorem in the context of each of these examples. The reader should also verify axioms (i) and (ii) in each example.

(1.2.1) Let H be a subgroup of a group G. Then we will say that H acts on G by left translation. This action will be specified by

$$h * g = hg$$

where $h \in H$, $g \in G$, and hg is the group operation. In this case,

$$or(g) = Hg$$
$$St(g) = \{e\}$$

for each $g \in G$.

(1.2.2) A group G acts on itself by conjugation. In this case, the action is specified by

$$h * g = hgh^{-1}$$

where $g, h \in G$. The orbit of an element g of G is given by

$$or(g) = \{hgh^{-1}: h \in G\}$$

The elements of or(g) are called conjugates of g. In this case, we denote

$$c(g) = or(g)$$

Note that for $g, h \in G$, $g = hgh^{-1}$ if and only if $gh = hg$. Thus the stabilizer of g is simply the centralizer of g in G. In this case, we denote

$$C_G(g) = St(g)$$

[cf. Exercise II, (4.22)]

(1.2.3) Let G be a group and let S(G) be the set of all subgroups of G. Then G acts on S(G) by conjugation. This action is specified by

$$g * H = gHg^{-1}$$

where $g \in G$ and H is a subgroup of G. (Note that if H is a subgroup of G and $g \in G$, then gHg^{-1} is also a subgroup of G) In this case,

$$or(H) = \{gHg^{-1}: g \in G\}$$

where H is a subgroup of G. The elements of or(H) are called <u>conjugates of H in G</u>. The stabilizer of H is called the <u>normalizer in G of H</u>, that is

$$N_G(H) = \{g: g \in G \text{ and } gH = Hg\} = St(H)$$

[cf. Exercise II, (4.21)]

(1.2.4) Let H be a subgroup of a group G. Then G acts on the set X of all left cosets of H in G by left translation. The action is specified by

$$g' * gH = (g'g)H$$

where $g, g' \in G$. In this case, the orbit and stabilizer of gH are given by

$$or(gH) = X$$
$$St(gH) = \{g': g' \in G \text{ and } g'gH = gH\}$$
$$= \{g': g' \in G \text{ and } g^{-1}g'g \in H\}$$

(1.2.5) Let G be a subgroup of B(X) for some set X. Recall that B(X) is the set of all bijections from X to X, and the operation is composition of functions. Then G acts on X by function action. This action is specified by

$$g * x = g(x)$$

where $g \in G$ and $x \in X$.

(1.2.6) Let G, H be groups and let f be a homomorphism from G to H. If H acts on a set X, then f induces an action of G on X. This action is specified by

$$g * x = f(g) * x$$

where $g \in G$ and $x \in X$. Here we use the symbol * ambiguously, but the ambient action is clear by the location of the elements involved.

(1.3) THEOREM Let G be a group and suppose that G acts on a set X.

(i) Define a binary relation ~ on X by, $x \sim y$ if there is a $g \in G$ such that $g * x = y$. Then ~ is an equivalence relation on X. Moreover, the equivalence class of $x \in X$ is or(x).

(ii) For each $x \in X$, St(x) is a subgroup of G.

(iii) If X is finite, then $|or(x)| = (G: St(x))$.

(iv) Let $\{x_i: i \in I\}$ be a complete residue system for ~ and let X be finite. Then

$$|X| = \sum_{i \in I} |or(x_i)| = \sum_{i \in I} (G: St(x_i))$$

PROOF (i) The fact that ~ is reflexive is immediate by taking $g = e$. Suppose that $x \sim y$ and $g \in G$ is such that $g * x = y$. Then

$$\begin{aligned} g^{-1} * y &= g^{-1} * (g * x) \\ &= (g^{-1}g) * x \\ &= e * x \\ &= x \end{aligned}$$

Therefore, $y \sim x$ and ~ is symmetric. Finally, suppose that $x \sim y$ and $y \sim z$. Let $g, h \in G$ be such that $g * x = y$ and $h * y = z$. Then

$hg * x = z$. Hence $\sim$ is transitive. The fact that the equivalence class of x with respect to $\sim$ is or(x) is clear from the definition of $\sim$.

(ii) Let $g, h \in St(x)$. Then $g * x = x$ and $h * x = x$. If we repeat the computational arguement for symmetry of $\sim$, then it is easily seen that $g^{-1} * x = x$. Hence

$$(g^{-1}h) * x = g^{-1} * (h * x) = g^{-1} * x = x$$

Therefore, $g^{-1}h \in St(x)$ and St(x) is a subgroup of G.

(iii) Let $g, h \in G$. We will show that

$$g * x = h * x \text{ if and only if } gSt(x) = hSt(x)$$

To see this, note that the following statements are equivalent

(a) $g * x = h * x$

(b) $(h^{-1}g) * x = x$

(c) $h^{-1}g \in St(x)$

(d) $gSt(x) = hSt(x)$

It follows that the mapping from the set of left cosets in G of St(x) to or(x) whose action is specified by

$$gSt(x) \mapsto g * x$$

is well defined and a bijection.

(iv) This is just a restatement of a fundamental counting principle which was stated in I, (2.2).

We will now state a Corollary which follows from Example (1.2.2). The reader is urged to consider each example in a similar way.

(1.4) COROLLARY Let G be a finite group and let $g \in G$. Then $C_G(g)$ is a subgroup of G and the number of elements in G which are conjugate to g is the index of $C_G(g)$ in G.

We will see in the sequel that the simple fact given in (1.3) (iv) will have profound consequences. Refering to (1.2.2), we will state one of those consequences now. The result will be called The Class Equation.

(1.5) COROLLARY Let G be a finite group and let G act on itself by conjugation.

(i) For any $g \in G$, $|c(g)| = 1$ if and only if $g \in Z(G)$.

(ii) For any $g \in G$, $|c(g)| = (G: C_G(g))$.

(iii) Let $\{g_i: i \in I\}$ be a complete residue system for the equivalence relation "is conjugate to". Then

$$\text{(CE)} \quad |G| = |Z(G)| + \sum_{i \in I} |c(g_i)|$$

We will see that the class equation, (CE), has numerous applications. The first application is the non-abelian extension of Cauchy's Theorem. [cf. II, (4.11)]

(1.6) THEOREM Let G be a finite group and let p be a prime divisor of the order of G. Then there is an element g in G such that $o(g) = p$.

PROOF We will proceed by induction on the order of G. The result is vacuously true for $o(G) = 1$. Let G be a group and assume that the result is valid for all groups whose order is less than the order of G. If the center Z(G) has order k, and k is divisible by p, then, by Cauchy's Theorem for Abelian Groups, Z(G), and hence G, contains an element of order p. We therefore assume that k is not divisible by p. Consider the class equation in this case.

$$|G| = k + \sum_{i \in I} |c(g_i)|$$

Since the left side is divisible by p, so is the right side. But, if the right side is divisible by p, and k is not divisible by p, then for some $i \in I$, $c(g_i)$ is not divisible by p. Since p is a divisor of $|G|$ and

$$|c(g_i)| = |G|/|St(g_i)|$$

we must have p is a divisor of $|St(g_i)|$. Also, since $|St(g_i)| < |G|$, there is an element in $|St(g_i)|$, and hence in G, with order p.

(1.7) PROPOSITION Let G be a group and let G act on a set X.

(i) For each $g \in G$, the mapping $1_g: X \to X$, defined by $1_g(x) = g \star x$, is a bijection.

(ii) The mapping $\theta: G \to B(X)$, defined by $\theta(g) = 1_g$, is a homomorphism of groups.

(iii) The kernel of θ is given by

$$\ker(\theta) = \{g: g \in G \text{ and } g \star x = x \text{ for every } x \in X\}$$

PROOF (i) If $x, y \in X$ and $g \in G$, then $g \star x = g \star y$ implies $(g^{-1}g) \star x = (g^{-1}g) \star y$. Hence $x = y$, and 1_g is injective. If $x \in X$, then $1_g(g^{-1} \star x) = x$; hence 1_g is bijective.

(ii) Let $g, h \in G$ and let $x \in X$. Then

$$\begin{aligned} 1_{gh}(x) &= gh \star x \\ &= g \star (h \star x) \\ &= g \star 1_h(x) \\ &= 1_g(1_h(x)) \end{aligned}$$

Therefore, $\theta(gh) = \theta(g) \circ \theta(h)$ and θ is a homomorphism.

(iii) This fact is clear from the definitions.

We should note that by combining examples (1.2.5) and (1.2.6), we can see that each homomorphism from a group G to the group B(X) induces an action of G on X. Thus a group action of G on a nonempty set X could have been defined as a homomorphism from G to B(X).

A group G is said to be simple if the only normal subgroups of G are G and {e}. Much of finite group theory is concerned with the identification and classification of simple groups. Note that one consequence of (ii) is that if G is a simple group, then G cannot act on a set X which is too small. This fact is made precise in the following Corollary.

(1.8) COROLLARY Let G be a simple group and let X be a set such that G acts on X and $|X| = k > 1$. Then G is finite and o(G) is a divisor of k!.

PROOF Since ker(θ) is a normal subgroup of G and since $k > 1$, we must have θ is injective. Therefore, by Lagrange's theorem, o(G) is a divisor of k!.

(1.8) PROPOSITION Let H be a subgroup of a finite group G and let (G: H) = k. If o(G) is not a divisor of k!, then H contains a nontrivial normal subgroup of G.

PROOF Let G act on X, the set of left cosets of H in G, by left translation. [cf. (1.2.4)] By (1.7), θ is a homomorphism from G to B(X). Since o(G) is not a divisor of k!, θ is not injective. Also, by the previous proposition, ker (θ) is contained in St(gH) for every $g \in G$. By (1.2.4) St(H) = H. Hence, we have ker (θ) is a nontrivial normal subgroup of G which is contained in H.

The next corollary is an immediate consequence of the proof of the previous proposition. It is an extension of Cayley's Theorem II, (2.12.5). In some cases, this fact decreases the size of the symmetric group into which a group G can be embedded as a subgroup.

(1.9) COROLLARY Let G be a finite group and let H be a subgroup of G which contains no nontrivial normal subgroups of G. Let G_H denote the set of all left cosets of H in G. Then θ is an injective homomorphism from G into $B(G_H)$. In particular, G is isomorphic to a subgroup of $B(G_H)$.

(1.10) COROLLARY Let G be a group of order pn, where p is a prime and $p > n$. Then G has a normal subgroup of order p. In particular, G is not simple.

PROOF By Cauchy's Theorem, G has an element, and hence a subgroup H, of order p. Since (G: H) = n and pn is not a divisor of n!, H must contain a nontrivial normal subgroup of G. Since the only nontrivial subgroup of H is H, the result obtains.

We will now establish three classical results which culminate in the fact that every group of order p^2, p a prime, is abelian. Let p be a prime. Then a group G is said to be a p-group if every element of G has order p^k for some nonnegative integer k.

(1.11) PROPOSITION Let p be a prime and let G be a finite p-group. Then G has a nontrivial center.

PROOF Consider the class equation (CE) for G

$$|G| = |Z(G)| + \sum_{i \in I} |c(g_i)|$$

By Cauchy's Theorem, $|G| = p^k$ for some k. Otherwise, G would have an element of order q for some prime $q \neq p$. Hence, for each $i \in I$, $|c(g_i)|$ is divisible by p. It follows that $|Z(G)|$ is divisible by p.

(1.12) PROPOSITION Let G be a group such that G/Z(G) is cyclic. Then G is abelian.

PROOF Let H = Z(G) and let aH be a generator of G/H. Let $x, y \in G$. Then for some $n, m \in \mathbf{Z}$ and for some $z, w \in H$, we have

$$x = a^n z$$
$$y = b^m w$$

Hence $xy = a^n z a^m w = a^m w a^n z = yx$, since powers of a commute with each other and z, w commute with everything.

(1.13) PROPOSITION Let p be a prime and let G be a group of order p^2. Then G is abelian.

PROOF Since G is a p-group, G has a nontrivial center Z. If $o(Z) = p$, then G/Z is cyclic, contradicting (1.12). Therefore, $o(Z) = p^2$ and G is abelian.

(1.14) PROPOSITION Let G be a finite p-group and let G act on a set X. Define

$$X_0 = \{x : x \in X \text{ and } |or(x)| = 1\}$$

Then $|X_0| \equiv |X| \pmod{p}$.

PROOF Consider the equation given in (1.3) (iv) which we repeat here.

$$|X| = \sum_{i \in I} |or(x_i)| = \sum_{i \in I} (G: St(x_i))$$

Since each term on the right side which is not 1 is divisible by p, the result follows.

(1.15) REMARK We will now note some facts which follow from (1.3) and (1.2.3). Let G be a group and let H be a subgroup of G, then the normalizer of H in G, i.e.

$$\begin{aligned} N_G(H) &= \{g: g \in G \text{ and } gH = Hg\} \\ &= \{g: g \in G \text{ and } gHg^{-1} = H\} \end{aligned}$$

is a subgroup of G. Moreover, $N_G(H)$ is the largest subgroup of G which contains H as a normal subgroup. In particular, H is normal in G if and only if $N_G(H) = G$. Finally note that, if G is finite, then the number of conjugates of H in G is $(G: N_G(H))$. In particular, the number of conjugates of H in G is a factor of the order of G.

(1.16) COROLLARY If G is a finite group and H is a subgroup of G with order p^k for some prime p and some $k > 0$, then

$$o(N_G(H))/o(H) \equiv o(G)/o(H) \pmod p$$

PROOF Let $X = \{gH: g \in G\}$ and let G act on X by left translation. Then note that $|or(gH)| = 1$ if and only if $g^{-1}hg \in H$. Hence, $|or(gH)| = 1$ if and only if $g \in N_G(H)$.

We will now establish the Sylow theorems. These results are very important structure theorems in the theory of finite groups.

(1.17) THEOREM <u>Sylow's First Theorem</u> Let G be a finite group and p be a prime number such that for some $n \in \mathbf{Z}$, $o(G) = p^k n$, $k > 0$ and p is not a divisor of n. Then there is an ascending sequence of p-subgroups in G

$$H_1 < H_2 < \cdots < H_k$$

such that

(i) $o(H_i) = p^i$ for $i = 1, 2, \ldots, k$.

(ii) H_i is a normal subgroup of H_{i+1} for i = 1, 2, ..., k-1.

PROOF By Cauchy's Theorem, G has a subgroup of order p. We proceed by induction and assume that H is a subgroup of order p^i where $i < k$. We will show that there is a subgroup K of G such that $o(K) = p^{i+1}$ and H is a normal subgroup of K. The result will then follow.

By (1.16),

$$o(N_G(H))/o(H) \equiv o(G)/o(H) \equiv 0 \pmod{p}$$

Hence, by Cauchy's Theorem, $N_G(H)/H$ has an element of order p, say $gN_G(H)$. Let K be the subgroup of G generated by H and g. Since $(K: H) = p$, we have $o(K) = p^{i+1}$. Since K is a subgroup of $N_G(H)$, we have H is a normal subgroup of K. The induction is now complete.

If G, p and k are as above and H is a subgroup of G with order p^k, then we will call H a p-Sylow subgroup of G.

(1.18) THEOREM Sylow's Second Theorem Let G be a finite group and p a prime divisor of the order of G. Let P be a p-Sylow subgroup of G and let H be any p-subgroup of G. Then, there is an $x \in G$ such that $H \subseteq xPx^{-1}$. Hence, H is a p-Sylow subgroup of G if and only if H is a conjugate of P in G.

PROOF Let $X = \{gP: g \in G\}$ and let H act on G by left translation. Define X_0 as in (1.14). Then

$$|X_0| \equiv (G: P) \pmod{p}$$

Since p is not a divisor of (G: P),

$$|X_0| \not\equiv 0 \pmod{p}$$

Hence, there is a $g \in G$ such that $gP \in X_0$. It follows that $hgP = gP$ for all $h \in H$. Therefore, $g^{-1}Hg \subseteq P$, or equivalently, $H \subseteq gPg^{-1}$.

If H is a p-Sylow subgroup of G, then $o(H) = o(P) = o(gPg^{-1})$. Hence $H = gPg^{-1}$. Clearly, any conjugate of P is a p-Sylow subgroup of G.

(1.19) THEOREM Sylow's Third Theorem. Let G be a finite group and let p be a prime divisor of the order of G. Let n be the number of p-Sylow subgroups of G. Then n is a divisor of the order of G and

$n \equiv 1 \pmod p$

PROOF Let X be the set of all p-Sylow subgroups of G. By Sylow's First Theorem, $X \neq \emptyset$. Let P be a p-Sylow subgroup of G and let P act on X by conjugation. Let

$$X_0 = \{H: H \in X \text{ and } |or(H)| = 1\}$$

Then, by (1.14), $|X_0| \equiv |X| \pmod p$. We claim that $X_0 = \{P\}$. Certainly $P \in X_0$. Let $H \in X_0$. Then $P \subseteq N_G(H)$. Hence, H and P are both p-Sylow subgroups of $N_G(H)$. By Sylow's Second Theorem, there is a $g \in N_G(H)$ such that $gHg^{-1} = P$. But $gHg^{-1} = H$, therefore, $H = P$ and our claim is established. It follows that $n \equiv 1 \pmod p$.

The fact that n is a divisor of the order of G has already been established. [(cf. (1.17)]

(1.20) EXAMPLES We will now give some examples which will illustrate the application of Sylows theorems.

(1.20.1) A_4 is not simple. Let H be the 2-Sylow subgroup of A_4. Then $o(H) = 4$ and hence, $(A_4: H) = 3$. Since 12 is not a divisor of 3!, then by (1.9), H must contain a nontrivial normal subgroup of G. Since H is a proper subgroup of G, N is a proper subgroup of G.

(1.20.2) Let p, q be primes such that $p > q$ and $p \not\equiv 1 \pmod q$. Then we claim that any group with order pq is abelian. To see this, let G be a group of order pq and let H, K be the p, q-Sylow subgroups of G, respectively. Since $p > q$, by (1.10), H is a normal subgroup of G. Since the number of conjugates of K is a divisor of pq and is congruent to 1 mod q, K has only one conjugate. Therefore, K is a normal subgroup of G. Certainly, $H \cap K = \{e\}$ and $HK = G$. Therefore, by II, Exercise (4.21), G is abelian. In fact, G is cyclic. The proof of this fact will be left as an exercise.

EXERCISES

(1.1) List the subgroups of D_4. Let D_4 act on the set of subgroups of D_4 by conjugation. List the orbits and stabilizers of all the subgroups of D_4.

(1.2) List the subgroups of S_3. Let S_3 act on this set by conjugation. List the orbits and stabilizers of each of the subgroups of S_3. In each case confirm Theorem (1.3).

(1.3) Let Q be the quaternion group, that is the groups whose elements are

$$Q = \{1, -1, i, -i, j, -j, k, -k\}$$

and whose operation is multiplication of quaternions. List all subgroups of Q. Let Q act on its set of subgroups by conjugation. Confirm that every subgroup of Q is normal by listing the orbits of each subgroup of Q.

(1.4) Let H, K be subgroups of a group G, and let H be a normal subgroup of K. Show that $K \subseteq N_G(H)$.

(1.5) Let $G = D_4$ and $H = \{e, st\}$. Find $N_G(H)$.

(1.6) Let D_4 act on D_4 by conjugation. List the orbits of each element of D_4. What is the center of D_4? Show that

$$\{e, s^2, s, t, st\}$$

is a complete residue system with respect to the equivalence relation 'conjugacy'. Verify the class equation for this case.

(1.7) List all 2-Sylow and 3-Sylow subgroups of A_4. Let A_4 act on its set of subgroups by conjugation. Compute this action on each of the Sylow subgroups, thus confirming Theorem (1.19).

(1.8) Let H be a normal subgroup of a finite group G and let $o(H) = p^k$. Show that H is a subgroup of every p-Sylow subgroup of G.

(1.9) Let G be a group of order pq, where p and q are primes and $p \not\equiv 1 \pmod{q}$. Show that G is cyclic. [cf. (1.21.2)]

(1.10) Let $f \in S_n$. Show that

$$f(a_1, a_2, \ldots, a_k)f^{-1} = (f(a_1), f(a_2), \ldots, f(a_k))$$

Deduce that two elements of S_n are conjugate in S_n if and only if they have similar decompositions as products of disjoint cycles.

(1.11) Let n be a positive integer. A partition of n is a sequence of positive integers $k_1, \ldots, k_m$ such that

$$k_1 + k_2 + \cdots + k_m = n$$
$$k_1 \le k_2 \le \cdots \le k_m$$

Let $f \in S_n$.

(i) Suppose that f is a product of disjoint cycles with orders $k_1, k_2, \ldots, k_m$. If we write the cycles in ascending order and if we include 1-cycles, then $k_1, \ldots, k_m$ is a partition of n which we will call the partition corresponding to f. Show that if $g \in S_n$, then g is a conjugate of f in S_n if and only if the partition corresponding to g is the same as the partition corresponding to f.

(ii) Show that the number of conjugacy classes in S_n is the number of partitions of n.

(iii) Find the number of elements in S_n which commute with f.

(1.12) Let $n \le 3$. Show that A_n is generated by the 3-cycles in A_n.

(1.13) Let H be a normal subgroup of A_n such that H contains a 3-cycle. Show that $H = A_n$.

(1.14) Show that A_5 is simple.

(1.15) Let G be a finite group and let $x \in G$ be such that x has two conjugates. Show that G is not simple.

(1.16) Show that every group of order 28 has a normal subgroup of order 7. Thus conclude that if a group of order 28 has a normal subgroup of order 4 then that group is abelian.

(1.17) Let G be a group of order 30. Show that the 3-Sylow and the 5-Sylow subgroups of G are normal subgroups of G. Conclude that G has a normal subgroup of order 15.

(1.18) Let G be a group of order p^2q, where p and q are primes. Show that either a p-Sylow or a q-Sylow subgroup of G must be normal in G.

(1.19) Show that every group of order 385 has a normal subgroup of order 77.

(1.20) Show that every group of order 1000 is not simple.

(1.21) Let G be a group with order 175. Show that G has normal subgroups of order 7 and 25. Thus show that G is abelian.

(1.22) Let H be a p-subgroup of a finite group G such that H is contained in exactly one p-Sylow subgroup P of G. Show that $N_G(H) \subseteq N_G(P)$.

(1.23) Let P be a p-Sylow subgroup of a finite group G and let H be a subgroup of G such that $N_G(P) \subseteq H$. Show that $N_G(H) = H$.

(1.24) Let N be a normal subgroup of a finite group G and let P be a p-Sylow subgroup of G. Show that $N \cap P$ is a p-Sylow subgroup of N.

(1.25) In this exercise, we will show that every group of order 112 is not simple. We will start by assuming that G is a simple group with order 112. Show that each of the following hold.

(a) The number of 2-Sylow subgroups of G is 7.

(b) Let G act of the set of 2-Sylow subgroups of G by conjugation and conclude from (1.8) that G is isomorphic to a subgroup G' of S_7.

(c) Show that G' is not a subgroup of A_7.

(d) Since G' has elements of order 7, deduce that $G' \cap A_7 \neq \{e\}$. Thus a contradiction and G is not simple.

2. SOLVABLE GROUPS

In this section we will develop the theory of solvable groups which will be needed to study Galois theory. We will see later that there is a very strong connection between solvable groups and polynomial equations which are solvable by radicals. Indeed, it is solvable groups that allow us to show that the general polynomial equation with degree larger than four is unsolvable by radicals

(2.1) DEFINITION Let G be a group. Then the set of commutators of G, which we will denote Com(G) is defined by

$$\text{Com}(G) = \{ghg^{-1}h^{-1}: g, h \in G\}$$

The subgroup of G which is generated by Com(G) will be called the commutator subgroup of G and will be denoted G'.

We will show that the commutator subgroup of a group G is in fact a normal subgroup of G.

(2.2) PROPOSITION Let H be a subgroup of a group G. Then H' is a subgroup of G'.

PROOF Certainly, Com(H) $\subseteq$ Com(G), hence H' is a subgroup of G'.

(2.3) PROPOSITION Let G be a group and let $g \in G$. Then

(i) $g\text{Com}(G)g^{-1} \subseteq \text{Com}(G)$;

(ii) if N is a normal subgroup of G, then N' is a normal subgroup of G; in particular, G' is a normal subgroup of G;

(iii) G/G' is abelian;

(iv) The Universal Property of Commutator Quotients. Let G be a group, let H be an abelian group and let $f: G \to H$ be a homomorphism. Then $G' \subseteq \ker(f)$ and f factors uniquely through k; that is, there is a unique homomorphism $f': G/G' \to H$ which makes the following diagram commutative.

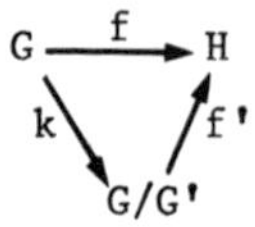

PROOF (i) Note that

$$gxyx^{-1}y^{-1}g^{-1} = gxg^{-1}gyg^{-1}gx^{-1}g^{-1}gy^{-1}g^{-1} \in \text{Com}(G)$$

Hence

$$(*) \quad g\text{Com}(G)g^{-1} \subseteq \text{Com}(G)$$

(ii) Let N be a normal subgroup of G and let $g \in G$. Then $gNg^{-1} = N$. Hence, $g(\text{Com}(N))g^{-1} \subseteq \text{Com}(N)$. [See the proof of (i)] It is easy to see that for every $x \in \text{Com}(N)$ we have $x^{-1} \in \text{Com}(N)$, therefore, N' is just the set of all finite products of elements in Com(N). By (*), and a computation similar to that in the proof of (i), we have $gN'g^{-1} = N'$. Hence N' is a normal subgroup of G.

(iii) Note that $gG'hG' = hG'gG'$ if and only if $(hg)^{-1}gh \in G'$. But, $(hg)^{-1}gh = g^{-1}h^{-1}gh \in \text{Com}(G')$. Hence, G/G' is abelian.

(iv) Let H be an abelian group and let $f: G \to H$ be a homomorphism. If $z = ghg^{-1}h^{-1} \in \text{Com}(G)$ then $f(z) = e_H$ (since H is abelian); hence $G' \subseteq \ker(f)$. The existence of f' is thus guaranteed by I, (1.12). For clarity, the action of f' is given by

$$f'(gG') = f(g)$$

(2.4) COROLLARY Let N be a normal subgroup of a group G such that G/N is abelian, then G' is a subgroup of N.

(2.5) DEFINITION Let G be a group. Then we define inductively the n-th derived subgroup of G as follows

(i) $G^{(1)} = G'$

(ii) $G^{(k+1)} = [G^{(k)}]'$

$G^{(n)}$ is called the <u>n-th derived subgroup of G</u>.

If G is a group, then G is said to be <u>solvable</u> if there is an $n \in \mathbf{N}$ such that $G^{(n)} = \{e\}$.

Note that, if G is solvable, then the n-th derived subgroups form a finite chain of subgroups of G

$$\{e\} < \cdots < G^{(i)} < G^{(i-1)} < \cdots < G^{(1)} < G$$

where $G^{(i)}$ is a normal subgroup of $G^{(i-1)}$ and $G^{(i-1)}/G^{(i)}$ is abelian. We will show that the existence of such a chain of subgroups of a group G is characteristic of solvable groups.

(2.6) THEOREM Let G be a group. Then, G is solvable if and only if there is a sequence of subgroups

(S) $\{e\} = G_n < G_{n-1} < \cdots < G_0 = G$

such that for each $i = 0, 1, \ldots, n-1$

(i) G_{i+1} is a normal subgroup of G_i,

(ii) G_i/G_{i+1} is abelian.

PROOF We have already observed that if G is solvable then a sequence of the form (S) which satisfies (i) and (ii) exists.

Conversely, assume that G is a group with a sequence (S) of subgroups which satisfies (i) and (ii). We will establish by induction that for each $i = 1, 2, \ldots n$, $G^{(i)} < G_i$. Since $G/G_1 = G_0/G_1$ is abelian, by (2.4), we have

$$G^{(1)} = G' < G_1$$

Assume that $G^{(i)} < G_i$; then by (2.2) we have

$$G^{(i+1)} = G^{(i)\prime} < G_i'$$

Since G_i/G_{i+1} is abelian, by (2.4), $G_i' < G_{i+1}$. Therefore,

$$G^{(i+1)} < G_{i+1}$$

Hence, by induction, $G^{(n)} < G_n = \{e\}$. Therefore, G is solvable.

A sequence of the form (S) which satisfies (i) and (ii) is called a solvable series for G. Hence, G is solvable if and only if G has a solvable series. This fact will be useful in some of the following examples.

(2.7) EXAMPLES

(2.7.1) Any abelian group is solvable. In this case, $G' = \{e\}$.

(2.7.2) S_3 is solvable since

$$\{e\} < A_3 < S_3$$

is a solvable series for S_3.

The symmetric group S_4 is also solvable. Let

$$V = \{e, (12)(34), (13)(24), (14)(23)\}$$

Then V is a normal subgroup of A_4 by Exercise (1.10). It follows that

$$\{e\} < V < A_4 < S_4$$

is a solvable series for S_4.

We will see later that S_n is a solvable group if and only if $n < 5$.

(2.7.3) Any finite p-group is solvable. Let G be a finite p-group. Then, by Sylows first theorem, there is a chain of subgroups

$$\{e\} = G_0 < G_1 < \cdots < G_n = G$$

such that G_{i-1} is a normal subgroup of G_i and $o(G_i) = p^i$. It follows that $o(G_i/G_{i-1}) = p$, and hence, G_i/G_{i-1} is abelian. Therefore, G has a solvable series, and by (2.6), G is solvable.

(2.7.4) Let p, q be primes such that $p^2 \not\equiv 1 \pmod{q}$. Then any group G of order p^2q is solvable. For let H be a p-Sylow subgroup and let K be a q-Sylow subgroup of G. If $p > q$, then it follows from (1.8), that H is a normal subgroup of G. Thus

$$\{e\} < H < G$$

is a solvable series for G. [H is abelian by (1.13)] If $p < q$, then let k be the number of q-Sylow subgroups of G. If $k = 1$, then K is a normal subgroup of G and

$$\{e\} < K < G$$

is a solvable series for G. By (1.19), k is a divisor of p^2q and $k \equiv 1 \pmod{q}$. If $k \neq 1$, then it follows that $k = p^2$ and

$p^2 \equiv 1 \pmod{q}$. Since this is contrary to our assumption, $k = 1$ and G is solvable.

(2.8) PROPOSITION Let G be a group and let $f: G \to H$ be a surjective homomorphism of groups.

(i) $f(\mathrm{Com}(G)) = \mathrm{Com}(H)$.

(ii) For each n, $f(G^{(n)}) = H^{(n)}$.

Hence, if G is solvable, then H is solvable.

PROOF To prove (i), let $h_1, h_2 \in H$. Then, there are $g_1, g_2 \in G$ such that

$f(g_1) = h_1$ and $f(g_2) = h_2$. Since

$$h_1 h_2 h_1^{-1} h_2^{-1} = f(g_1 g_2 g_1^{-1} g_2^{-1})$$

it follows that $\mathrm{Com}(H) \subseteq f(\mathrm{Com}(G))$. The reverse inclusion is clear.

We will now establish (ii). Certainly,

$$f(G') = f(\langle \mathrm{Com}(G) \rangle) = \langle f(\mathrm{Com}(G)) \rangle = \langle \mathrm{Com}(H) \rangle = H'$$

Hence, by induction, (ii) follows.

(2.9) PROPOSITION Let G be a group.

(i) If G is solvable and H is a subgroup of G, then H is solvable.

(ii) If N is a normal subgroup of G, then G is solvable if and only if N and G/N are solvable.

PROOF (i) Suppose that $G^{(n)} = \{e\}$. Then, since $H^{(n)} < G^{(n)}$, H is also solvable.

(ii) Assume first that G is solvable; then by (i), N is solvable. Let $k: G \to G/N$ be the canonical map. Since k is surjective, it follows by (2.8) (ii) that G/N is solvable.

Assume now that N and G/N are solvable. Let $k: G \to G/N$ be the canonical map. Since G/N is solvable, there is an m such that

$$(G/N)^{(m)} = \{e\}$$

Since k is surjective,

$$k(G^{(m)}) = (G/N)^{(m)} = \{e\}$$

Therefore, $G^{(m)} < N$. Since N is solvable, $N^{(n)} = \{e\}$ for some n. It follows that

$$G^{(n+m)} < N^{(n)} = \{e\}$$

Hence G is solvable.

(2.10) COROLLARY Let

$$1 \to N \to G \to H \to 1$$

be an exact sequence of groups. Then G is solvable if and only if N and H are solvable.

We will now show that the symmetric group S_n is solvable if and only if $n < 5$.

(2.11) PROPOSITION (i) Let $n \geq 5$. If N is a normal subgroup of S_n which contains every 3-cycle, then N' contains every 3-cycle.

(ii) S_n is solvable if and only if $n < 5$.

PROOF (i) Since N contains every 3-cycle and $n \geq 5$, $f, g \in N$ where

$$f = (123)$$
$$g = (145)$$

A simple computation shows that

$$f^{-1}g^{-1}fg = (135)$$

Hence, $(135) \in N'$. By (2.3)(ii), N' is a normal subgroup of G, therefore, $h(135)h^{-1} \in N'$ for all $h \in G$. Choose h such that $h(x) = 1$, $h(y) = 3$ and $h(z) = 5$ (for arbitrary x,y,z). Then

$$(xyz) = h^{-1}(135)h \in N'$$

[cf. Exercise (1.10)] and the proof of (i) is complete.

(ii) For $n < 5$, S_n is solvable. [See (2.7.2)] Let $n \geq 5$. We will show that for all i, $G^{(i)}$ contains every 3-cycle. By (i), with

$N = G$, G' contains every 3-cycle. By induction, $G^{(i)}$ contains every 3-cycle and hence, G is not solvable.

EXERCISES

(2.1) Show that D_4 is solvable by producing a solvable series.

(2.2) Compute $(D_4)^{(n)}$ for each positive integer n.

(2.3) Show that any group of order 12 is solvable.

(2.4) Show that any group of order 36 is solvable.

(2.5) Let G be a finite group which is not simple. Suppose that for any proper divisor, k, of the order of G, all groups with order k are solvable. Show that G is solvable.

(2.6) Let p, q be primes and let G be a group with order pq. Show that G is solvable.

(2.7) Show that the finite direct product of solvable groups is solvable.

(2.8) Show that all groups with order less than 60 are solvable.

(2.9) Show that all groups of order less than 120, except possibly a group of order 60, are solvable. Note that the order of A_5 is 60 and A_5 is not solvable. [See Exercises (1.25) and (2.5)]

(2.10) Let G be a group and let f be an endomorphism of G. Show that

(a) $f(G') \subseteq G'$.
(b) $f(G^{(i)}) \subseteq G^{(i)}$.
(c) $G^{(i)}$ is a normal subgroup of G.

(2.11) Let a, b, c, d be positive integers and note that

$$(a, b, c) = (a, b)(b, c)$$

Show that (a, b)(c, d) is a product of 3-cycles. Thus conclude that A_n is generated by 3-cycles.

(2.12) Let $n \geq 5$ and let N be a normal subgroup of A_n such that N contains a 3-cycle.

(a) Show that N contains every 3-cycle. [cf. (2.11)]

(b) If N is a nontrivial normal subgroup of A_n, show that N must contain a 3-cycle.

(c) Deduce that A_n is simple for $n \geq 5$.

4
Rings

In this chapter we consider rings - algebraic systems consisting of a set together with two binary operations. Our analysis of rings will follow the pattern set forth in Chapter II. We define rings and subrings in Section 1 and discuss their basic properties. In Sections 2 and 3 we study ring homomorphisms, their kernels, and quotient rings. In Section 4 we generalize the familiar construction of the field of rational numbers from the ring of integers.

1. RINGS AND SUBRINGS

(1.1) DEFINITION A ring $\langle R, +, *\rangle$ is a nonempty set R together with two binary operations $+$ and $*$ on R (generally called addition and multiplication) satisfying the following properties.

(i) $\langle R, +\rangle$ is an additive group.

(ii) Multiplication is associative: for all $s,t,u \in R$, $(s * t) * u = s * (t * u)$.

(iii) Multiplication is left distributive over addition: for all $s,t,u \in R$, $s * (t + u) = (s * t) + (s * u)$.

(iv) Multiplication is right distributive over addition: for all $s,t,u \in R$, $(s + t) * u = (s * u) + (t * u)$.

When the operations are implicit in the context, we write simply R for $\langle R, +, *\rangle$ and st for $s * t$.

We recall the following facts from group theory.

(i) The additive identity is called the zero element and is denoted 0.

(ii) If $s,t \in R$, then $s - t = s + (-t)$ where $-t$ is the additive inverse of t.

(iii) Let $s \in R$ and $n \in \mathbf{N} \cup \{0\}$. Then as in II, (1.5), $ns = 0$ if $n = 0$, $ns = s + s + \cdots + s$ (n times) if $n > 0$, and $(-n)s = -(ns)$.

(iv) For all $n,m \in \mathbf{Z}$, and $s,t \in R$, $(n + m)s = ns + ms$ and $n(s + t) = ns + nt$.

Since multiplication is associative, we generally omit parenthesis, writing stu for the common value of $s(tu)$ and $(st)u$. If n is a positive integer, s^n is defined inductively as follows: $s^1 = s$, and $s^n = ss^{n-1}$ $(n > 1)$. More informally, $s^n = ss \cdots s$ (n times). Then, for all $n,m \in \mathbf{N}$, $s^n s^m = s^{n+m}$.

If there is a nonzero element $1 \in R$ such that $1s = s = s1$ for all $s \in R$, then 1 is called an identity for R and R is said to be a ring with identity (or, when more convenient, a ring with 1). In this case, we define $s^0 = 1$ for all $s \in R$.

The following proposition lists some elementary properties of a ring.

(1.2) PROPOSITION Let R be a ring. Then the following assertions hold.

(i) For all $s \in R$, $s0 = 0 = 0s$.

(ii) For all $s,t \in R$, $(-s)t = -(st) = s(-t)$.

(iii) For all $s,t \in R$, $(-s)(-t) = st$.

(iv) For all $s,t \in R$, $n \in \mathbf{Z}$, $(ns)t = n(st) = s(nt)$.

PROOF (i) $s0 = s(0 + 0)$ (0 is an additive identity)

$= s0 + s0$ (left distributive property)

Hence $s0$ is an idempotent in the group $\langle R, + \rangle$. Since 0 is the unique additive idempotent by II, (1.4), it follows that $s0 = 0$. One proves similarly that $0s = 0$.

(ii) $st + (-s)t = [s + (-s)]t$ (right distributive property)

$= 0t$

$= 0$ [by part (i)]

Hence $(-s)t = -(st)$ by the uniqueness of inverses. The proof that $s(-t) = -(st)$ is similar.

(iii) $(-s)(-t) = -[s(-t)] = -[-(st)] = st$.

(iv) The proof of (iv) is left to the student - induction and the distributive properties must be used.

The ring $R = \{0\}$ with $0 + 0 = 0$ and $0 \cdot 0 = 0$ is called the trivial ring and is denoted by (0). If $R \neq (0)$, R is called a nontrivial ring.

A ring R is called <u>commutative</u> if $st = ts$ for all $s,t \in R$. If R is commutative, then $(st)^n = s^n t^n$ holds for all $s,t \in R$ and $n \in \mathbf{N}$.

(1.3) DEFINITION A subset S of a ring R is a subring of R if S is itself a ring under the operations of R.

(1.4) PROPOSITION A nonempty subset S of a ring R is a subring of R if and only if it satisfies the following properties.

(i) If $s,t \in S$, then $s - t \in S$.

(ii) If $s,t \in S$, then $st \in S$.

PROOF. If S is a subring of R, it clearly satisfies properties (i) and (ii). Let S be a nonempty subset of R satisfying (i) and (ii). Since S satisfies (i), S is an additive subgroup of R by II, (1.10). By (ii), multiplication is a binary operation on S. The associative and distributive properties are inherited from R under restriction.

(1.5) EXAMPLES At this point we present some elementary examples. More examples are given in (1.9).

(1.5.1) The ring $\mathbf{C}$ of complex numbers under ordinary addition and multiplication admits the following chain of subrings.

$$\mathbf{Z} \subseteq \mathbf{Q} \subseteq \mathbf{R} \subseteq \mathbf{C}$$

(1.5.2) By II, (1.19) the only additive subgroups of $\mathbf{Z}$ are of the form $n\mathbf{Z}$ with $n \in \mathbf{N} \cup \{0\}$. It is easily verified [using (1.4)] that in fact $n\mathbf{Z}$ is a subring of $\mathbf{Z}$ for all $n \in \mathbf{N} \cup \{0\}$.

(1.5.3) $\langle \mathbf{R}, +, *\rangle$ where + is ordinary addition and $x * y = x + y - xy$ is not a ring since the distributive properties fail (although all the other properties hold).

(1.5.4) $\langle \mathbf{Z}, +, *\rangle$ where + is ordinary addition and $m * n = -(mn)$ is a ring.

(1.6) DEFINITION Let R be a ring with identity. An element $s \in R$ is a <u>unit</u> if s has an inverse with respect to multiplication; that is, if there is an element $t \in R$ such that $ts = st = 1$.

If R is a ring with identity, we define a subset $R^\times$ of R by

$R^\times = \{s \in R: s \text{ is a unit}\}$

(1.7) PROPOSITION Let R be a ring with identity. Then $R^\times$ is a group under multiplication.

PROOF Let $s,t \in R^\times$. Then if s has an inverse u and t has an inverse v, the product st has inverse vu and hence is an element of $R^\times$. Thus multiplication restricts to a binary operation on $R^\times$. The associativity of multiplication is inherited from R. Since 1 is a unit, $1 \in R^\times$ and is in fact the identity for $R^\times$. Finally, the elements of $R^\times$ are a priori invertible.

$R^\times$ is called the group of units of R. Since $R^\times$ is a group, we have the following facts from group theory [cf. II, (1.4)].

(i) If $s \in R^\times$, the inverse of s is unique and is denoted by s^{-1}.

(ii) For all $s \in R^\times$, $(s^{-1})^{-1} = s$.

(iii) If $s \in R^\times$ and $n \in \mathbf{N}$, then $s^n \in R^\times$ and $(s^n)^{-1} = s^{-n} = (s^{-1})^n$.

(iv) For all $s,t \in R^\times$, $(st)^{-1} = t^{-1}s^{-1}$.

If R is a commutative ring with identity, then $R^\times$ is an abelian group.

(1.8) DEFINITION A field is a commutative ring R with identity in which every nonzero element has an inverse. Thus, if R is a field,

$$R^* = \{s \in R: s \neq 0\} = R^\times$$

is an abelian group under multiplication.

(1.9) EXAMPLES

(1.9.1) **Q**, **R** and **C** are fields. Notice that while **Z** is a subring of the field **Q**, it is not itself a field. In fact $\mathbf{Z}^\times = \{-1, 1\}$.

(1.9.2) The Ring $\mathbf{Z}_n$. Let n be a positive integer. Then $\langle \mathbf{Z}_n, +, \cdot \rangle$ is a commutative ring with identity called the ring of integers modulo n.

As an example, the Cayley tables for addition and multiplication in the ring $\mathbf{Z}_4$ are given below.

+	[0]	[1]	[2]	[3]
[0]	[0]	[1]	[2]	[3]
[1]	[1]	[2]	[3]	[0]
[2]	[2]	[3]	[0]	[1]
[3]	[3]	[0]	[1]	[2]

•	[0]	[1]	[2]	[3]
[0]	[0]	[0]	[0]	[0]
[1]	[0]	[1]	[2]	[3]
[2]	[0]	[2]	[0]	[2]
[3]	[0]	[3]	[2]	[1]

The reader is urged to use these Cayley tables to verify the following facts.

(i) $\mathbf{Z}_4$ contains zero divisors - that is, nonzero elements s and t (not necessarily distinct) such that st = 0.

(ii) $\mathbf{Z}_4$ does not satisfy the cancellation law - that is, there are elements $r, s, t \in \mathbf{Z}_4$, with $r \neq 0$ and $s \neq t$, such that rs = rt.

(iii) $\mathbf{Z}_4^\times = \{[1], [3]\}$.

By Exercise I, (4.3), $\mathbf{Z}_n^\times = \{[m] \in \mathbf{Z}_n: (m, n) = 1\}$. Thus $\mathbf{Z}_n$ is a field if and only if n is prime. If p is a prime, we will sometimes denote the field $\mathbf{Z}_p$ by F_p.

(1.9.3) The Ring $M_2(R)$. In I, (3.4.5) we defined addition and multiplication in the set $M_2(\mathbf{R})$ of 2×2 matrices with real entries. $M_2(\mathbf{R})$ is a noncommutative ring with identity I_2 (the distributive laws may be verified by direct computation). Then

$$[M_2(\mathbf{R})]^\times = GL_2(\mathbf{R}) = \{A \in M_2(\mathbf{R}): \det(A) \neq 0\}$$

(1.9.4) The Ring $M_2(F)$. Let F be any field. Generalizing the definition of $M_2(\mathbf{R})$ we define $M_2(F)$ to be the set of 2×2 matrices with entries from the field F. Addition, multiplication, and the definition of determinant are the same as in II, (1.6.9). Then $M_2(F)$ is a noncommutative ring with identity I_2 and, again,

$$[M_2(F)]^\times = \{A \in M_2(F): \det(A) \neq 0\}$$

In Exercise (1.20) the student is asked to verify this using the fact that F is a field. As in (1.9.3) we denote the group $[M_2(F)]^\times$ by $GL_2(F)$. In particular, if p is prime, $M_2(F_p)$ is a finite noncommutative ring with identity and $GL_2(F_p)$ is a finite nonabelian group.

(1.9.5) The Ring $\mathbf{Z}[\alpha]$. Let a be a squarefree integer - that is, if $m \in \mathbf{N}$, then $m^2 | a$ if and only if $m = 1$. (This is equivalent to saying that a has no repeated primes in its prime decomposition.) Let $\alpha = \sqrt{a}$. (Then α is an irrational complex number which is real if and only if a is positive.)

We define a subset $\mathbf{Z}[\alpha]$ of $\mathbf{C}$ by

$$\mathbf{Z}[\alpha] = \{c + d\alpha: c,d \in \mathbf{Z}\}$$

PROPOSITION (i) Let $a,b,c,d \in \mathbf{Z}$. Then

$$c + d\alpha = e + f\alpha \iff c = e \text{ and } d = f$$

Hence the representation of $c + d\alpha$ $(c,d \in \mathbf{Z})$ of a number in $\mathbf{Z}[\alpha]$ is unique.

(ii) $\mathbf{Z}[\alpha]$ is a subring of $\mathbf{C}$ containing $\mathbf{Z}$. Moreover, if $a > 0$, then $\mathbf{Z}[\alpha] \subseteq \mathbf{R}$.

PROOF (i) Suppose that $c + d\alpha = e + f\alpha$ $(c,d,e,f \in \mathbf{Z})$. Then $c - e = (d - f)\alpha$. If $d \neq f$, then $\alpha = (c - e)/(d - f) \in \mathbf{Q}$. This contradiction implies that d = f. Then c - e = 0, and it follows that c = e.

The proof of (ii) is left to the reader [use (1.4)]. The key fact in showing closure under multiplication is that $\alpha^2 = a \in \mathbf{Z}$.

The ring $\mathbf{Z}[\alpha]$ is in fact the smallest subring of **C** containing both **Z** and α and is called the subring of **C** generated by **Z** and α.

(1.9.6) The Rings R^X, $F(R)$ and $C^n(R)$. Let R be a ring and X a nonempty set. Let

$$R^X = \{f: f \text{ is a function from } X \text{ to } R\}$$

If f and g are elements of R^X we define their sum and product by

$$(f + g)(x) = f(x) + g(x) \quad \text{and} \quad (fg)(x) = f(x)g(x) \quad (x \in X)$$

Then R^X is a ring. If R is commutative, so is R^X. If R has an identity, then the constant function f(x) = 1 is an identity for R^X.

In the particular case that X = R, we shall use the notation F(R) in place of R^X - that is, [as in I, (3.4.6)],

$$F(R) = \{f: f \text{ is a function from } R \text{ to } R\}$$

Then, considering the ring **R** of real numbers, the set $C^0(\mathbf{R})$ of all continuous functions from **R** to **R** is a subring of $F(\mathbf{R})$.

Let n be a positive integer. By II, (1.20.5), the set $C^n(\mathbf{R})$ of all n-times differentiable functions from **R** to **R** with continuous n-th derivative is an additive group. We recall from calculus Leibnitz' formula for the derivative of a product: $D(fg) = f(Dg) + g(Df)$. Hence, if $f,g \in C^1(\mathbf{R})$, $fg \in C^1(\mathbf{R})$ (observe that in this case all the functions on the right hand side of the equation are continuous and hence fg has continuous first derivative).

More generally, one may show that if $f,g \in C^n(\mathbf{R})$, $fg \in C^n(\mathbf{R})$ (this follows from the generalized Leibnitz formula for the n-th derivative of a product).

Thus we have the following chain of subrings

$$\cdots \subseteq C^n(\mathbf{R}) \subseteq \cdots \subseteq C^1(\mathbf{R}) \subseteq C^0(\mathbf{R}) \subseteq F(\mathbf{R})$$

The set of polynomial functions on **R**, usually denoted by **R**[x], is a subring of $C^n(\mathbf{R})$ for all $n \in \mathbf{N}$. An element of **R**[x] is a function of the form

$$p(x) = a_0 + a_1x + a_2x^2 + \cdots + a_nx^n$$

for some $n \in \mathbf{N} \cup \{0\}$ and $a_0, a_1, \ldots, a_n \in \mathbf{R}$. The ring **R**[x] will be studied in greater detail in Chapter V.

(1.9.7) The Ring of Quaternions. Let the set K consist of all symbols $a + bi + cj + dk$ with $a,b,c,d \in \mathbf{R}$. We define $a + bi + cj + dk = a' + b'i + c'j + d'k$ if and only if $a = a'$, $b = b'$, $c = c'$ and $d = d'$.

Addition is defined as follows:

$$(a + bi + cj + dk) + (a' + b'i + c'j + d'k)$$
$$= (a + a') + (b + b')i + (c + c')j + (d + d')k$$

Multiplication is defined by multiplying the two symbols formally and collecting terms using the relations

$$i^2 = j^2 = k^2 = ijk = -1$$
$$ij = k,\ ji = -k,\ jk = i,\ kj = -i,\ ki = j \text{ and } ik = -j$$

Verify by direct computation that K is a noncommutative ring with zero element $0 = 0 + 0i + 0j + 0k$ and identity $1 = 1 + 0i + 0j + 0k$.

If $a + bi + cj + dk$ is a nonzero element of K, then it is a unit of K with inverse $(a/f) - (b/f)i - (c/f)j - (d/f)k$ where $f = a^2 + b^2 + c^2 + d^2$. A ring R with 1 in which every nonzero element is a unit is called a <u>division ring</u>. Thus a commutative division ring is a field. Exercise (2.14) shows that K may also be interpreted as a subring of the ring $M_2(\mathbf{C})$ of all 2×2 matrices with complex entries.

(1.9.8) The Ring End(A). Let A be an abelian group (written additively). As in II, (2.6), we define

$$\text{End}(A) = \{f\text{: } f \text{ is a group homomorphism from A to A}\}$$

We define addition and composition of functions as usual - that is, if $f, g \in End(A)$ and $a \in A$, then

$$(f + g)(a) = f(a) + g(a) \text{ and } (f \circ g)(a) = f(g(a))$$

Exercise (1.10) shows that $\langle End(A), +, \circ \rangle$ is a ring with identity and that $[End(A)]^\times = Aut(A)$.

(1.10) GENERAL EXAMPLES

(1.10.1) Direct Products. Let R and R' be two given rings. Then the direct product $R \times R'$ is a ring under addition and multiplication defined as follows:

addition: $(s, s') + (t, t') = (s + t, s' + t')$

multiplication: $(s, s')(t, t') = (st, s't')$

for $s,t \in R$ and $s',t' \in R'$. Note that the binary operations are those of R in the first coordinate and R' in the second coordinate. If R and R' are commutative, so is $R \times R'$. If R and R' are rings with identities 1 and 1' respectively, then (1, 1') is an identity for $R \times R'$ and $(R \times R')^\times = R^\times \times (R')^\times$.

(1.10.2) The Center of a Ring. Let R be a ring. Define

$$Z(R) = \{r \in R: rs = sr \text{ for all } s \in R\}$$

Then Z(R) is a subring of R called the center of R [cf. Exercise (1.7)].

EXERCISES

(1.1) Determine if the given set together with the indicated operations is a ring.

(a) $\langle \mathbf{Z}, -, \bullet \rangle$

(b) $\langle \mathbf{Z}, +, * \rangle$ where $a * b = 2ab$.

(c) The set $B(\mathbf{Z})$ of bijections of $\mathbf{Z}$ under function addition and multiplication.

(1.2) (a) Prove that a subset S of $\mathbf{Z}$ is a subring of $\mathbf{Z}$ if and only if it is an additive subgroup of **Z**.

(b) Show that the set $S = \{x \in \mathbf{R}: x = a\sqrt{2}, a \in \mathbf{Z}\}$ is an additive subgroup of **R** but is not a subring of **R**.

(1.3) Determine if the given subset S of **C** is a subring of **C**.

(a) $S = \{x \in \mathbf{C}: x = a + b\sqrt{3}, a,b \in \mathbf{Z}\}$.

(b) $S = \{x \in \mathbf{C}: x = a + b\sqrt[3]{2}, a,b \in \mathbf{Z}\}$.

(c) $S = \{x \in \mathbf{C}: x^2 - 2x = 0\}$.

(1.4) Determine all binary operations $*$ on **R** such that $x * y = P(x, y)$ is a polynomial function in x and y and $\langle \mathbf{R}, +, * \rangle$ is a ring.

(1.5) Let R be a ring and $x \in R$. Prove that the set $I(x) = \{r \in R: rx = 0\}$ is a subring of R.

(1.6) Prove that if $\alpha = \sqrt{a}$ where a is a squarefree integer, then $\mathbf{Z}[\alpha]$ is a subring of **C** containing **Z** [cf. (1.9.5), (ii)].

(1.7) Prove that Z(R) is a subring of R for any ring R.

(1.8) Let $T_2(\mathbf{R}) = \{A \in M_2(\mathbf{R}): A \text{ is upper triangular}\}$.

(a) Prove that $T_2(\mathbf{R})$ is a subring of $M_2(\mathbf{R})$ containing I_2.

(b) Find the center of $T_2(\mathbf{R})$.

(c) Find $[T_2(\mathbf{R})]^\times$.

(1.9) Let R be a ring and $x \in R$. Let $Z(x) = \{r \in R: rx = xr\}$. Prove that Z(x) is a subring of R.

(1.10) Let A be an abelian group written additively. Prove that $\langle End(A), +, \circ \rangle$ is a ring with identity and that $[End(A)]^\times = Aut(A)$ [cf. (1.9.7)].

(1.11) Let A be the abelian group $\mathbf{Z} \times \mathbf{Z}$. Prove that End(A) is a noncommutative ring.

(1.12) Show that a ring R is commutative if and only if $(a + b)^2 = a^2 + 2ab + b^2$ for all $a,b \in R$.

(1.13) Let R be a ring and let $T = \{r \in R: r$ has finite additive order}. Prove that T is a subring of R.

(1.14) Let R be a ring. An element $r \in R$ is said to be nilpotent if $r^n = 0$ for some $n \in \mathbf{N}$.

(a) Prove that if R is commutative and $N = \{r \in R: r \text{ is nilpotent}\}$, then N is a subring of R.

(b) Show that the word commutative is essential in (a) by exhibiting two nilpotent elements of $M_2(\mathbf{R})$ whose sum is not nilpotent.

(1.15) Find all nilpotent elements of the ring $\mathbf{Z}_{20}$.

(1.16) Let R be a ring. An element $r \in R$ is said to be idempotent if $r^2 = r$.

(a) Show that $A = \begin{pmatrix} 1 & 0 \\ 0 & 0 \end{pmatrix} \in M_2(\mathbf{R})$ is idempotent.

(b) Find all idempotent elements of $\mathbf{Z}_{20}$.

(c) For what $n \in \mathbf{N}$ does the ring $\mathbf{Z}_n$ have no idempotents other than zero and the identity?

(1.17) Let R be a ring with no nonzero nilpotent elements and let r be an idempotent element of R. Prove that $r \in Z(R)$.

(1.18) Prove that $(\mathbf{Z}[i])^{\times} = \{1, -1, i, -i\}$.

(1.19) Let G be the following subset of the ring K of quaternions: $G = \{1, -1, i, -i, j, -j, k, -k\}$. Show that G is a nonabelian group under multiplication in which every proper subgroup is normal.

(1.20) Let F be a field. Show that $M_2(F)$ is a ring and that

$[M_2(F)]^{\times} = \{A \in M_2(F): \det(A) \neq 0\}$ [cf. (1.9.4)].

(1.21) Find the inverse of $A = \begin{pmatrix} [1] & [7] \\ [3] & [9] \end{pmatrix}$ in $GL_2(F_{11})$.

(1.22) Find the number of elements in $GL_2(F_p)$.

(1.23) Let X be a nonempty set and let $S = \mathbf{P}(X)$ [cf. I, (1.4.2)]. Define binary operations + and $*$ on S by

$$A + B = A \triangledown B \qquad [A,B \in \mathbf{P}(S)]$$
$$A * B = A \cap B \qquad [A,B \in \mathbf{P}(S)]$$

(a) Show that S is a commutative ring with zero element the empty set and identity the set X.

(b) Show that $A^2 = A$ for all $A \in S$.

(c) Show that $2A = 0$ for all $A \in S$.

(1.24) Prove that if R and R' are rings then $R \times R'$ is a ring with the properties stated in (1.10.1).

(1.25) Verify that the inverse of a nonzero element of the ring K of real quaternions is as given in (1.9.7).

(1.26) Let $R = \mathbf{Z} \times \mathbf{Z}$. Show that the element (1, 0) is an identity for the subring $S = \mathbf{Z} \times (0)$ of R. Conclude that a subring of a ring R with identity 1_R may have an identity different than 1_R.

2. HOMOMORPHISMS, IDEALS, AND QUOTIENT RINGS

(2.1) DEFINITION Let R and S be rings. A function $f: R \to S$ is called a homomorphism of rings if it satisfies the following properties.

(i) For all $s,t \in R$, $f(s + t) = f(s) + f(t)$; that is, f is a homomorphism of groups from $\langle R, +\rangle$ to $\langle S, +\rangle$.

(ii) For all $s,t \in R$, $f(st) = f(s)f(t)$.

If f is also bijective, we say that f is an isomorphism of rings. In this case, we call R and S isomorphic and write $R \cong S$. Given rings R and S, the map $R \to S$ defined by $s \mapsto 0$ for all $s \in R$ is a homomorphism of rings which we call the zero homomorphism.

Let $f: R \to S$ be a homomorphism of rings. Then, in particular, f is a homomorphism of additive groups. We recall the following facts from group theory II, (2.2).

(i) $f(0) = 0$.

(ii) $f(-s) = -[f(s)]$ for all $s \in R$.

(iii) $f(ns) = nf(s)$ for all $s \in R$, $n \in \mathbf{N}$.

(2.2) PROPOSITION (i) The identity map $1_R: R \to R$ is an isomorphism of rings for all rings R.

(ii) If $f: R \to S$ and $g: S \to T$ are ring homomorphisms, then so is $g \circ f: R \to T$.

(iii) If $f: R \to S$ is an isomorphism of rings, then so is $f^{-1}: S \to R$.

PROOF The proof is routine and is left to the student.

(2.3) COROLLARY (i) $R \cong R$ for all rings R.

(ii) If $R \cong S$, then $S \cong R$.

(iii) If $R \cong S$ and $S \cong T$, then $R \cong T$.

Thus isomorphism is an equivalence relation on the class of all rings.

(2.4) EXAMPLES We present some elementary examples at this point. More examples will be given in (2.10).

(2.4.1) If $n \in \mathbf{N}$, then the map $f: \mathbf{Z} \to \mathbf{Z}_n$ by $f(m) = [m]$ is a homomorphism of rings.

(2.4.2) Define $f: M_2(\mathbf{R}) \to \mathbf{R}$ by

$$\begin{pmatrix} a & b \\ c & d \end{pmatrix} \mapsto a$$

Then f is not a homomorphism of rings (Why?). However, if f' is the map f restricted to the subring $T_2(\mathbf{R})$ of all upper triangular 2×2 matrices with entries from $\mathbf{R}$, then $f': T_2(\mathbf{R}) \to \mathbf{R}$ is a homomorphism of rings.

(2.5) NOTATION AND TERMINOLOGY Let R and S be rings. Then Hom(R, S) denotes the set of all homomorphisms of rings from R to S. Hom(R, S) contains at least one element, namely the zero homomorphism. In the special case R = S, Hom(R, R) is denoted End(R) and elements of End(R) are called endomorphisms. Bijective endomorphisms - that is, isomorphisms from R onto R - are called automorphisms of R. The set of all such automorphisms is denoted Aut(R) and is a group under composition.

(2.6) PROPOSITION Let $f \in \mathrm{Hom}(R, R')$ and S be a subring of R. Then $S' = f(S)$ is a subring of R'.

PROOF We will use (1.4). Since $f|_S: S \to R'$ is, in particular, a homomorphism of additive groups, S' is an additive subgroup of R by II, (2.9). Let $s', t' \in S'$. Then there are elements $s, t \in S$ such that $f(s) = s'$ and $f(t) = t'$. Since S is a subring of R, $st \in S$. Hence $s't' = f(s)f(t) = f(st) \in S'$. It follows by (1.4) that S' is a subring of R'.

Let R and S be rings with identities 1 and 1' respectively. An element $f \in \mathrm{Hom}(R, S)$ is called a <u>unital</u> homomorphism if $f(1) = 1'$.

(2.7) PROPOSITION (i) Let R be a ring with 1. Then the identity map $1_R: R \to R$ is a unital homomorphism.

(ii) If R, S, and T are rings with identity and $f: R \to S$, $g: S \to T$ are unital homomorphisms, then $g \circ f: R \to T$ is a unital homomorphism.

(iii) Let R and S be rings with identity and $f: R \to S$ a unital homomorphism. Then $f(R^\times) \subseteq S^\times$ [in particular, $[f(s)]^{-1} = f(s^{-1})$ for all $s \in R^\times$]. Hence, if f' is the restriction of f to $R^\times$, $f': R^\times \to S^\times$ is a homomorphism of groups. If f is an isomorphism of rings, then f' is an isomorphism of groups.

PROOF The proofs of (i) and (ii) are routine and are left to the student.

To prove (iii), let 1 and 1' be the identities of R and S respectively and let $s \in R^\times$. Then

$$\begin{aligned} f(s)f(s^{-1}) &= f(ss^{-1}) \quad \text{(since f is a homomorphism)} \\ &= f(1) \\ &= 1' \qquad \text{(since f is unital)} \end{aligned}$$

Similarly, $f(s^{-1})f(s) = 1'$. Hence $f(s)$ is an element of $S^\times$ with inverse $f(s^{-1})$. It follows that $f(R^\times) \subseteq S^\times$. Now, for all $s, t \in R^\times$,

$f(st) = f(s)f(t)$; hence f' is a homomorphism of groups.

Suppose now that f is an isomorphism. Then f' is an injective group homomorphism. Since f and f^{-1} are both homomorphisms,

$$S^{\times} = ff^{-1}(S^{\times}) \subseteq f(R^{\times}) \subseteq S^{\times}$$

and hence $S^{\times} = f(R^{\times})$ and f' is a surjective map.

For the interested reader we point out that the preceding proposition gives us an example of a functor from one category to another (cf. Appendix B). Let C be the category whose objects are rings with identity and whose morphisms are unital ring homomorphisms and let D be the category whose objects are groups and whose morphisms are group homomorphisms. For each object R in C we assign the object $T(R) = R^{\times}$ in D and to each unital homomorphism of rings $f: R \to S$ we assign the homomorphism of groups $T(f): R^{\times} \to S^{\times}$, where $T(f)$ is the restriction of f to $R^{\times}$.

(2.8) DEFINITION Let $f \in \text{Hom}(R, S)$. Then the kernel of f, ker f, is the kernel of f as a homomorphism of additive groups. That is,

$$\ker f = \{s \in R: f(s) = 0\}$$

We shall see later that the kernel of f is a special type of subring of R called an ideal. The following proposition follows from group theory [cf. II, (2.8)].

(2.9) PROPOSITION A homomorphism $f: R \to S$ of rings is injective if and only if $\ker f = (0)$.

(2.10) EXAMPLES

(2.10.1) The map $f: \mathbf{Z} \to \mathbf{Z}_n$ defined by $f(m) = [m]$ is a surjective homomorphism of rings with $\ker f = n\mathbf{Z}$.

(2.10.2) The complex conjugation map

$f: \mathbf{C} \to \mathbf{C}$ defined by $x + iy \mapsto x - iy$

is an automorphism of $\mathbf{C}$ which fixes every element of the subring $\mathbf{R}$. We shall see in Chapter IX that in fact f and the identity map are the only two automorphisms of $\mathbf{C}$ which fix all elements of $\mathbf{R}$.

(2.10.3) Let a be a squarefree integer and let $\alpha = \sqrt{a}$. Define a map $f: \mathbf{Z}[\alpha] \to \mathbf{Z}[\alpha]$ by $c + d\alpha \mapsto c - d\alpha$ [cf. (1.9.5)]. Since the representation of $c + d\alpha$ ($c, d \in \mathbf{Z}$) of a number in $\mathbf{Z}[\alpha]$ is unique, f is a well-defined map. In fact, f is an automorphism of the ring $\mathbf{Z}[\alpha]$.

(2.10.4) Let X be the closed interval [0, 1] in $\mathbf{R}$ and let

$$C(X) = \{f: f \text{ is a continuous function from } X \text{ to } \mathbf{R}\}$$

We define the restriction map

$$\tau_X: C^0(\mathbf{R}) \to C(X) \quad \text{by} \quad f \mapsto f|_X$$

Then τ_X is a surjective homomorphism of rings with

$$\ker \tau_X = \{f \in C^0(\mathbf{R}): f(x) = 0 \text{ for all } x \in X\}$$

We shall use the notation I(X) for the ker of τ_X. Thus I(X) is the set of all continuous functions from $\mathbf{R}$ to $\mathbf{R}$ which vanish on the interval X.

(2.10.5) The Evaluation Map. Let R be a ring and X a nonempty set. We recall from (1.9.6) that

$$R^X = \{f: f \text{ is a function from } X \text{ to } R\}$$

is a ring under function addition and multiplication. Let $b \in X$. We define

$$v_b: R^X \to R \quad \text{by} \quad f \mapsto f(b)$$

The map v_b is called evaluation at b. Then v_b is a homomorphism of rings and

$$\ker v_b = \{f \in R^X: f(b) = 0\}$$

(2.10.6) We now consider a particular case of the preceding

example. Let τ be the restriction of the evaluation map v_0: $F(\mathbf{R}) \to \mathbf{R}$ to the subring $\mathbf{R}[x]$. Then

$$\tau: \mathbf{R}[x] \to \mathbf{R} \text{ by } p \mapsto p(0)$$

Observe that if $p = a_0 + a_1x + \cdots + a_nx^n$, then $\tau(p) = a_0$. Thus ker τ is the set of all polynomial functions with zero constant term.

(2.11) Some General Examples of Homomorphisms.

(2.11.1) Inclusion of Subrings. Let S be a subring of R. Then the inclusion map from S to R defined by $s \mapsto s$ for all $s \in S$ is a homomorphism of rings.

(2.11.2) Projections and Injections. Let $R \times R'$ be a direct products of rings. Then the projection maps

$$p_1: R \times R' \to R \text{ by } (r, r') \mapsto r \quad (r \in R, r' \in R')$$
$$p_2: R \times R' \to R' \text{ by } (r, r') \mapsto r' \quad (r \in R, r' \in R')$$

and the injection maps

$$i_1: R \to R \times R' \text{ by } r \mapsto (r, 0) \quad (r \in R)$$
$$i_2: R' \to R \times R' \text{ by } r' \mapsto (0, r') \quad (r' \in R')$$

are homomorphisms of rings.

(2.11.3) The Diagonal Map. Let R be a ring. Then the diagonal map $d: R \to R \times R$ defined by $s \mapsto (s, s)$ is an injective homomorphism of rings. Hence by (2.6) $d(R) = \{(s, s): s \in R\}$ is a subring of $R \times R$.

Let S be a subring of a ring R. Since S is in particular a subgroup of the additive group $\langle R, +\rangle$, the relation congruence modulo S defined by

$$r \equiv t \pmod{S} \text{ if } r - t \in S \quad (r, t \in R)$$

is a congruence relation on the set R with respect to the binary operation + and, if $r \in R$, then

$$[r] = S + r = \{t \in R: t = s + r \text{ for some } s \in S\}$$

The quotient set

$$R/S = \{[r]: r \in R\} = \{S + r: r \in R\}$$

is then an additive group with addition defined by

$[r] + [t] = [r + t]$ or, equivalently,

$$(S + r) + (S + t) = S + r + t \quad (r,t \in R)$$

We wish to find conditions on the subring S which make the relation congruence modulo S a congruence relation with respect to the operation of multiplication on R. The following conditions are required for all $r,t \in R$:

(i) If $r \equiv t \pmod S$, then $ru \equiv tu \pmod S$ for all $u \in R$.

(ii) If $r \equiv t \pmod S$, then $ur \equiv ut \pmod S$ for all $u \in R$.

It is therefore necessary that whenever $r,t \in R$ are such that $r - t \in S$, then $(r - t)u$ and $u(r - t) \in S$ for all $u \in R$. This condition will clearly be satisfied if whenever $s \in S$ and $u \in R$, su and us are elements of S. We are thus motivated to make the following definition.

(2.12) DEFINITION Let I be a nonempty subset of a ring R. Then I is an _ideal_ of R if the following conditions are satisfied.

(i) If $s,t \in I$, then $s - t \in I$ (thus I is an additive subgroup of R).

(ii) If $s \in I$, then for all $r \in R$, $rs \in I$ and $sr \in I$.

Observe that properties (i) and (ii) guarantee that every ideal of R is in particular a subring of R. If R is a ring then R itself and the trivial subring (0) are clearly ideals of R.

(2.13) EXAMPLE Let R be a commutative ring and $x \in R$. Let

$$Rx = \{s \in R: s = rx \text{ for some } r \in R\}$$

Then Rx is an ideal of R called the principal ideal generated by x and is sometimes denoted by (x).

If R has an identity then (x) is the smallest ideal of R containing x and (x) = R if and only if x is a unit of R [cf. Exercise (2.12)].

For example, in the ring **Z** of integers, the principal ideal generated by 2 is the ideal (2) = 2**Z**. Observe that in fact every additive subgroup of **Z** is an ideal of **Z**. This is not generally true in rings.

(2.14) PROPOSITION Let I be an ideal of the ring R.

(i) If addition and multiplication are defined on R/I by

$$[r] + [s] = [r + s] \text{ and } [r][s] = [rs]$$

then $\langle R/I, +, \bullet \rangle$ is a ring with zero element [0]. If R has an identity 1, then [1] is an identity for R/I. If R is commutative, so is R/I. R/I is called the quotient ring of R by I.

(ii) The canonical map $k_I\colon R \to R/I$ defined by $k_I(r) = [r]$ $(r \in R)$ is a surjective homomorphism of rings with ker $k_I = I$.

PROOF (i) By II, (4.2), $\langle R/I, +\rangle$ is an additive group. By I, (3.10), multiplication is an associative binary operation on R/I (which is commutative if multiplication is commutative in R) and, if 1 is an identity for R, then [1] is an identity for R/I. We verify the left distributive property. Let $r,s,t \in R$.

$$\begin{aligned} [r]([s] + [t]) &= [r][s + t] && \text{(definition of + in R/I)} \\ &= [r(s + t)] && \text{(definition of} \cdot \text{in R/I)} \\ &= [rs + rt] && \text{(distributive property in R)} \\ &= [rs] + [rt] && \text{(definition of + in R/I)} \end{aligned}$$

The right distributive property is proved in a similar manner.

(ii) By II, (4.2), the canonical map $k_I\colon R \to R/I$ is a surjective additive group homomorphism. Let $r,s \in R$. Then

$$k_I(rs) = [rs] = [r][s] = k_I(r)k_I(s)$$

and it follows that k_I is a homomorphism of rings.

Since [r] = I + r, the above theorem, written in coset notation, states that if I is an ideal of R, then the quotient set R/I is a ring under the operations

$$(I + r) + (I + s) = I + (r + s)$$
$$(I + r)(I + s) = I + rs$$

The zero element of R/I is the coset $I + 0 = I$ and, if R has an identity, $I + 1$ is an identity for R/I. Finally, the canonical map $k_I: R \to R/I$ is defined by $k_I(r) = I + r$.

(2.15) PROPOSITION Let $f: R \to R'$ be a homomorphism of rings. Then the following assertions hold.

(i) If I' is an ideal of R', then $I = f^{-1}(I')$ is an ideal of R.

(ii) Ker f is an ideal of R.

(iii) If f is surjective and I is an ideal of R, then $I' = f(I)$ is an ideal of R'.

PROOF By II, (4.3), all of the subsets in question are additive subgroups; thus we need only verify property (ii) of (2.12).

(i) Let $s \in I$ and $r \in R$. We must show that rs and $sr \in I$, or, equivalently, that f(rs) and $f(sr) \in I$. Let $r' = f(r)$ and $s' = f(s)$ (so that $s' \in I'$). Then $f(rs) = f(r)f(s) = r's' \in I'$ (since I' is an ideal of R). Hence $rs \in f^{-1}(I') = I$. Similarly, $sr \in I$.

(ii) Ker f is the inverse image under f of the trivial ideal of R' and hence is an ideal by (i).

(iii) Let $r' \in R'$ and $s' \in I'$. Then $s' = f(s)$ for some $s \in I$ and, since f is surjective, $r' = f(r)$ for some $r \in R$. Since I is an ideal of R, rs and sr are elements of I. Hence $r's' = f(r)f(s) = f(rs) \in I'$ and similarly $s'r' \in I'$.

(2.16) THEOREM (Fundamental Theorem of Ring Homomorphisms) Let $f: R \to R'$ be a surjective homomorphism of rings and $K = \ker f$. Then the following assertions hold.

(i) $R/K \cong R'$ via the map $f': [r] \mapsto f(r)$ $(r \in R)$.

(ii) The mappings whose actions are specified below

$$I \to f(I) \quad \text{and} \quad I' \to f^{-1}(I')$$

are a pair of inverse maps between the set of all ideals I of R containing K and the set of all ideals I' of R'.

PROOF (i) By II, (4.5), the map $f'\colon R/K \to R'$ by $f'([r]) = f(r)$ is an isomorphism of additive groups. Let $r,s \in R$. Then

$$\begin{aligned} f'([r][s]) &= f'([rs]) && \text{(multiplication in } R/I\text{)} \\ &= f(rs) && \text{(definition of } f'\text{)} \\ &= f(r)f(s) && \text{(since } f \text{ is a homomorphism)} \\ &= f'([r])f'([s]) && \text{(definition of } f'\text{)} \end{aligned}$$

Thus f' is an isomorphism of rings.

(ii) This statement follows from (2.15) and II, (4.5).

As in groups, the first statement of the previous theorem is generally called the <u>first isomorphism theorem</u>.

(2.17) PROPOSITION Let I and J be ideals of a ring R. Then $I + J$ is an ideal of R.

PROOF $I + J$ is an additive subgroup of R by II, (4.6). If $s = u + v \in I + J$ (with $u \in I$ and $v \in J$) and $r \in R$, then $ru \in I$ and $rv \in J$. Hence $rs = r(u + v) = ru + rv \in I + J$. In a similar manner, $sr \in I + J$.

(2.18) THEOREM (The Second Isomorphism Theorem) Let I and J be ideals of R. Then $I \cap J$ is an ideal of I, J is an ideal of $I + J$, and

$$I/(I \cap J) \cong (I + J)/J$$

PROOF We have seen in II, (4.7), that the map $f = k_J|_I\colon I \to R/J$ is an injective additive group homomorphism with im $f = (I + J)/J$. Since f is the restriction of a ring homomorphism to a subring, f is a homomorphism of rings and the result therefore follows.

(2.19) THEOREM Let $f\colon R \to R'$ be a surjective homomorphism of rings and suppose that I is an ideal of R containing ker f. Let $I' = f(I)$. Then $R/I \cong R'/I'$.

PROOF. The map $k_I \circ f: R \to R'/I'$ is a surjective homomorphism of rings with kernel I [cf. II, (4.8)] and the result therefore follows from the first isomorphism theorem.

(2.20) COROLLARY (The Fundamental Theorem of Quotient Rings) Let I be an ideal of R.

(i) The mappings whose actions are as specified below

$$J \mapsto J/I \quad \text{and} \quad J' \mapsto k_I^{-1}(J')$$

are a pair of inverse maps between the set of all ideals J of R containing I and the set of all ideals J' of the quotient ring R/I.

(ii) Under the correspondence given in (i), J/I is an ideal of the quotient ring R/I and $(R/I)/(J/I) \cong R/J$.

PROOF We need only apply the preceding theorem using the surjective homomorphism $k_I: R \to R/I$.

(2.21) EXAMPLES

(2.21.1) Let n be a positive integer. Then the map $f: \mathbf{Z} \to \mathbf{Z}_n$ defined by $f(m) = [m]$ is a surjective homomorphism of rings with $\ker f = n\mathbf{Z}$. Hence the ring $\mathbf{Z}/n\mathbf{Z}$ is isomorphic to the ring $\mathbf{Z}_n$.

(2.21.2) Let R be a ring with identity. Define a map

$$f: \mathbf{Z} \to R \text{ by } m \mapsto m1$$

Then f is a homomorphism of rings with

$$\ker f = \{m \in \mathbf{Z}: m1 = 0_R\}$$

If f is injective, then R contains a copy of $\mathbf{Z}$. If f is not injective, then $\ker f = n\mathbf{Z}$ for some $n \in \mathbf{N}$ with $n > 1$ and, since $\mathbf{Z}_n \cong \mathbf{Z}/n\mathbf{Z}$, R contains a copy of $\mathbf{Z}_n$.

(2.21.3) Let X be the closed interval [0, 1] in $\mathbf{R}$ and let

$$C(X) = \{f: f \text{ is a continuous function from } X \text{ to } \mathbf{R}\}$$

We recall from (2.10.4) that the restriction map

$$\tau_X: C^0(\mathbf{R}) \to C(X) \text{ by } f \mapsto f|_X$$

is a surjective homomorphism of rings with kernel

$$I(X) = \{f \in C^0(\mathbf{R}): f(x) = 0 \text{ for all } x \in X\}$$

Hence $I(X)$, the set of all continuous functions from R to R which vanish on X, is an ideal of $C^0(\mathbf{R})$ and by (2.16) τ_X induces an isomorphism $C^0(\mathbf{R})/I(X) \cong C(X)$.

(2.21.4) Let R be a ring and X a nonempty set. Let $b \in X$. By (2.10.5) the evaluation map

$$v_b: R^X \to R \text{ by } f \mapsto f(b)$$

is a surjective homomorphism of rings with

$$\ker v_b = \{f \in R^X: f(b) = 0\}$$

Hence $\ker v_b$ is an ideal of R^X and v_b induces an isomorphism

$$R^X/\ker v_b \cong R.$$

(2.21.5) By (2.10.6) the map

$$\tau: \mathbf{R}[x] \to \mathbf{R} \text{ by } p \mapsto p(0) \text{ for } p \in \mathbf{R}[x]$$

is a surjective homomorphism of rings with

$$\ker \tau = \{p \in \mathbf{R}[x]: p \text{ has zero constant term}\}$$

We may then observe that in fact $\ker \tau = (x)$, the principal ideal generated by the polynomial $p(x) = x$ in $\mathbf{R}[x]$ (that is, a polynomial $q \in \mathbf{R}[x]$ has zero constant term if and only if x is a factor of q). We may therefore conclude from the first isomorphism theorem that

$$\mathbf{R}[x]/(x) \cong \mathbf{R}$$

(2.21.6) Let $I_1, \ldots, I_n$ be a collection of ideals of a ring R. We define a map f from R to the direct product of the quotient rings of R by I_i as follows

$$f: R \to \prod_{i=1}^{n} R/I_i \text{ by } r \mapsto (I_1 + r, \ldots, I_n + r)$$

The following facts may be verified.

(i) f is a homomorphism of rings.

(ii) ker $\tau = \bigcap_{i=1}^{n} I_i$.

(iii) $R/\bigcap_{i=1}^{n} I_i$ is isomorphic to a subring of $\prod_{i=1}^{n} R/I_i$.

In Section 3 we will discuss conditions under which the map f given above is surjective. Examples are given in the exercises to show that this is not always the case.

EXERCISES

(2.1) Decide if the given map is a homomorphism of rings.

(a) $f: \mathbf{R} \to \mathbf{R}$ by $f(x) = 2x$.

(b) $f: \mathbf{R} \to \mathbf{R}$ by $f(x) = 2^x$.

(c) $\tau: C^1(\mathbf{R}) \to C^0(\mathbf{R})$ by $\tau(f) = Df$.

(d) $f: M_2(\mathbf{R}) \to \mathbf{R}$ by $f(A) = \det A$.

(2.2) Let R be a ring.

(a) Is $\langle End(R), +, \circ\rangle$ a ring?

(b) Is $\langle End(R), +, *\rangle$ where $*$ is multiplication of functions a ring?

(2.3) Let R be a ring and $a \in R$. Let $I(a) = \{r \in R: ra = 0\}$.

(a) Prove that if R is commutative then I(a) is an ideal.

(b) Show that the commutative hypothesis in part (a) is necessary by exhibiting elements $A,B,C \in M_2(\mathbf{R})$ such that $B \in I(A)$ but $BC \notin I(A)$.

(2.4) Let I be an ideal of the ring R. Prove that the sets T and S defined below are ideals of R.

$T = \{r \in R: rx = 0 \text{ for every } x \in I\}$

$S = \{r \in R: rx \in I \text{ for every } x \in I\}$

(2.5) Let R be a commutative ring and, as in Exercise (1.14), $N = \{x \in R: x \text{ is nilpotent}\}$.

(a) Prove that N is an ideal of R.

(b) Prove that R/N is a ring with no nonzero nilpotent elements.

(2.6) Let R, S, R' and S' be rings. Show that if $f \in \text{Hom}(R, R')$ and $g \in \text{Hom}(S, S')$ then the map $f \times g: R \times S \to R' \times S'$ defined by $(r, s) \mapsto (f(r), g(s))$ is an element of $\text{Hom}(R \times S, R' \times S')$.

(2.7) Prove that the ring $M_2(\mathbf{R})$ contains no proper nontrivial ideals.

(2.8) Let S be a ring and $f: M_2(\mathbf{R}) \to S$ be a nonzero homomorphism. Prove that f is injective.

(2.9) Find all ideals of the ring $T_2(\mathbf{R})$ of all upper triangular 2×2 matrices with entries from $\mathbf{R}$.

(2.10) Prove that if S is a subring of a ring R then the equivalence relation congruence mod S is a congruence relation with respect to multiplication if and only if S is an ideal of R.

(2.11) Decide if the given homomorphism of rings is surjective [cf. (2.21.6)].

(a) $f: \mathbf{Z} \to \mathbf{Z}_{10} \times \mathbf{Z}_2$ by $r \mapsto ([r], [r])$.

(b) $f: \mathbf{Z} \to \mathbf{Z}_6 \times \mathbf{Z}_5$ by $r \mapsto ([r], [r])$.

(2.12) Let R be a commutative ring and $x \in R$. As in (2.13), let $(x) = Rx = \{rx: r \in R\}$.

(a) Prove that (x) is an ideal of R.

(b) Prove that if R has an identity then (x) contains x and $(x) = R$ if and only if x is a unit.

(2.13) Show that the map $f: \mathbf{C} \to M_2(\mathbf{R})$ defined by

$$f(a + bi) = \begin{pmatrix} a & b \\ -b & a \end{pmatrix}$$

is an injective unital homomorphism of rings. Conclude that $\mathbf{C}$ may be identified with a subring of $M_2(\mathbf{R})$.

(2.14) Let K be the ring of real quaternions. Show that the map $f: K \to M_2(\mathbf{C})$ defined by

$$f(a + bi + cj + dk) = \begin{pmatrix} a + bi & c + di \\ -c + di & a - bi \end{pmatrix}$$

is an injective unital homomorphism of rings. Conclude that K may be identified with a subring of $M_2(\mathbf{C})$.

(2.15) Let f: $\mathbf{Z} \to \mathbf{Z}$ be a nonzero ring homomorphism. Prove that f is the identity map.

(2.16) Let $T = T_2(\mathbf{R})$ and

$$I = \left\{ \begin{pmatrix} 0 & b \\ 0 & 0 \end{pmatrix} : b \in \mathbf{R} \right\}$$

Show that I is an ideal of T and that $T/I \cong \mathbf{R} \times \mathbf{R}$. What ideals of T containing I correspond to the ideals $\mathbf{R} \times (0)$ and $(0) \times \mathbf{R}$?

(2.17) Illustrate the fundamental theorem of quotient rings (2.20) by constructing Hasse diagrams for the ideals of $\mathbf{Z}$ containing $12\mathbf{Z}$ and the ideals of $\mathbf{Z}_{12}$.

(2.18) Let R be a ring and I, J ideals of R. Let

$$IJ = \{\sum_{i=1}^{n} r_i s_i : r_i \in I,\ s_i \in I,\ n \in \mathbf{N}\}$$

Show that IJ is an ideal of R and that $IJ \subseteq I \cap J$.

(2.19) Give an example of rings R and R' with identity and a homomorphism f: $R \to R'$ which is not unital.

(2.20) Let I and J be ideals of R. Then R is said to be a <u>direct sum</u> of I and J if $R = I + J$ and $I \cap J = (0)$. Prove that R is a direct sum of I and J if and only if for every $x \in R$, there are unique elements $r \in I$ and $s \in J$ such that $x = r + s$.

(2.21) Show that if R is a commutative ring with identity and e is a central idempotent of R [that is, $e \in Z(R)$ and $e^2 = e$], then $1 - e$ is also a central idempotent and R is a direct sum of the ideals Re and $R(1 - e)$.

(2.22) Let R be an arbitrary ring and let S be the set $R \times \mathbf{Z}$. Define addition and multiplication on S as follows.

$$(a, m) + (b, n) = (a + b, m + n)$$
$$(a, m)(b, n) = (ab + mb + na, mn)$$

(a) Show that under these operations S is a ring with identity (0, 1).

(b) Show that the map $f: R \to S$ defined by $f(a) = (a, 0)$ is an injective homomorphism of rings and hence R may be identified with the subring $S' = \{(a, 0): a \in R\}$ of S. Thus every ring may be embedded in a ring with identity.

(2.23) An additive subgroup I of a ring R is said to be a left ideal of R if $ru \in I$ for all $r \in R$ and $u \in I$. Prove that if R is a ring and $x \in R$, each of the following subsets of R is a left ideal of R.

(a) $I = Rx = \{rx: r \in R\}$.

(b) $I = (0: x) = \{r \in R: rx = 0\}$.

3. MAXIMAL IDEALS AND THE CHINESE REMAINDER THEOREM

(3.1) PROPOSITION Let R be a ring with identity and let I be an ideal of R. Then the following statements are equivalent.

(i) I contains a unit.

(ii) $1 \in I$.

(iii) $I = R$.

PROOF (i)$\Longrightarrow$(ii) Suppose that the unit s is an element of I. Then, since I is an ideal of R, $1 = s^{-1}s \in I$.

(ii)$\Longrightarrow$(iii) Suppose that $1 \in I$. Then for all $r \in R$, $r = r1 \in I$ and hence $I = R$.

(iii)$\Longrightarrow$(i) If $I = R$ then I contains the unit 1.

(3.2) PROPOSITION Let R be a commutative ring with identity. Then the following statements are equivalent.

(i) R is a field.

(ii) R contains no proper nontrivial ideals; that is, if I is an ideal of R, then either $I = R$ or $I = (0)$.

PROOF (i)$\Longrightarrow$(ii) Let R be a field and I a nontrivial ideal of R. Then, since $I \neq (0)$, there is a nonzero element s in I. Since R is a field, s is a unit. Hence, by the previous proposition, $I = R$.

(ii)$\Longrightarrow$(i) Suppose that R contains no proper nontrivial ideals. Let s be a nonzero element of R and let Rs = (s) be the principal ideal generated by s. Then Rs is an ideal of R containing the nonzero element s. Hence, by hypothesis, Rs = R and it follows that s is a unit of R.

(3.3) THEOREM Let I be a proper ideal of R such that R/I, the quotient ring of R by I, is a commutative ring with identity (this occurs, for example, if R is a commutative ring with identity). Then I is a maximal ideal of R if and only if R/I is a field.

PROOF By the fundamental theorem of quotient rings, (2.20), the ideals of the ring R/I are in one-to-one correspondence with the ideals of R containing I. Then I is a maximal ideal of R if and only if there are no proper ideals of R strictly containing I; that is, if and only if the ring R/I contains no proper nontrivial ideals. By the preceding proposition this occurs if and only if R/I is a field.

(3.4) EXAMPLES

(3.4.1) By (1.9.2) $\mathbf{Z}_n$ is a field if and only if n is prime. Since $\mathbf{Z}_n \cong \mathbf{Z}/n\mathbf{Z}$, it follows by (3.3) that $n\mathbf{Z}$ is a maximal ideal of $\mathbf{Z}$ if and only if n is prime. This fact may also be verified directly using the ideal structure of $\mathbf{Z}$ and the definition of prime [see Exercise (3.1)].

(3.4.2) We recall from (2.21.4) that if $b \in \mathbf{R}$ and $I(b) = \{p \in \mathbf{R}[x]: p(b) = 0\}$, then $\mathbf{R}[x]/I(b) \cong \mathbf{R}$. Since $\mathbf{R}$ is a field, it follows by (3.3) that I(b) is a maximal ideal of $\mathbf{R}[x]$.

As a particular example, we observed in (2.21.5) that I(0) = (x) (the principal ideal generated by x, or equivalently, the set of polynomials with zero constant term). Thus (x) is a maximal ideal of $\mathbf{R}[x]$. We shall prove in Chapter V that in fact for all $b \in \mathbf{R}$, I(b) = (x - b), the principal ideal of $\mathbf{R}[x]$ generated by the polynomial x - b.

A pair of ideals I, J of a ring R is said to be <u>comaximal</u> if

$I + J = R$. A collection C of ideals of R is said to be pairwise comaximal if the pair I, J is comaximal for all $I \neq J \in C$. We observe from (3.1) that if R has an identity, then the pair I, J of ideals of R is comaximal if and only if $1 \in I + J$.

(3.5) LEMMA Let $I \neq J$ be maximal ideals of R. Then the pair I, J is comaximal.

PROOF Since $I + J$ is an ideal of R which properly contains the maximal ideal I, $I + J = R$.

(3.6) PROPOSITION Let R be a ring with identity and let $\{I_1, \ldots, I_n\}$ be a pairwise comaximal collection of ideals of R. Then for $k = 1, \ldots, n$, the pair of ideals I_k, $\bigcap_{i\neq k} I_i$ is comaximal.

PROOF We shall prove the result for $k = 1$; that is, we shall show that $I_1 + \bigcap_{i\neq 1} I_i = R$. The proof for general k is similar.

We proceed by induction on n, the number of ideals in the collection. If $n = 2$, then $\bigcap_{i\neq 1} I_i = I_2$ and $I_1 + \bigcap_{i\neq 1} I_i = I_1 + I_2 = R$ by hypothesis.

Suppose that $n > 2$ and the result is true for all pairwise comaximal collections of ideals with $n - 1$ elements. Since the pair I_1, I_n is comaximal there are elements $a \in I_1$, $b \in I_n$ such that $1 = a + b$. By the induction hypothesis, the pair I_1, $\bigcap_{i=2}^{n-1} I_i$ is comaximal; hence there are elements $c \in I_1$ and $d \in \bigcap_{i=2}^{n-1} I_i$ such that $1 = c + d$. Then

$$1 = (c + d)(a + b) = ca + cb + da + db$$

Since $a, c \in I_1$, $(ca + cb + da) \in I_1$. Since $d \in \bigcap_{i=2}^{n-1} I_i$ and $b \in I_n$, $db \in \bigcap_{i=2}^{n} I_i$. Hence $1 \in I_1 + \bigcap_{i=2}^{n} I_i$ and the result now follows by induction.

(3.7) THEOREM. (The Chinese Remainder Theorem) Let R be a ring with identity and let $C = \{I_1, \ldots, I_n\}$ be a pairwise comaximal collection of ideals of R. Let $r_1, \ldots, r_n$ be arbitrary elements of R. Then the system of congruences

$$x \equiv r_1 \pmod{I_1}$$
$$\vdots$$
$$x \equiv r_n \pmod{I_n}$$

has a solution - that is, there is an element $r \in R$ such that $r \equiv r_i \pmod{I_i}$ $(i = 1, \ldots, n)$.

PROOF By the previous proposition, for each $k = 1, \ldots, n$, $R = I_k + \bigcap_{i \neq k} I_i$. Hence for each k there are elements $s_k \in I_k$ and $t_k \in \bigcap_{i \neq k} I_i$ such that $r_k = s_k + t_k$. Then

(*) $r_k \equiv t_k \pmod{I_k}$ for each k

Observe that if $j \neq k$, then $t_j \in \bigcap_{i \neq j} I_i \subseteq I_k$ and hence

(**) $t_j \equiv 0 \pmod{I_k}$ if $j \neq k$

Let $r = t_1 + \cdots + t_n$. Then, for $k = 1, \ldots, n$,

$$\begin{aligned} r &\equiv (t_1 + \cdots + t_n) \pmod{I_k} \\ &\equiv t_k \pmod{I_k} && \text{[by (**)]} \\ &\equiv r_k \pmod{I_k} && \text{[by (*)]} \end{aligned}$$

(3.8) COROLLARY Let R be a ring with identity and let $C = \{I_1, \ldots, I_n\}$ be a collection of ideals of R. Define

$$f: R \to \prod_{i=1}^{n} R/I_i \quad \text{by} \quad r \mapsto (I_1 + r, \ldots, I_n + r)$$

Then f is a homomorphism of rings with $\ker f = \bigcap_{i=1}^{n} I_i$. If the collection C is pairwise comaximal, then f is surjective and hence induces an isomorphism

$$R/(\bigcap_{i=1}^{n} I_i) \cong \prod_{i=1}^{n} R/I_i$$

We first discuss applications of the preceding theorems in the ring of integers **Z**.

(3.9) LEMMA Let $m, n \in \mathbf{Z}$. Then the pair of ideals $m\mathbf{Z}$, $n\mathbf{Z}$ is comaximal if and only if m and n are relatively prime.

PROOF The integers m and n are relatively prime if and only if there are integers x and y such that $mx + ny = 1$; that is, if and only if $1 \in m\mathbf{Z} + n\mathbf{Z}$. By (3.1) this occurs if and only if the pair $m\mathbf{Z}$, $n\mathbf{Z}$ is comaximal.

(3.10) THEOREM (The Chinese Remainder Theorem for Integers) Let $m_1, \ldots, m_n$ be positive integers which are relatively prime in pairs. Then if $a_1, \ldots, a_n$ are any integers, the system of congruences

$$\begin{aligned} x &\equiv a_1 \pmod{m_1} \\ &\vdots \\ x &\equiv a_n \pmod{m_n} \end{aligned}$$

has a solution.

PROOF We recall that if $a, b \in \mathbf{Z}$ and m is a positive integer, then $a \equiv b \pmod{m}$ if and only if $a \equiv b \pmod{m\mathbf{Z}}$. The result now follows from the preceding lemma and the general Chinese Remainder Theorem.

We recall from II, (1.19) that if $m_1, \ldots, m_n$ are elements of $\mathbf{Z}$ which are relatively prime in pairs and $m = m_1 \cdots m_n$, then $m = \text{lcm}(m_1, \ldots, m_n)$ and $m\mathbf{Z} = \bigcap_{i=1}^{n} m_i\mathbf{Z}$.

(3.11) PROPOSITION Let $m_1, \ldots, m_n$ be elements of $\mathbf{Z}$ which are relatively prime in pairs and let $m = m_1 \cdots m_n$. Then the map

$$f: \mathbf{Z} \to \prod_{i=1}^{n} \mathbf{Z}/m_i\mathbf{Z} \text{ by } a \mapsto (m_1\mathbf{Z} + a, \ldots, m_n\mathbf{Z} + a)$$

is a surjective homomorphism of rings with $\ker f = m\mathbf{Z}$. Hence

$$\mathbf{Z}_m \cong \mathbf{Z}/m\mathbf{Z} \cong \prod_{i=1}^{n} \mathbf{Z}/m_i\mathbf{Z}$$

PROOF The result follows from Corollary (3.8).

(3.12) COROLLARY Let n be an integer greater than 1 with prime factorization

$$n = m_1 \cdots m_n \quad \text{where } m_i = p_i^{r_i}$$

Then $\mathbf{Z}_n \cong \prod_{i=1}^{n} \mathbf{Z}/m_i\mathbf{Z}$.

We have the following examples of the above Corollary.

$\mathbf{Z}_{20} \cong \mathbf{Z}_4 \times \mathbf{Z}_5$ since $20 = 2^2\ 5$
$\mathbf{Z}_{60} \cong \mathbf{Z}_4 \times \mathbf{Z}_3 \times \mathbf{Z}_5$ since $60 = 2^2\ 3\ 5$

(3.13) EXAMPLE Let $(b_1, c_1), \ldots, (b_n, c_n)$ be points in $\mathbf{R}^2$ with distinct first coordinates. We shall use the Chinese Remainder Theorem to prove that there is a polynomial function $p \in \mathbf{R}[x]$ which contains these points [so that $p(b_i) = c_i$ for $i = 1, \ldots, n$].

By (3.4.2), for each i the set

$$I(b_i) = \{p \in \mathbf{R}[x]: p(b_i) = 0\}$$

is a maximal ideal of $\mathbf{R}[x]$. These ideals are distinct since if $i \neq j$, $b_i \neq b_j$ and hence the polynomial $x - b_i$ is an element of $I(b_i)$ but not of $I(b_j)$. Thus, by (3.5), the collection $C = \{I(b_1), \ldots, I(b_n)\}$ is pairwise comaximal.

For each i, let $p_i \in \mathbf{R}[x]$ be the constant polynomial $p_i(x) = c_i$. By the Chinese Remainder Theorem there is a polynomial $p \in \mathbf{R}[x]$ such that

$$p \equiv p_i \pmod{I(b_i)} \qquad (i = 1, \ldots, n)$$

Then, for each i, $p - p_i \in I(b_i)$; that is, $(p - p_i)(b_i) = 0$. Thus $p(b_i) = p_i(b_i) = c_i$ as desired.

EXERCISES

(3.1) Use the definition of prime number and the ideal structure of $\mathbf{Z}$ to show that if $n \in \mathbf{Z}$, then the ideal $n\mathbf{Z}$ is maximal if and only if n is prime.

(3.2) Let R be a commutative ring with identity. Show that the following statements are equivalent.

(i) R is a field.

(ii) (0) is a maximal ideal of R.

(iii) Every nonzero homomorphism of rings $f: R \to S$ is injective.

(3.3) Let R be a commutative ring with identity and I an ideal of R. Show that I is a maximal ideal if and only if for all $r \in R - I$ there exists an $x \in R$ such that $1 - rx \in I$.

(3.4) A commutative ring with identity which has a unique maximal ideal is said to be a <u>local</u> ring. Prove that if R is a commutative ring with identity then the following statements are equivalent.

(i) R is local.

(ii) All nonunits of R are contained in some ideal $M \neq R$.

(iii) The set of nonunits of R is an ideal.

(3.5) Prove that every nonzero homomorphic image of a local ring is local.

(3.6) Let p be a prime and $R = \{a/b \in \mathbf{Q}: p \nmid b\}$. Prove that R is a local ring.

(3.7) Prove that a commutative ring with identity is local if and only if for all $r,s \in R$, if $r + s = 1$ then either r or s is a unit.

(3.8) Use the methods of the Chinese Remainder Theorem to solve the following system of congruences.

$x \equiv 4 \pmod{7}$

$x \equiv 5 \pmod{12}$

$x \equiv 8 \pmod{25}$

(3.9) Decide which of the following subsets of $T_2(\mathbf{R})$ are maximal ideals.

(a) $\left\{\begin{pmatrix} a & b \\ 0 & 0 \end{pmatrix} : a,b \in \mathbf{R}\right\}$

(b) $\left\{\begin{pmatrix} 0 & b \\ 0 & c \end{pmatrix} : b,c \in \mathbf{R}\right\}$

(c) $\left\{\begin{pmatrix} 0 & b \\ 0 & 0 \end{pmatrix} : b \in \mathbf{R}\right\}$

(3.10) If $f(x) = a_0 + a_1x + \cdots + a_nx^n \in \mathbf{R}[x]$ with $a_n \neq 0$, then we define the degree of f to be n.

(a) Prove that if f and g are nonzero elements of **R**[x], then deg fg = deg f + deg g.

(b) Prove that if f(x) and g(x) are nonzero elements of **R**[x], then their product is nonzero.

(3.11) Prove the division algorithm for **R**[x]: given f(x), g(x) nonzero elements of **R**[x], there exist q(x), r(x) $\in$ **R**[x] such that g(x) = f(x)q(x) + r(x) and either r(x) = 0 or deg r(x) < deg f(x) [cf. I, (4.9)].

(3.12) If f(x), g(x) $\in$ **R**[x], we say that f(x) divides g(x), and write f(x)|g(x), if g(x) = f(x)q(x) for some q(x) $\in$ **R**[x]. Let f(x) $\in$ **R**[x] and a $\in$ **R**. Prove that (x - a)|f(x) if and only if f(a) = 0.

(3.13) If d,f,g $\in$ **R**[x], then d is said to be a greatest common divisor of f and g [and we write d = gcd(f, g)] if d satisfies the following properties.

(i) d|f and d|g.

(ii) If q $\in$ **R**[x] is such that q|f and q|g, then q|d.

Prove that if f,g are nonzero elements of **R**[x] then there is an element d $\in$ **R**[x] such that d = gcd(f, g) and, for some k,l $\in$ **R**[x], d = fk + gl [cf. I, (4.13)].

(3.14) Prove that if I is a nonzero ideal of **R**[x] then I = (f) where f is a polynomial in I of minimal degree. Thus R[x] is a principal ideal domain; that is, an integral domain in which every ideal is principal.

(3.15) If I = (f) and J = (g) are ideals of **R**[x], show that I $\subseteq$ J if and only if g|f.

(3.16) An element f $\in$ **R**[x] with nonzero degree is said to be irreducible if whenever g $\in$ **R**[x] is such that deg g $\neq$ 0 and g|f then g = cf for some c $\in$ **R**. Show that an ideal I = (f) of **R**[x] (with f $\in$ **R**[x]) is maximal if and only if f is irreducible. Conclude that if b $\in$ **R** then I = (x - b) is a maximal ideal of **R**[x].

(3.17) Use the methods of the Chinese Remainder Theorem to find an element $f \in \mathbf{R}[x]$ which passes through the following points: $(-1, 2)$, $(0, 4)$, $(1, 5)$ and $(3, -2)$.

(3.18) Prove that $\mathbf{R}[x]/(x^2 + 1) \cong \mathbf{C}$.

4. PRIME IDEALS, INTEGRAL DOMAINS, AND THE FRACTION FIELD

In this section A will denote a commutative ring with identity.

A nonzero element $a \in A$ is said to be a zero divisor if $ab = 0$ for some nonzero element $b \in A$. The ring A is then said to be an integral domain if it contains no zero divisors. Thus A is an integral domain if and only if whenever $ab = 0$ for $a,b \in A$, either $a = 0$ or $b = 0$.

(4.1) PROPOSITION Let $s \in A^\times$. Then s is not a zero divisor of A.
PROOF Suppose that $sb = 0$ for some $b \in A$. Since $s \in A^\times$, s has an inverse. Then $b = s^{-1}sb = s^{-1}0 = 0$ and the result now follows.

(4.2) COROLLARY Let A be a field. Then A is an integral domain.

(4.3) PROPOSITION Let s be a nonzero element of A and $\iota_s: A \to A$ by $\iota_s(a) = sa$ $(a \in A)$. Then the following assertions hold.

(i) ι_s is injective if and only if s is not a zero divisor of A.

(ii) ι_s is surjective if and only if $s \in A^\times$.

PROOF (i) Since ι_s is an additive group homomorphism, ι_s is injective if and only if $\ker \iota_s = (0)$. This occurs if and only if, whenever $sa = 0$, $a = 0$; that is, if and only if s is not a zero divisor of A.

(ii) Since $\operatorname{im} \iota_s = sA = (s)$, the principal ideal of A generated by s [cf. (2.13)], ι_s is surjective if and only if $sA = A$. Since A has an identity, this occurs if and only if $s \in A^\times$.

(4.4) COROLLARY Let A be an integral domain. Then A satisfies the cancellation property; that is, if $a,b,c \in A$ with $ab = ac$ and $a \neq 0$, then $b = c$.

(4.5) THEOREM Let A be a finite integral domain. Then A is a field.
PROOF By hypothesis, A is a commutative ring with identity. Let s be a nonzero element of A. Since A is an integral domain, s is not a zero divisor. Hence by the preceding proposition the map $\iota_s: A \to A$ is injective. Since A is finite ι_s is also surjective [cf. I, exercise (1.24)]. Thus, again by the preceding proposition, $s \in A^{\times}$ and it therefore follows that A is a field.

A proper ideal P of A is said to be a <u>prime</u> ideal if, whenever $a,b \in A$ with $ab \in P$, then either $a \in P$ or $b \in P$. Thus A is an integral domain if and only if (0) is a prime ideal of A. The following theorem generalizes this result.

(4.6) THEOREM Let I be a proper ideal of A. Then I is a prime ideal if and only if A/I, the quotient ring of A by I, is an integral domain.
PROOF Since A is a commutative ring with identity, so is A/I. We recall that an element [a] of A/I is nonzero if and only if a is not an element of I. Thus, since $[a][b] = [ab]$ for all $a,b \in A$, A/I is an integral domain if and only if whenever a and b are not elements of I, their product ab is not an element of I; that is, if and only if I is a prime ideal.

(4.7) PROPOSITION Let M be a maximal ideal of A. Then M is a prime ideal.
PROOF By (3.3) if M is a maximal ideal of A, the quotient ring A/M is a field. Then by (4.2), A/M is an integral domain. Hence M is a prime ideal by the preceding theorem.

In the ring **Z** of integers, the ideal (0) is a prime ideal which is not maximal. Since every other proper ideal of **Z** is of the form n**Z** for some $n > 1$, and the quotient ring $\mathbf{Z}/n\mathbf{Z} \cong \mathbf{Z}_n$ is finite, it follows by (4.5) and (4.6) that for $n > 1$, $n\mathbf{Z}$ is a prime ideal if and only if it is a maximal ideal. By (3.4.1) this occurs if and only if n is prime. The student is urged to construct a direct proof that $n\mathbf{Z}$ is a prime ideal if and only if n is prime.

The prime ideals of the polynomial ring $\mathbf{R}[x]$ will be studied in Chapter V.

We saw in (2.21.2) that the map $f: \mathbf{Z} \to A$ defined by $m \mapsto m1$ is a homomorphism of rings with $\ker f = m\mathbf{Z}$ for some $m \in \mathbf{Z}$. We then say that the integral domain A has <u>characteristic</u> m. Since $\mathbf{Z}/m\mathbf{Z} \cong \text{im } f$ which is an integral domain, $m\mathbf{Z}$ is a prime ideal of $\mathbf{Z}$ and therefore either $m = 0$ or $m = p$ for some prime p.

(4.8) PROPOSITION Let A be an integral domain. Then the following assertions hold.

(i) If A has characteristic 0, then A contains an isomorphic copy of $\mathbf{Z}$. In this case $na = 0$ for $n \in \mathbf{Z}$ and $a \in A$ if and only if either $n = 0$ or $a = 0$.

(ii) If A has characteristic p, then A contains an isomorphic copy of the field F_p. In this case $pa = 0$ for all $a \in A$.

PROOF (i) If A has characteristic 0, and $f: \mathbf{Z} \to A$ by $n \to n1$ as in the preceding paragraph then $\ker f = (0)$ and f is an injective map. Then the subring $B = \text{im } f$ of A is isomorphic to $\mathbf{Z}$. If $na = 0$ for n a nonzero element of $\mathbf{Z}$ and $a \in A$, then $0 = na = n(1a) = (n1)a = f(n)a$. Since $f(n) \neq 0$ and A is an integral domain, $a = 0$.

(ii) Suppose that A has characteristic p. Then A contains an isomorphic copy of the field $F_p \cong \mathbf{Z}/p\mathbf{Z}$. If $a \in A$, then $pa = p(1a) = (p1)a = f(p)a = 0a = 0$.

We conclude this section by generalizing the construction of the field of rational numbers from the ring of integers.

(4.9) PROPOSITION Let $S = A^* = \{s \in A: s \neq 0\}$. Define a relation $\sim$ on $A \times S$ by

$$(a, s) \sim (b, t) \Leftrightarrow at = bs \quad (a,b \in A,\ s,t \in S)$$

Then ~ is an equivalence relation on $A \times S$ and ~ satisfies the following properties.

(i) $(a, s) \sim (0, t)$ if and only if $a = 0$ ($a \in A$, $s,t \in S$).

(ii) $(at, st) \sim (a, s)$ ($a \in A$, $s,t \in S$).

(iii) $(a, s) \sim (1, 1)$ if and only if $a = s$ ($a \in A$, $s \in S$).

PROOF It is easily seen that ~ is reflexive and symmetric. To establish transitivity let $a,b,c \in A$ and $s,t,u \in S$ be such that $(a, s) \sim (b, t)$ and $(b, t) \sim (c, u)$.

$(a, s) \sim (b, t)$ implies $at = bs$

$(b, t) \sim (c, u)$ implies $bu = ct$

Then multiplying the first equation by u and the second by s (and using the fact that A is commutative) we get $aut = cst$. Since A is an integral domain and $t \neq 0$, $au = ct$ and hence $(a, s) \sim (c, u)$.

(i) If $a \in A$ and $s,t \in S$, then $(a, s) \sim (0, t)$ if and only if $at = 0s = 0$. Since A is an integral domain and $t \neq 0$, this occurs if and only if $a = 0$.

(ii) Since A^* is closed under multiplication, if $s,t \in A^*$, $st \in A^*$. Then $(at, st) \sim (a,s)$ since $ats = ast$.

(iii) If $a \in A$, $s \in S$, $(a, s) \sim (1, 1)$ if and only if $a1 = 1s$; that is, if and only if $a = s$.

(4.10) THEOREM Let $S = A^*$ and ~ be defined on $A \times S$ as in (4.9). Define

$$a/s = [(a, s)] \quad (a \in A,\ s \in S) \text{ and let}$$

$$K = A/\sim \ = \{a/s\colon a \in A,\ s \in S\}$$

Define operations of addition and multiplication on K as follows.

$$a/s + b/t = (at + bs)/st \quad (a,b \in A,\ s,t \in S)$$
$$(a/s)(b/t) = ab/st \quad (a,b \in A,\ s,t \in S)$$

Then K is a field with zero element 0/1 and identity 1/1.

PROOF Since K consists of equivalence classes, we must first show that the operations defined above are well-defined (that is, independent of the equivalence class representative). To this end, let $a, b, c, d \in A$ and $s, t, u, v \in S$ be such that

(*) $a/s = b/t$ and $c/u = d/v$ (so that $at = bs$ and $cv = du$)

To show that $a/s + c/u = b/t + d/v$, we must show that $(au + cs)/su = (bv + dt)/tv$ [that is, $(au + cs)tv = (bv + dt)su$]. But

$$\begin{aligned}(au + cs)tv &= autv + cvst \\ &= uvbs + dust \qquad \text{[by (*)]} \\ &= (bv + dt)su\end{aligned}$$

and it follows that addition is a well-defined operation on K. We leave it to the student to verify that multiplication is also well-defined.

The verification that addition and multiplication are commutative and associative and that the distributive laws hold are routine and are left to the student [(4.9) must be used].

Since for all $a \in A$, $s \in S$, $0/1 + a/s = (0s + a1)/1s = a/s$, $0/1$ is a zero element for K. The additive inverse of a/s is $(-a)/s$ since

$$a/s + (-a)/s = (as - as)/ss = 0/ss = 0/s = 0/1$$

For all $a \in A$ and $s \in S$, $(1/1)(a/s) = a/s$ and hence $1/1$ is an identity for K.

If $a/s \neq 0/1$, then $a \neq 0$ and hence $a \in S$. Then $s/a \in K$ and $(a/s)(s/a) = 1/1$ so that s/a is an inverse for a/s. Thus K is a field.

The field K given in the previous proposition is generally called the <u>fraction field of A</u>.

(4.11) THEOREM Let K be the fraction field of A. Define

$$k: A \to K \quad \text{by} \quad k(a) = a/1$$

Then the following assertions hold.

(i) k is a (unital) injective homomorphism of rings and hence A is embedded as a subring of the field K.

(ii) (Universal Property) Let B be a commutative ring with identity and $g: A \to B$ a unital homomorphism of rings such that $g(S) \subseteq B^{\times}$ (where $S = A^{*}$). Then there is a unique unital homomorphism of rings $g': K \to B$ making the following diagram commutative.

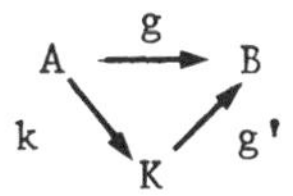

PROOF (i) If $a,b \in A$, then $k(a + b) = (a + b)/1 = a/1 + b/1 = k(a) + k(b)$ and $k(ab) = ab/1 = (a/1)(b/1) = k(a)k(b)$ and hence k is a homomorphism. Since $k(1) = 1/1$, the identity of K, k is unital. If $a \in \ker k$, then $0/1 = k(a) = a/1$, and by (4.9), $a = 0$. Hence k is injective.

(ii) Let B be a commutative ring with identity and $g: A \to B$ be a unital homomorphism of rings such that $g(S) \subseteq B^{\times}$. Then for all $s \in S$, $g(s)$ is a unit of B. We define

$$g': K \to B \quad \text{by} \quad a/s \mapsto g(a)g(s)^{-1}$$

We must first show that g' is a well-defined map. Let $a,b \in A$ and $s,t \in S$ be such that $a/s = b/t$ (so that $at = bs$). Then

$$g(a)g(t) = g(at) = g(bs) = g(b)g(s)$$

and hence, multiplying both sides of the above equation by $g(s)^{-1}g(t)^{-1}$, $g(a)g(s)^{-1} = g(b)g(t)^{-1}$. It follows that $g'(a/s) = g'(b/s)$.

We next verify that g' is a homomorphism. Let a/s, $b/t \in K$ ($a,b \in A$, $s,t \in S$).

$$\begin{aligned} g'((a/s)(b/t)) &= g'(ab/st) \\ &= g(ab)g(st)^{-1} \\ &= g(a)g(b)g(t)^{-1}g(s)^{-1} \\ &= g(a)g(s)^{-1}g(b)g(t)^{-1} \\ &= g'(a/s)g'(b/t) \end{aligned}$$

The verification that $g'((a/s) + (b/t)) = g'(a/s) + g'(b/t)$ is similar and is left to the student.

If $a \in A$, $g' \circ k(a) = g'(k(a)) = g'(a/1) = g(a)g(1)^{-1} = g(a)$ since g is unital. Thus $g' \circ k = g$.

Finally, we show that g' is unique. Let $h: K \to B$ be a unital homomorphism of rings such that $h \circ k = g$. Since $k(S) \subseteq K^{\times}$ and $h: K \to B$ is unital, $h[k(S)] \subseteq B^{\times}$ by (2.7), (iii), and for all $s \in S$

(*) $h(k(s)^{-1}) = [hk(s)]^{-1}$

If $a/s \in K$ with $a \in A$ and $s \in S$, then

$$\begin{aligned} h(a/s) &= h(k(a)k(s)^{-1}) \\ &= hk(a)h(k(s)^{-1}) && \text{(since h is a homomorphism)} \\ &= hk(a)[hk(s)]^{-1} && \text{[by (*)]} \\ &= g(a)g(s)^{-1} && (\text{since } h \circ k = g) \\ &= g'(a/s) && (\text{definition of } g') \end{aligned}$$

Hence $h = g'$.

Since the map $k: A \to K$ by $a \mapsto a/1$ of the preceding proposition is injective, we will often identify a with its image a/1 in K and hence consider A to be a subring of K.

(4.12) COROLLARY Let A be an integral domain with fraction field K and consider A to be a subring of K. Let E be a field and $g: A \to E$ be an injective homomorphism of rings. Then there is a unique (injective) homomorphism $g': K \to E$ such that $g'|_A = g$. In particular, any field E containing A contains an isomorphic copy K' of K with $A \subseteq K' \subseteq E$.

PROOF Since g is injective, $g(s) \neq 0$ for all $s \in S$. Hence, since E is a field, $g(s) \in E^{\times}$. The preceding proposition and exercise (4.11) then guarantee the existence and uniqueness of the map g' [where g' is defined by $g'(a/s) = g(a)g(s)^{-1}$]. Since ker g' is a proper ideal of the field K, ker g' = (0) and it follows that g' is injective.

Finally, if A is contained in the field E, let $g: A \to E$ be the inclusion map and let $K' = g'(K)$. The result then follows.

(4.13) COROLLARY Let A and A' be integral domains with fraction fields K and K' respectively. Then if $\alpha: A \to A'$ is an injective unital ring homomorphism there is a unique homomorphism $\tau: K \to K'$ such that $\tau|_A = \alpha$ (we say in this case that the map τ extends α). If α is an isomorphism, τ is also an isomorphism.

(4.14) EXAMPLES

(4.14.1) The fraction field of the ring of integers **Z** is the field of rational numbers **Q**. By (4.8) any field of characteristic 0 contains an isomorphic copy of **Z**. Hence by the preceding corollary, any field of characteristic 0 contains an isomorphic copy of **Q**.

(4.14.2) Let A = **R**[x], the ring of polynomials with real coeffecients. It is shown in Chapter V that A is an integral domain. The fraction field K of A is the set of all rational functions; that is, $K = \{p(x)/q(x): p(x), q(x) \in A \text{ and } q(x) \neq 0\}$.

REMARK The preceding corollary gives us another example of a functor from one category to another.

Let C be the category whose objects are integral domains and whose morphisms are injective unital homomorphisms. Let D be the category whose objects are fields and whose morphisms are unital homomorphisms. We may define a functor T from C to D as follows.

To each object A in C we assign the object T(A) = K in D where K is the fraction field of A. Let $g: A \to B$ be an injective unital homomorphism of integral domains and let A and B have fraction fields K and L respectively. By (4.13) there is a unique homomorphism $g': K \to L$ such that $g'|_A = g$. We then let $T(g) = g'$.

EXERCISES

(4.1) Let R be a ring with identity and A an integral domain. Prove that if $f: R \to A$ is a nonzero homomorphism of rings, then f is unital.

(4.2) Let R be a commutative ring with identity and I an ideal of R. Show that the following statements are equivalent.

(i) I is a prime ideal of R.

(ii) If J and K are ideals of R with $JK \subseteq I$, then either $J \subseteq I$ or $K \subseteq I$ [cf. exercise (2.18) for the definition of JK].

(4.3) Let $f: R \to R'$ be a surjective homomorphism of rings and $K = \ker f$. Prove that the mappings

$$P \mapsto f(P) \quad \text{and} \quad P' \mapsto f^{-1}(P')$$

are a pair of inverse maps between the set of all prime ideals P of R containing K and the set of all prime ideals P' of R'.

(4.4) Show that if I is an ideal of R then the mapping $P \mapsto P/I$ is a bijection from the set of all prime ideals P containing I to the set of all prime ideals of the quotient ring R/I.

Let R be a commutative ring (not necessarily an integral domain). A nonempty subset S of R is said to be <u>multiplicative</u> if $0 \notin S$ and whenever $s,t \in S$, $st \in S$.

(4.5) Let A be an integral domain. Prove that each of the following subsets of A is multiplicative.

(i) $S = A - P$ where P is a prime ideal of A.

(ii) $S = A^*$.

(iii) $S = A^\times$.

(4.6) Let A be an integral domain and S a multiplicative subset of A. Define $\sim$ on $A \times S$ by

$$(a, s) \sim (b, t) \text{ if and only if } at = bs \quad (a,b \in A,\ s,t \in S)$$

Prove that $\sim$ is an equivalence relation on $A \times S$.

(4.7) Let A be an integral domain, S a multiplicative subset of A and $\sim$ be defined as in the preceding exercise. For $(a, s) \in A \times S$, let $a/s = [(a, s)]$ and $AS^{-1} = A/\sim = \{a/s: a \in A, s \in S\}$.

(a) Define addition and multiplication as follows.

$$a/s + b/t = (at + bs)/st \quad \text{and} \quad (a/s)(b/t) = ab/st$$

Show that $\langle AS^{-1}, +, \ \rangle$ is an integral domain with identity s/s (where s is any element of S).

(b) Show that the map f_S: $A \to AS^{-1}$ defined by $a \mapsto as/s$ ($a \in A$ and s any fixed element of S) is a unital injective homomorphism of rings satisfying $f_S(S) \subseteq (AS^{-1})^{\times}$.

(c) (Universal Property of f_S) Prove that if B is a commutative ring with identity and g: $A \to B$ is a unital homomorphism satisfying $g(S) \subseteq B^{\times}$, then there is a unique unital homomorphism of rings g': $AS^{-1} \to B$ such that $g' \circ f_S = g$.

(4.8) Let S be a multiplicative subset of an integral domain A.

(a) Prove that if I is an ideal of A, then

$$IS^{-1} = \{a/s\colon a \in I,\ s \in S\}$$

is an ideal of AS^{-1}.

(b) Prove that if J is an ideal of AS^{-1}, then $f_S^{-1}(J)$ is an ideal of A.

(c) Prove that the mapping $P \mapsto PS^{-1}$ defines a bijection from the set of prime ideals disjoint from S to the set of prime ideals of AS^{-1}.

(4.9) Let p be a prime and $S = \mathbf{Z} - (p)$. Show that $\mathbf{Z}S^{-1} = \{a/s\colon a, s \in \mathbf{Z} \text{ and } p \nmid s\}$ and that $\mathbf{Z}S^{-1}$ is a local ring [cf. exercise (3.4)].

(4.10) Let S be a multiplicative subset of a commutative ring R (not necessarily an integral domain). Define a relation $\sim$ on $R \times S$ by

$$(r, s) \sim (r', s') \qquad t(rs' - r's) = 0 \text{ for some } t \in S$$

(a) Show that $\sim$ is an equivalence relation and, if R is an integral domain, then $\sim$ is the same as in exercise (4.6).

(b) If $(r, s) \in R \times S$, denote $[(r, s)]$ by r/s. Let

$$RS^{-1} = R/{\sim} = \{r/s\colon r \in R,\ s \in S\}$$

Define addition and multiplication on RS^{-1} by $(r/s + r'/s') = (rs' + r's)/ss'$ and $(r/s)(r'/s') = rr'/ss'$. Show that RS^{-1} is a commutative ring with identity (called the ring of fractions of R by S).

(c) Show that the map f_S: $R \to RS^{-1}$ defined by $r \mapsto rs/s$ (for any $s \in S$) is a homomorphism of rings such that $f_S(S) \subseteq (RS^{-1})^{\times}$.

(d) (Universal Property of f_S) Prove that if T is any commutative ring with identity and g: $R \to T$ is a unital homomorphism of rings such that $g(S) \subseteq T^{\times}$, then there is a unique unital homomorphism g': $RS^{-1} \to T$ such that $g' \circ f_S = g$.

5
Factorization in Commutative Rings

In this chapter we continue our study of rings with a major restriction - All rings will be commutative with unity. We will see that, by further restriction to rings in which all ideals are principal, many of the familiar theorems about integers will obtain. In section 1 we define Euclidean rings and principal ideal rings and develop some of their properties. In section 2 we discuss primes and unique factorization domains. In section 3 we study prime factorization in Noetherian domains and establish a very general form of the Fundamental Theorem of Arithmetic.

1. EUCLIDEAN RINGS AND PRINCIPAL IDEAL RINGS

Recall that for a ring R, R^* is the set of all nonzero elements of R.

(1.1) DEFINITION Let R be a ring. A function $d: R^* \to \mathbf{Q}$ is said to be a Euclidean function on R if the following conditions are satisfied.

(i) For every $r \in R^*$, $d(r) \geq 0$.

(ii) For every $a, b \in R^*$, there are q, r such that $a = bq + r$ and either $d(r) < d(b)$ or $r = 0$.

A ring R is said to be a Euclidean ring with respect to d if d is a Euclidean function on R. A Euclidean ring is a ring (commutative with unity) which admits a Euclidean function. If, in addition, R is an integral domain, then we will call R a Euclidean domain with respect to d, or simply a Euclidean domain. As we will see there are many natural examples of Euclidean rings.

(1.2) EXAMPLES

(1.2.1) Define d: $\mathbf{Z}^* \to \mathbf{Z}$ by $d(n) = |n|$ for each $n \in \mathbf{Z}^*$. Then, by the division algorithm, d is a Euclidean function on **Z**. Hence, **Z** is a Euclidean domain with respect to d. This example is a primary motivation for Definition (1.1).

(1.2.2) The ring of Gaussian integers. Let

$$R = \{a + bi: a,b \in \mathbf{Z}\}$$

be equipped with the usual multiplication and addition in **C**. Then R is a ring which is called the ring of Gaussian integers. Define $d: R^* \to \mathbf{Z}$ by,

$$d(a + bi) = a^2 + b^2$$

To establish that d is a Euclidean function on R, we will need only show that (ii) of (1.1) is valid. We proceed to show this by first noting that for each $z,w \in R$ we have

$$d(zw) = d(z)d(w)$$

The verification of this fact will be left as an exercise to the student.

We will first establish the validity of (ii) in the case b is a positive integer. Let $a = a_1 + a_2 i \in R$ and use the division algorithm to choose q_1, q_2, r_1 and r_2 in **Z** so that

$$a_j = q_j b + r_j \qquad (j = 1, 2)$$

where $|r_j| \le b/2$. (This is more than just a direct application of the division algorithm.) Then

$$\begin{aligned} a &= a_1 + a_2 i \\ &= q_1 b + r_1 + (q_2 b + r_2)i \\ &= (q_1 + q_2 i)b + r_1 + r_2 i \end{aligned}$$

Furthermore, either $r_1 + r_2 i = 0$ or

$$d(r_1 + r_2 i) = r_1^2 + r_2^2 \leq b^2/2 < d(b)$$

Hence, (ii) is valid if b is a positive integer.

We return to the general case. Let

$$b = b_1 + b_2 i$$
$$a = a_1 + a_2 i$$

be nonzero elements of R. Denote by $\bar{b} = b_1 - b_2$. Since $b\bar{b}$ is a positive integer, we can find, q,r such that

$$a\bar{b} = qb\bar{b} + r$$

and either $r = 0$ or $d(r) < d(b\bar{b})$. It follows that either

$$d(b)d(\bar{b}) = d(b\bar{b}) > d(r) = d[(a + qb)\bar{b}] = d(a - qb)d(\bar{b})$$

or $r = 0$. Let $r_1 = a - qb$. Then, since $d(\bar{b}) > o$, we have either $d(r_1) < d(b)$ or $r_1 = 0$. Thus (ii) of (1.1) obtains in the general case. It follows that R is a Euclidean domain with respect to d.

(1.2.3) Let

$$R = \{a + bi\sqrt{2} : a, b \in \mathbf{Z}\}$$

Define $d(a + bi\sqrt{2}) = a^2 + 2b^2$. The verification that d is a Euclidean function on R will be left as an exercise. With this verification complete, R is a Euclidean domain with respect to d.

(1.2.4) We will see later that $\mathbf{R}[x]$ is a Euclidean domain with respect to the restricted degree function.

If R is a ring and $a \in R$, then recall that the set Ra is an ideal in R which we call the principal ideal generated by a. Since R is a commutative ring with unity, Ra is the smallest ideal of R which contains a. Thus we will sometimes write Ra = (a).

(1.3) DEFINITION A ring R will be called a principal ideal ring if every ideal in R is principal. More precisely, if I is an ideal in

R, then there is an $a \in R$ such that $I = (a)$. If, in addition, R is an integral domain, then we will say that R is a principal ideal domain.

(1.4) THEOREM Let R be a Euclidean ring. Then R is a principal ideal ring.

PROOF Let d be a Euclidean function on R and let I be a nonzero ideal in R. Choose $b \in I$ so that $b \neq 0$ and $d(b) \leq d(r)$ for each nonzero $r \in I$. We will show that $(b) = I$.

Clearly, $(b) \subseteq I$, hence we need only show that $I \subseteq (b)$. Let $a \in I$ and $a \neq 0$. Choose $q, r \in R$ so that $a = bq + r$ and either $r = 0$ or $d(r) < d(b)$. Since $r = a - bq \in I$, $d(r) < d(b)$ is impossible by our choice of b. Therefore, $r = 0$ and $a = bq \in (b)$. Hence, R is a principal ideal ring.

(1.5) COROLLARY Let R be a Euclidean ring and let I be an ideal in R. If $b \in I$ is such that $d(b) \leq d(r)$ for every $r \in I$, then $(b) = I$.

POLYNOMIALS IN ONE VARIABLE

(1.6) DEFINITION In Chapter IV we defined the ring of polynomials $\mathbf{R}[x]$ in the indeterminate x with coefficients in $\mathbf{R}$, the field of real numbers. We now give an informal extension of this definition to the case of an arbitrary commutative ring with unity. The formal definition shall be given in Chapter VI as a special case of the multivariable construction.

Let R be a ring (commutative with unity). Then a polynomial f(x) (or more simply f) in the indeterminate x with coefficients in R is an expression of the form

$$a_n x^n + a_{n-1} x^{n-1} + \cdots + a_1 x^1 + a_0 x^0$$

where $a_i \in R$ for $i = 0, 1, \ldots, n$ and a_i is called the coefficient of x^i. The term $a_i x^i$ is called the i-th order term, and it is an implicit restriction that only finitely many terms have nonzero coefficient. As a matter of notational convenience, we will write x for x^1 and a_0 for a_0x^0, omit terms with coefficient 0, and omit coefficients which are 1. Moreover, we ignore the typographical ordering of the terms. Thus we could just as well write

$$f(x) = a_0 + a_1x + \cdots + a_nx^n$$

In general, two polynomials are defined to be equal if corresponding coefficients are equal. The set of all polynomials in the indeterminate x with coefficients in R is denoted R[x].

Among the more distinguished elements of R[x] are the so called constant polynomials -- polynomials of the form a_0 -- which under our notational convention become indistinguishable from elements a_0 in R. Thus R is identified with a subset of R[x]. We leave to the student to show that this identification is an injective ring homomorphism. The constant polynomial 0 is called the zero polynomial.

Using the summation notation, define polynomials f and g as follows:

$$f(x) = \sum_i a_ix^i$$

$$g(x) = \sum_i b_ix^i$$

Where it is understood that only finitely many of the coefficients are nonzero. We will define the sum and product of f and g as follows.

$$(f + g)(x) = \sum_i (a_i + b_i)x^i$$

$$(fg)(x) = \sum_i c_ix^i$$

where

$$c_i = \sum_{j+k=i} a_jb_k$$

The last sum being taken over only those pairs of nonnegative integers whose sum is i. Note that the definition of the product gives $x^i\ x^j = x^{i+j}$ and restores our usual interpretation of the superscript as an exponent. Just as in the case of the real numbers, R[x] is a commutative ring with unity under these operations.

The degree of a nonzero polynomial f in R[x], which is written deg(f), is naturally defined to be the order of the highest order term with nonzero coefficient. Thus if

$$f(x) = \sum_i a_i x^i$$

and if deg(f) = n, then $a_n \neq 0$ and $a_m = 0$ for every m > n. For convenience, we will define the degree of the zero polynomial to be $-\infty$. The only properties of this symbol to keep in mind are $-\infty < m$ and $-\infty + m = -\infty$ for all integers m. Two basic properties of degree to be used throughout are

(i) $\deg(f + g) \leq \max\{\deg(f), \deg(g)\}$

(ii) $\deg(fg) \leq \deg(f) + \deg(g)$

for all $f,g \in R[x]$. Statement (ii) becomes an equality in general if and only if R is an integral domain.

Finally, if $f \in R[x]$ is defined by

$$f(x) = \sum_i a_i x^i$$

and $r \in R$, then we can define the notion of evaluation of f at r by writing

$$f(r) = \sum_i a_i r^i$$

Here the sum on the right is simply a finite ring sum in R. Thus f defines a function from R to R. In general, this function is not a homomorphism. We can also define the evaluation map at r as a function from R[x] to R whose action is given by

$$f \longmapsto f(r)$$

The fact that this function is a ring homomorphism will be left as an exercise to the student.

(1.6) THEOREM Let F be a field. Then F[x] is a Euclidean ring with respect to the restricted degree function, that is the degree function restricted to the nonzero elements of F[x].

PROOF Let $f, g \in F[x]$ be such that $\deg(f) = n$ and $\deg(g) = m$. More precisely, let

$$f(x) = a_0 + a_1x + \cdots + a_nx^n$$
$$g(x) = b_0 + b_1x + \cdots + b_mx^m$$

We will establish (ii) of (1.1) by induction on n. First we note two important special cases.

Case 1: If $n < m$, then we can take $q = 0$ and $r = f$. Then it is easily seen that $f = qg + r$ and $\deg(r) = n < m = \deg(g)$.

Case 2: If $n = m = 0$, then we can take $q = a_0/b_0$ and $r = 0$.

We now begin our induction on n by assuming that (ii) is valid for any polynomial f in F[x] with degree less than n. By Case 1, we can assume that $n \geq m$. Let

$$q_1(x) = (a_n/b_m)x^{n-m}$$
$$r_1(x) = f(x) - q_1(x)g(x)$$

Then either $r_1 = 0$, in which case we are done, or $\deg(r_1) < n$. In the latter case, by our induction assumption, there exist polynomials q_2 and r_2 in F[x] such that $r_1 = q_2g + r_2$ and either $\deg(r_2) < m$ or $r_2 = 0$. Now

$$f = q_1g + r_1 = q_1g + q_2g + r_2 = (q_1 + q_2)g + r_2$$

Hence, by induction, (ii) is valid for all f and g in F[x]. Thus, F[x] is a Euclidean ring with respect to restricted deg.

The careful reader will note that the well known algorithm for dividing polynomials is contained in the proof of (1.6). We illustrate this with an example.

(1.7) EXAMPLE In **Q**[x], let

$$f(x) = x^4 + 3x^3 - x + 1$$
$$g(x) = 2x^2 + 1$$

To construct q, r we first take

$$q_1(x) = (1/2)x^2$$
$$r_1(x) = 3x^3 - (1/2)x^2 - x + 1$$

Then $f(x) = q_1(x)g(x) + r_1(x)$ and $\deg(g) < \deg r_1$, hence we continue with

$$q_2(x) = (3/2)x$$
$$r_2(x) = -(1/2)x^2 - (5/2)x + 1$$

Then $r_1(x) = q_2(x)g(x) + r_2(x)$ and $\deg(r_2) > \deg(g)$, hence we continue with

$$q_3(x) = -1/4$$
$$r_3(x) = -(5/2)x + 5/4$$

Since $\deg(r_3) < \deg(g)$, we are done. Set

$$q(x) = (1/2)x^2 + (3/2)x - 1/4$$
$$r(x) = r_3(x) = -(5/2)x + 5/4$$

By back substitution, we can now get

$$f(x) = q(x)g(x) + r(x)$$

and $\deg(r) < \deg(g)$.

The process which is illustrated in Example (1.7) may continue through several iterations. The crucial fact being $\deg(r_k) < \deg(r_{k-1})$ for each k. Hence, we will eventually get $\deg(r_k) < \deg(g)$ or $r_k = 0$. Once this has been accomplished, we set

$$q = q_1 + q_2 + \cdots + q_k$$
$$r = r_k$$

(1.8) NOTATION AND TERMINOLOGY Previously, we defined the notation $m|n$ for integers m and n. We will now take r, s to be in some ring R and define $r|s$ if there is a $t \in R$ such that $rt = s$. In this case, we will say that 'r divides s in the ring R.'

(1.9) THEOREM <u>The Factor Theorem</u> Let F be a field and $f \in F[x]$. If

$a \in F$, then $(x - a) | f(x)$ in the ring $F[x]$ if and only if $f(a) = 0$.
PROOF Assume first that $(x - a) | f(x)$; then there is a $q \in F[x]$ such that

$$f(x) = q(x)(x - a)$$

But then, since the evaluation map is a homomorphism, $f(a) = q(a)(a - a) = 0$.

Now assume that $a \in F$ is such that $f(a) = 0$. Since $F[x]$ is a Euclidean ring, there are $q, r \in F[x]$ such that $f(x) = q(x)(x - a) + r$ and $\deg(r) < 1$ or $r = 0$. Since $f(a) = 0$, by evaluation, we have $r = 0$.

(1.10) COROLLARY <u>The Remainder Theorem</u> Let F be a field, $f \in F[x]$, and $a \in F$. Choose $q, r \in F[x]$ such that $f(x) = q(x)(x - a) + r$ and $\deg(r) = 0$. Then $r = f(a)$.
PROOF See the proof of the second part of (1.10).

(1.11) REMARK Note that one consequence of (1.10) is the Horner algorithm for computing $f(a)$. More specifically, if f is divided synthetically by $x - a$, the resulting remainder is $f(a)$.

(1.12) COROLLARY Let F be a field and $f \in F[x]$. Let $\{a_1, \ldots, a_k\} \subseteq F$ be such that a_i are distinct and $f(a_i) = 0$ for $i = 1, \ldots, k$. Then $k \leq \deg(f)$.

We should note that the previous corollary fails in a division ring. For in the ring of real quaternions, there are in fact infinitely many solutions to the equation $x^3 = -1$.

EXERCISES

(1.1) Show that $(2x - 2) | (x^3 - 1)$ in $\mathbf{Q}[x]$. Does $(2x - 2) | (x^3 - 1)$ in $\mathbf{Z}[x]$?

(1.2) Does $(1 - i\sqrt{2}) | (2 + 3i\sqrt{2})$ in $\mathbf{Z}[i\sqrt{2}]$?

(1.3) Does $[8]|[4]$ in $\mathbf{Z}_{12}$?

(1.4) In the ring $\mathbf{Z}$, find k so that

(a) $(k) = (12) + (8)$

(b) $(k) = (6)(15)$

(c) $(k) = (m) + (n) \qquad (m,n \in \mathbf{Z})$

(1.5) Find $f \in \mathbf{Z}[x]$ so that $(f) = (x^2 - 4) + (x^2 - x - 2)$.

(1.6) Let R be a ring and let s be a zero divisor in R. Let $t \in R$ be such that $t|s$. Show that t is a zero divisor in R.

(1.7) Let R be a ring. Show that R[x] is an integral domain if and only if

$$\deg(fg) = \deg(f) + \deg(g) \qquad (f,g \in R[x])$$

Thus deduce that R[x] is an integral domain if and only if R is an integral domain.

(1.8) Let $R = \mathbf{Z}[i\sqrt{2}]$ and define $d: R^* \to \mathbf{Z}$ by

$$d(a + b\sqrt{2}) = a^2 + 2b^2$$

Show that d is a Euclidean function on R, and hence, R is a Euclidean domain.

(1.9) Let $R = \mathbf{Z}[\sqrt{2}]$ and define $d: R^* \to \mathbf{Z}$ by

$$d(a + b\sqrt{2}) = |a^2 - 2b^2|$$

Show that d is a Euclidean function on R, and hence, R is a Euclidean domain.

(1.10) In some texts, a Euclidean function d is sometimes assumed to satisfy the additional axiom

(iii) If $a,b,ab \in R^*$, then $d(ab) \geq d(a)$.

Let R be a ring and let d be a Euclidean function on R. Define $d': R^* \to \mathbf{Z}$ by

$$d'(r) = \min \{d(s): s \in R, Rs = Rr\}$$

for any $r \in R$. Show that d' is a (well defined) Euclidean function on R which satisfies (iii).

(1.11) Let R be a ring and let d be a Euclidean function on R which satisfies (iii) of the previous exercise. Let $r \in R^*$. Show that r is a unit if and only if $d(r) \leq d(s)$ for each $s \in R$.

(1.12) Show that any field is a Euclidean domain.

(1.13) Show that the evaluation map from R[x] to R is a ring homomorphism.

(1.14) Let R be a Euclidean domain with Euclidean function d. Assume that d also satisfies

$$d(rs) = d(r)d(s) \qquad (r,s \in R)$$

$$d(r + s) = \max \{d(r), d(s)\} \qquad (r,s \in R)$$

Show that either R is a field, or there is a field F such that $R \cong F[x]$.

(1.15) Let F be a field. Show that F[x, y] is not a principal ideal ring.

(1.16) Let R be a principal ideal ring and $f: R \to S$ be a surjective ring homomorphism. Show that S is a principal ideal ring.

(1.17) Show that $\{f: f \in \mathbf{Z}[x], f(0) \text{ is even}\}$ is an ideal in $\mathbf{Z}[x]$ which is not principal.

(1.18) Show that $\{m/n: m,n \in \mathbf{Z}, n \text{ odd}\}$ is a principal ideal domain.

(1.19) Let R be a principal ideal domain and let S be a multiplicative system in R. [See IV, (4.9)] Show that RS^{-1} is a principal ideal domain.

(1.20) Show that in the ring of real quaternions, there are infinitely many solutions to the equation $x^2 - 1 = 0$. Deduce that commutativity is essential in (1.13).

2. PRIMES AND UNIQUE FACTORIZATION

In this section we continue our study of the relation 'divides in the ring R.' Our purpose is to introduce the notion of prime element and prime factorization in a ring R. A reminder--All rings are commutative with unity.

(2.1) PROPOSITION Let R be a ring and let $a,b,c \in R$. Then

(i) $a|b$ if and only if $Rb \subseteq Ra$

(ii) $a|b$ and $b|c$ implies $a|c$

(iii) $a|b$ and $a|c$ implies $a|(b + c)$ and $a|(b - c)$

(iv) $a|b$ implies $a|(bc)$

PROOF (i) If $a|b$, then there is an $r \in R$ such that $ra = b$. Let $s \in R$, then

$$sb = s(ra) \in Ra$$

Hence, $Rb \subseteq Ra$.

Conversely, if $Rb \subseteq Ra$, then $b \in Ra$. Therefore, there is an $r \in R$ such that $ra = b$. Hence, $a|b$.

The remaining facts follow easily and will be left as exercises.

(2.2) NOTATION AND TERMINOLOGY Let R be a ring. Then we define

$$R^{\times} = \{s: s \in R, st = 1 \text{ for some } t \in R\}$$

It is easily verified that $R^{\times}$ is a group under multiplication which we will call the group of units of R. It is equally easy to verify that $s \in R^{\times}$ if and only if $Rs = R$.

(2.3) DEFINITION Let R be a ring and let $a,b \in R$. Then we will say that a and b are associates in R, and we will write $a \sim b$, if there is a $u \in R^{\times}$ such that $a = ub$.

Clearly, the relation $\sim$ is an equivalence relation on R. Denote the equivalence class of $s \in R$ by $[s]$. Then, $[0] = \{0\}$ and $[1] = R^{\times}$.

(2.4) EXAMPLES We will now give some examples to illustrate the action of the relation $\sim$ in some familiar rings.

(2.4.1) Note that $\mathbf{Z}^\times = \{1, -1\}$. Hence, for each $n \in \mathbf{Z}$, we have $[n] = \{n, -n\}$.

(2.4.2) Recall that $\mathbf{Z}[i] = \{a + bi: a,b \in \mathbf{Z}\}$. It is easy to see that $\mathbf{Z}[i]^\times = \{1, -1, i, -i\}$. Therefore,

$$[a + bi] = \{a + bi, -a - bi, ai - b, -ai + b\}$$

(2.4.3) Let R be an integral domain. Then we claim that $R[x]^\times = R^\times$. Let $p,q \in R[x]$ be such that $pq = 1$. Then $0 = \deg(pq) = \deg(p) + \deg(q)$. Hence, $\deg(p) = \deg(q) = 0$. Therefore, $p,q \in R^\times$. Since $R^\times \subseteq R[x]^\times$, our claim is established.

Consequently, if $p \in R[x]$, then

$$[p] = \{ap: a \in R^\times\}$$

We will see in Exercise (2.9) that if R is not an integral domain then $R^\times$ is not necessarily equal to $R[x]^\times$.

(2.5) PROPOSITION Let R be an integral domain and $a,b \in R$. Then the following statements are equivalent.

(i) $Ra = Rb$

(ii) $a \sim b$

(iii) $a|b$ and $b|a$

PROOF To establish (i) implies (ii), assume that $Ra = Rb$. Then there are $r,s \in R$ such that $a = rb$ and $b = sa$. Now $a = rb = rsa$ implies $(1 - rs)a = 0$. Since R is an integral domain, either $a = 0$ or $rs = 1$. In case $a = 0$, then $b = 0$ and we are done. If $rs = 1$, then $r,s \in R^\times$ and $a \sim b$.

To show that (ii) implies (iii), assume that $a \sim b$. Let $u \in R^\times$ be such that $a = ub$. Then $b = u^{-1}a$ and hence $a|b$ and $b|a$ as required.

The fact that (iii) implies (i) follows from Proposition (2.1). Thus our cycle is complete and the proposition is verified.

In Exercise (2.8) we will see that statements (i) and (ii) are not necessarily equivalent if R is not an integral domain.

(2.6) PROPOSITION Let R be an integral domain and $0 \neq p \in R$. Then, the following statements are equivalent.

(i) If $a \in R$ and $a|p$, then either $a \sim p$ or $a \in R^\times$.

(ii) If $a,b \in R$ and $p = ab$, then either $p \sim a$ or $p \sim b$.

(iii) If $a,b \in R$ and $p = ab$, then either $a \in R^\times$ or $b \in R^\times$.

PROOF We begin by showing that (i) implies (ii). Let $a,b \in R$ be such that $p = ab$. Then $a|p$; hence, by (i), either $a \sim b$ or $a \in R^\times$. If $a \in R^\times$, then $p \sim b$ as required.

To show that (ii) implies (iii), let $a,b \in R$ be such that $p = ab$. Then by (ii), $p \sim a$ or $p \sim b$. If $p \sim a$, then there is a $u \in R^\times$ such that $p = ua$. Since $p = ab$, and R is an integral domain, we have either $a = 0$ or $u = b$. The former case being impossible, we must have $b = u \in R^\times$. Similarly, if we first assume that $p \sim b$, then we can show that $a \in R^\times$.

Finally, we establish (iii) implies (i). Let $a \in R$ be such that $a|p$. Then there is a $b \in R$ such that $p = ab$. By (iii), either $a \in R^\times$ or $b \in R^\times$. If $b \in R^\times$, then $a \sim p$. Thus (iii) implies (i) and our cycle is complete.

(2.7) DEFINITION Let R be an integral domain and p a nonunit in R. Then p will be called prime in R, or a prime element in R, if any, and hence all, of the three conditions in the previous proposition are satisfied.

(2.8) EXAMPLES We now give some examples of prime elements in rings. The reader is urged to work through every example completely and supply any missing details.

(2.8.1) In the ring of integers, p is prime in our sense if and only if p is prime in the usual sense.

(2.8.2) Let $R = \mathbf{R}[x]$. Then $x^2 + 1$ is prime in R. Assume that $f,g \in R$ are such that $x^2 + 1 = f(x)g(x)$ and $f \notin R^\times$. Then $\deg(f) = \deg(g) = 1$. We can assume that $f(x) = x - a$ for some $a \in \mathbf{R}$. By the factor theorem, a is a root of $x^2 + 1 = 0$, which is impossible. Hence, $x^2 + 1$ is prime in $\mathbf{R}[x]$.

(2.8.3) Let $R = \mathbf{Z}[i\sqrt{5}] = \{a + bi\sqrt{5}\colon a,b \in \mathbf{Z}\}$. Recall that for any complex number $a + bi$, the absolute value of $a + bi$ is defined by $|a + bi| = \sqrt{a^2 + b^2}$. The following facts about absolute value of complex numbers are easily verified.

(a) If $z,w \in \mathbf{C}$, then $|zw| = |z||w|$.

(b) If $z \in \mathbf{C}$ and $z \neq 0$, then $|z^{-1}| = |z|^{-1}$.

Returning to R, note that for any $0 \neq s \in R$, $|s| \geq 1$. Hence, by (b), $R^{\times} = \{1, -1\}$. Assume that $s \in R$ admits a factorization $s = tr$ in R. Then $|s| = |t||r|$. Hence, either $|t| \leq \sqrt{|s|}$ or $|r| \leq \sqrt{|s|}$. Therefore, $p \in R$ is prime in R if and only if for any $s \in R$ such that $1 < |s| \leq \sqrt{|p|}$, we have s is not a divisor of p in R. Using this fact, we can easily see that 2, 3, $1 + i\sqrt{5}$, and $1 - i\sqrt{5}$ are primes in R.

In Chapter IV, we defined an ideal I in a ring R to be prime if for any $a,b \in R$ such that $ab \in I$, we have either $a \in I$ or $b \in I$. We will now examine the connection between prime ideals and prime elements in an integral domain.

(2.9) PROPOSITION Let R be an integral domain and let $0 \neq p \in R$ be such that Rp is a prime ideal in R. Then p is a prime in R.

PROOF Let $a,b \in R$ be such that $p = ab$. Since Rp is a prime ideal and $ab \in Rp$, then either $a \in Rp$ or $b \in Rp$. It suffices to assume that $a \in Rp$. Let $s \in R$ be such that $a = sp$, then $p = ab = spb$. Since R is an integral domain. $sb = 1$ and $b \in R^{\times}$. Hence, p satisfies (2.6)(iii) and is prime as required.

WARNING The converse of (2.9) fails. Take $R = \mathbf{Z}[i\sqrt{5}]$. Then

$$(2)(3) = 6 = (1 + i\sqrt{5})(1 - i\sqrt{5})$$

Now, $1 + i\sqrt{5}$ is prime, $6 \in R(1 + i\sqrt{5})$, but $2 \notin R(1 + i\sqrt{5})$ and $3 \notin R(1 + i\sqrt{5})$.

(2.10) DEFINITION Let R be an integral domain. Then R is said to be a unique factorization domain, or a UFD, if both of the following conditions are satisfied.

(F) Every nonzero element of R is either a unit or a product of primes.

(U) Each factorization of an element of R as a product of primes is unique up to associates.

We can state (U) more explicitly. If $p_1p_2 \cdots p_n = q_1q_2 \cdots q_m$ are both prime factorizations, then $m = n$ and with the appropriate reordering, $p_i \sim q_i$ for $i = 1, 2, \ldots, n$.

(2.11) PROPOSITION Let R be an integral domain in which (F) holds. Then R is a unique factorization domain if and only if

Rp is a prime ideal if and only if p is a prime element in R

PROOF Assume that Rp is a prime ideal in R whenever p is a prime element of R. Suppose that we have two equal prime factorizations

(1) $p_1p_2 \cdots p_n = q_1q_2 \cdots q_m$

We must show that $m = n$ and up to a possible reordering, $p_i \sim q_i$ for $i = 1, 2, \ldots, n$. We argue by induction on n.

The result being clear for $n = 1$, we assume that the assertion is true for all products of primes with length less that n. Since $p_1p_2 \cdots p_n \in Rq_1$, we must have $p_i \in Rq_1$ for some $i = 1, 2, \ldots, n$. Without loss of generality, we may assume that $p_1 \in Rq_1$. Let $u \in R^\times$ be such that $p_1 = uq_1$. After substituting into (1) and cancellation, we have

$$up_2p_3 \cdots p_n = q_2q_3 \cdots q_m$$

The assertion now follows by induction.

The converse will be left as an exercise to the student.

EXERCISES

(2.1) If R is a ring, $u \in R^\times$, and $s|u$, then show that $s \in R^\times$.

(2.2) If R is a ring and $u \in R$, then show that $u \in R^\times$ if and only if $R = Ru$.

(2.3) If R is a ring, $u \in R^{\times}$, $t \in R$, and $t \sim u$, then show that $t \in R^{\times}$.

(2.4) If R is a ring, $s \in R$ is a zero divisor in R and $t \in R$, then show that $t \sim s$ implies that t is a zero divisor in R.

(2.5) Find all divisors of $x^7 - x$ in $\mathbf{Q}[x]$.

(2.6) Let $R = \mathbf{Z}_5$ and show that
(a) $x^2 + 1$ is not prime in $R[x]$.
(b) $x^3 + x + 1$ is prime in $R[x]$.

(2.7) Show that in $\mathbf{Z}_6$, [2] and [4] are not associates. Show that $[2]\mathbf{Z}_6 = [4]\mathbf{Z}_6$.

(2.8) Show that (x) is a prime ideal which is not maximal in $\mathbf{Z}[x]$.

(2.9) Show that in $\mathbf{Z}_4[x]$, there are invertible elements with nonzero degree.

(2.10) Show that (x) is a prime ideal which is not maximal in $\mathbf{Q}[x, y]$.

(2.11) Show that $\mathbf{Q}[x, y]$ is not a principal ideal domain.

(2.12) If R is a principal ideal domain and $I \neq 0$ is an ideal in R, then show that I is maximal if and only if I is prime.

(2.13) Complete the proof of (2.11).

(2.14) Let R be a ring and let S be a monoid with identity element e. Define $S^{\times} = \{t: t \in S \text{ and } ts = st = e \text{ for some } s \in S\}$. We will say that a mapping $f: R \to S$ is multiplicative if for every $s,t \in R$, we have $f(st) = f(s)f(t)$ and $f(1) = e$. If $f: R \to S$ is multiplicative, show that $R^{\times} \subseteq f^{-1}(S^{\times})$.

(2.15) Let $R = \mathbf{Z}[i\sqrt{10}]$ and define $f: R \to \mathbf{Z}$ by $f(a + bi\sqrt{10}) = a^2 + 10b^2$. Show that f is multiplicative and that $u \in R^{\times}$ if and only if $u = 1$ or $u = -1$.

(2.16) Let R be a ring and let $f: R \to \mathbf{Z}$ be multiplicative. Let $p \in R$ be such that

(D) $q \in R$ and $|f(q)| \le \sqrt{|f(p)|}$ implies q is not a divisor of p.

Show that p is prime in R.

(2.17) Let $R = \mathbf{Z}[i\sqrt{5}]$ and define $f: R \to \mathbf{Z}$ by $f(a + bi\sqrt{5}) = a^2 + 5b^2$. Show that f is multiplicative. Use this to show that 3, $2 + i\sqrt{5}$, and $2 - i\sqrt{5}$ are prime in R. Note that

$$3^2 = 9 = (2 + i\sqrt{5})(2 - i\sqrt{5})$$

(2.18) Let $R = \mathbf{Z}[\sqrt{10}]$ and define $f: R \to \mathbf{Z}$ by

$$f(a + b\sqrt{10}) = a^2 - 10b^2$$

Show that f is multiplicative and

(a) If $u \in R$, then $u \in R^\times$ if and only if f(u) is 1 or -1.

(b) 2, 3, $4 + \sqrt{10}$, and $4 - \sqrt{10}$ are primes in R.

(c) $(2)(3) = 6 = -(4 + \sqrt{10})(4 - \sqrt{10})$.

(d) (F) holds in R.

(2.19) Find all primes in $\mathbf{Z}[i]$.

(2.20) Find all units in $\mathbf{Z}[\sqrt{2}]$.

(2.21) If R is a UFD and $S \subseteq R$ is a multiplicative set, then show that RS^{-1} is a UFD.

(2.22) Let R be an integral domain. Show that the following conditions are equivalent.

(a) R is a UFD.

(b) If I is a prime ideal, then there is an $s \in R$ such that $Rs \subseteq I$ and Rs is a prime ideal.

(2.23) Complete the proof of (2.11)

3. FACTORIZATION AND NOETHERIAN DOMAINS

In this section we will study prime factorization in Noetherian domains. As before, all rings are commutative with unity. We begin with a definition.

(3.1) DEFINITION A ring R is called a Noetherian ring if any ideal I of R is finitely generated; that is, there are $a_1, a_2, \ldots, a_n \in R$ such that $I = \sum_i Ra_i$. If, in addition, R is an integral domain, then R will be called a Noetherian domain.

EXAMPLES Clearly, any principal ideal domain is a Noetherian domain. Hence **Z** is a Noetherian domain and if F is a field, then F[x] is a Noetherian domain.

(3.2) THEOREM Let R be a ring. Then the following conditions are equivalent.

(a) R is a Noetherian ring.

(b) If $I_1 \subseteq I_2 \subseteq \ldots \subseteq I_n \subseteq \ldots$ is a chain of ideals in R, then there is an $n \in \mathbf{Z}$ such that $I_m = I_n$ for all $m \geq n$.

(c) Any nonempty set of ideals of R has a maximal member.

PROOF We first establish (a) implies (b). If R is a Noetherian ring and $I_1 \subseteq I_2 \subseteq \ldots \subseteq I_n \subseteq \ldots$ is a chain of ideals in R, then $I = \bigcup_{i \in \mathbf{N}} I_i$ is an ideal in R. Hence, there are $a_1, \ldots, a_k \in R$ such that $I = \langle a_1, \ldots, a_k \rangle$. For some $n \in \mathbf{Z}$, $a_1, a_2, \ldots, a_k \in I_n$. Hence, $I_m = I_n$ for every $m \geq n$.

We next establish (b) implies (c). Let S be a nonempty set of ideals of R. Select $I_1 \in R$. If I_1 is maximal in S, we are done. Otherwise, select I_2 in S so that $I_1 \subseteq I_2$. Continue to construct a strictly increasing chain of ideals from S. Thus we contradict the condition in (b).

We will leave the proof that (c) implies (a) as an exercise to the student.

In our next theorem, we will extend our list of examples of Noetherian rings.

(3.3) THEOREM The Hilbert Basis Theorem. Let R be a Noetherian ring. Then R[x] is a Noetherian ring.

PROOF The proof of this theorem is somewhat technical but not difficult. Hence, we will leave many of the details as exercises. Let I be an ideal in R[x]. We will show that I is finitely generated. For k = 0, 1, 2, ... let I_k be the set consisting of 0 and all leading coefficients of polynomials in I with degree k. Then I_k is an ideal in R for each k. Furthermore, $I_0 \subseteq I_2 \subseteq \cdots$.

By (3.2), there is an $n \in \mathbf{Z}$ such that $I_n = I_m$ for all $m \geq n$. Since R is Noetherian, each of the ideals I_k are finitely generated. Hence, there is a finite set F of elements of $\cup I_i$ such that I_k is generated by $F \cap I_k$ for each k. Form a set P as follows: for each $a \in F$, choose one polynomial f_a in I with leading coefficient a. Set

$$P = \{f_a: a \in F\}$$

We leave as an exercise to show that I is generated by P. Hence R is Noetherian.

Note that among the consequences of (3.3), we have $\mathbf{Z}[x]$ is Noetherian and for any field F, $F[x_1, x_2, \ldots, x_n]$ is Noetherian.

(3.4) DEFINITION Let R be an integral domain and let $a,b \in R$. We will say that $d \in R$ is a gcd of a and b if

(a) $d|a$ and $d|b$.

(b) If $c \in R$ is such that $c|a$ and $c|b$, then $c|d$.

In this case, we will write d = gcd(a, b). Note that if d and d' are both gcd's of a and b in R, then d and d' are associates in R. Finally, if gcd(a, b) = 1, then we will say that a and b are relatively prime in R.

WARNING gcd's do not necessarily exist.

(3.5) PROPOSITION Let R be a principal ideal domain and let $a,b \in R$. Then there are $r,s \in R$ such that ra + sb = gcd(a, b).

PROOF Let I = Ra + Rb. Then I is an ideal in R and hence, there is a $d \in R$ such that I = Rd. Clearly $d|a$ and $d|b$. Let $c \in R$ be such that $c|a$ and $c|b$. Then it is easily seen that $c|d$. Hence d = gcd(a, b). Necessarily, d = ra + sb for some $r,s \in R$.

(3.6) COROLLARY Let R be a principal ideal domain and let $a,b,c \in R$ be such that $a|bc$ and $\gcd(a, b) = 1$. Then $a|c$.

PROOF Let $r,s \in R$ be such that $ra + sb = 1$. Then $rac + sbc = c$. Since $a|(rac)$ and $a|(sbc)$, we have $a|c$.

(3.7) COROLLARY Let p be a prime in a principal ideal domain R. Then for each $a,b \in R$, $p|(ab)$ implies $p|a$ or $p|b$.

(3.8) THEOREM Let R be a principal ideal domain and p a nonzero element of R. Then the following statements are equivalent.

(a) p is a prime in R.

(b) Rp is a prime ideal in R.

(c) Rp is a maximal ideal in R.

PROOF By IV (4.7), (c) implies (b) and by (2.9), (b) implies (a). To show that (a) implies (b), let p be a prime in R and let $ab \in Rp$. Then $p|(ab)$. By (3.7), either $p|a$ or $p|b$. Hence, either $a \in Rp$ or $b \in Rp$ and Rp is a prime ideal as required.

Finally, we show that (b) implies (c). Suppose that Rp is a prime ideal in R and that $Rp \subseteq I = Ra$. Then $a|p$, hence either $a \in R^{\times}$ or $a \sim p$. The former case gives $I = R$. The latter case gives $Rp = Ra$. Hence, Rp is a maximal ideal in R.

(3.9) THEOREM Let R be a Noetherian domain. Then R satisfies (F) of (2.10).

PROOF Suppose that (F) is not satisfied. Then there is an $s \in R$ such that $s \notin R^{\times}$ and s is not a product of primes. Let S denote the set of all such s. Denote by $T = \{Rs\colon s \in S\}$. Then T is a nonempty set of ideals of R. Let Rc be a maximal element of T. Then $c \notin R^{\times}$ and c is not a prime, hence there are c_1, c_2 such that $c = c_1c_2$ and neither c_1 nor c_2 is a unit. Since Rc is properly contained in both Rc_1 and Rc_2, neither c_1 nor c_2 is in S. But then c_1 and c_2 are products of primes and hence, so is $c = c_1c_2$. With this contradiction, we are left with the conclusion that R satisfies (F).

(3.10) COROLLARY A Noetherian domain is a unique factorization domain if and only if Rp is a prime ideal in R whenever p is a prime in R.

PROOF See (2.11).

(3.11) COROLLARY Every principal ideal domain is a unique factorization domain.

PROOF See (3.8)

(3.12) COROLLARY The Fundamental Theorem of Arithmetic. **Z** is a unique factorization domain.

(3.13) COROLLARY If F is a field, then F[x] is a unique factorization domain.

For the remainder of this section we will be considering polynomials with rational or integer coefficients. The principal aim will be to develop a criterion for a polynomial f to be irreducible, or prime, in **Z**[x]. Most of the following results obtain if **Z** is replaced by an arbitrary unique factorization domain and **Q** is replaced by the quotient field. The generalizations of these results will be left as an exercise to the student.

(3.13) DEFINITION Let $f(x) = a_0 + \dots + a_n x^n$ be a polynomial in **Z**[x]. Then the content of f, denoted c_f, will be the greatest common divisor of the set $\{a_0, \dots, a_n\}$. We will say that f is primitive if c_f is 1.

(3.14) PROPOSITION Let f and g be primitive polynomials. Then fg is a primitive polynomial.

PROOF Let

$$f(x) = a_0 + a_1 x + \cdots + a_n x^n$$
$$g(x) = b_0 + b_1 x + \cdots + b_m x^m$$
$$f(x)g(x) = c_0 + c_1 x + \cdots + c_k x^k$$

If fg is not a primitive polynomial, then there is a prime p which divides all of the c_i's. Let i be the first nonnegative integer such that p does not divide a_i. Similarly define j so that p does not divide b_j. Consider the coefficient c_{i+j}.

$$c_{i+j} = a_i b_j + \text{Sum}$$

Here the expression Sum contains sums of products of the form $a_q b_h$ such that $q + h = i + j$ and $q \neq i$, $h \neq j$. Necessarily, p does not divide $a_i b_j$. Yet $p|(\text{Sum})$ and $p|(c_{i+j})$, which is a contradiction. Hence fg is primitive.

(3.15) REMARK Let f be a polynomial in **Z**[x]. Then f can be written in the form

$$f = c_f g$$

Where g is a primitive polynomial in **Z**[x].

(3.16) PROPOSITION Let f be a primitive polynomial in **Z**[x]. Suppose that f = gh is a nontrivial factorization of f in **Q**[x] (i.e. neither g nor h is in **Q**). Then f admits a nontrival factorization in **Z**[x].

PROOF By multiplying by a suitable integer n, we can remove all denominators in the coefficients of g and h and thus represent

$$nf = mg_1h_1$$

Where g_1 and h_1 are primitive polynomials in **Z**[x] with the same degree as g and h, respectively, and (m, n) = 1. By the previous proposition, g_1h_1 is primitive. Hence, the content of the left side is n and the content of the right side is m. Thus m = n = 1. Therefore, f factors in **Z**[x].

(3.17) THEOREM The Eisenstein Criterion. Let

$$f(x) = a_0 + a_1x + \cdots + a_nx^n$$

be a polynomial in **Z**[x]. Let p be a prime and suppose that p does not divide a_n, p^2 does not divide a_0 and p divides all other coefficients of f. Then f is prime in **Z**[x].

PROOF We may assume that f is primitive. Suppose that f is reducible, then there are polynomials g,h $\in$ **Z**[x] say

$$g(x) = b_0 + \cdots + b_m x^m$$
$$h(x) = c_0 + \cdots + c_k x^k$$

such that f = gh. By definition of multiplication, $a_0 = b_0c_0$. Furthermore, since $p|a_0$ and p^2 does not divide a_0, we have p divides exactly one of the integers b_0 or c_0. The other case being similar, assume that $p|b_0$ and p does not divide c_0. If p divides b_i for each i, then $p|a_n$ which is contrary to assumption. Hence, let i be the first positive integer such that p does not divide b_i. Necessarily, $i < m \leq n$. Again, by definition of multiplication,

$$a_i = b_ic_0 + b_{i-1}c_1 + \cdots + b_0c_i$$

Since $p|a_i$ and p divides every term on the right side except maybe the first, we must have p divides b_i, since p does not divide c_0. Thus we reach a contradiction, and f is irreducible.

(3.18) EXAMPLE Let p be a prime, then the polynomial

$$x^{p-1} + x^{p-2} + \cdots + 1$$

is an irreducible polynomial in **Z**[x]. Note that the Eisenstein criterion does not apply immediately. However, if we consider the polynomial g(x) = f(x + p), we can see, after expansion that Eisenstein's criterion does apply to g. Thus g is irreducible. It is easy to see that g is irreducible if and only if f is irreducible.

EXERCISES

(3.1) Show that (c) implies (a) in (3.2).

(3.2) Show that the sets I_k defined in the proof of (3.3) are ideals in R.

(3.3) If P and I are as defined in the proof of (3.3), show that I is generated by P.

(3.4) Show that the polynomial g defined in Example (3.18) is irreducible if and only if f is irreducible. Show that Eisensteins's criterion does apply to the polynomial g.

(3.5) Show that gcd's do not necessarily exist.

(3.6) Show that there are prime ideals in the ring **Q**[x, y] which are not maximal.

(3.7) Show that $x^n - p$ is an irreducible polynomial in **Q**[x] for each prime p.

(3.8) Show that $x^4 + x^3 + x + 1$ is not irreducible over any field F.

(3.9) Show that $x^4 + 2x + 2$ is irreducible over **Q**.

(3.10) If I is an ideal in a Noetherian ring R, show that R/I is Noetherian.

(3.11) If R is an integral domain such that every prime ideal is finitely generated, show that R is Noetherian.

(3.12) If I is an ideal in a ring R, define

$$\text{rad}(I) = \{r: r \in R \text{ and } r^n \in I \text{ for some } n \in \mathbf{N}\}$$

Show that rad(I) is an ideal in R. This ideal is called the radical of I.

(3.13) If I is an ideal in a ring R, then I is said to be primary if for every $r, s \in R$

$$rs \in I,\ r \notin I \text{ implies } s^n \in I \text{ for some } n \in \mathbf{N}$$

If I is a primary ideal in a ring R, show that rad(I) is a prime ideal in R. See exercise (3.12).

(3.14) Show that a nonzero ideal I in **Z** is primary if and only if $I = (p^n)$ for some prime p and some $n \in \mathbf{N}$.

(3.15) If I is an ideal in a Noetherian ring R which is not

primary, show that there are ideals J and K such that $I \neq J$, $I \neq K$ and $I = J \cap K$.

(3.16) Show that every ideal in a Noetherian ring R is the finite intersection of primary ideals. Interpret this result for the ring **Z**. See exercise (3.14).

6
Algebras

We now introduce three ring constructions which are paramount: multi-variable polynomials, matrices, and (in the extended exercises) group rings. In fact, each of these partakes of more structure than that of just an abstract ring, and so we first introduce the formalism of an algebra.

Throughout this chapter, 'ring' means ring with one, 'ring homomorphism' means unital ring homomorphism, and $\mathbf{N}$ contains 0.

1. ALGEBRAS AND MORPHISMS

(1.1) DEFINITION Let k be a commutative ring. Then a <u>k-algebra</u> is an arbitrary ring A together with a homomorphism of rings $f: k \to A$ such that f(k) lies in Z(A), the center of A. We call A a <u>commutative k-algebra</u> if the ring A is commutative. Note that our definition is slightly restrictive: the term 'algebra' sometimes signifies a more general structure, one which we'll mention briefly in the exercises.

If A is a k-algebra with respect to $f: k \to A$, then for λ in k, a in A, the product $f(\lambda)\cdot a$ is defined as a ring product in A. In fact, by convention we write

$$\lambda \cdot a \quad \text{for} \quad f(\lambda) \cdot a$$

and thus suppress the homomorphism f entirely. This causes little confusion since there is scarcely any other interpretation for the product $\lambda\cdot a$ when λ and a are in different rings.

(1.2) PROPOSITION Let A be a k-algebra (with respect to the homomorphism $f: k \to A$).

(i)	$(\lambda \cdot \mu)a = \lambda(\mu \cdot a)$	$(\lambda,\mu \in k;\ a \in A)$
(ii)	$\lambda(a + b) = [\lambda \cdot a + \lambda \cdot b]$	$(\lambda \in k;\ a,b \in A)$
(iii)	$1 \cdot a = a$	$(1 = 1_k;\ a \in A)$
(iv)	$(\lambda + \mu)a = \lambda \cdot a + \mu \cdot a$	$(\lambda,\mu \in k;\ a \in A)$
(v)	$\lambda(a \cdot b) = (\lambda \cdot a)b = a(\lambda \cdot b)$	$(\lambda \in k;\ a,b \in A)$

PROOF With our notational convention in mind, each law follows from the ring properties of A and from the fact that f is a (unital) ring homomorphism into the center of A. For example, we prove (iv):

$$\begin{aligned}(\lambda + \mu)a &= f(\lambda + \mu) \cdot a && \text{(by convention)}\\ &= [f(\lambda) + f(\mu)] \cdot a && \text{(f is a homomorphism)}\\ &= f(\lambda) \cdot a + f(\mu) \cdot a && \text{(distributive law in A)}\\ &= \lambda \cdot a + \mu \cdot a && \text{(by convention)}\end{aligned}$$

The other verifications are similar; only (v) requires the centrality of f.

We shall see later that the properties (i)-(iv) qualify A as a k-module (Chapter VII).

Note that any homomorphism of commutative rings $f: k \to k'$ automatically endows k' with the structure of a k-algebra since in this case Im(f) certainly lies in the center of k'.

(1.3) MORPHISMS OF ALGEBRAS Let A be a k-algebra and A' be a k'-algebra, where k and k' are (possibly distinct) commutative rings. Observe that we suppress any explicit mention of the underlying homomorphism.

DEFINITION A morphism of algebras $A \to A'$ is a pair of ring homomorphisms $\phi: A \to A'$ and $\psi: k \to k'$ such that for all λ in k and a in A, we have

$$\phi(\lambda \cdot a) = \psi(\lambda) \cdot \phi(a)$$

In the special case $k = k'$ and $\psi = 1_k$, this condition reduces to

$$\phi(\lambda \cdot a) = \lambda \cdot \phi(a)$$

and we then speak of a homomorphism of k-algebras or a k-homomorphism.

The reader can verify easily that algebras and morphisms constitute a category: the only point of contention is that when we compose the pairs of maps which define a morphism, we again obtain a morphism. [See Exercise (1.4).] Similarly, if we fix a commutative ring k, k-algebras and k-homomorphisms also form a category.

(1.4) MONOID ALGEBRAS We now construct a type of algebra which is basic to the discussion of polynomials and group rings.

Let S be a nonempty set, and let k be a commutative ring. Then a formal sum in S over k is an expression of the form

$$\lambda_1 \cdot s_1 + \cdots + \lambda_n \cdot s_n$$

where $\lambda_1,\ldots,\lambda_n$ lie in k and $s_1,\ldots,s_n$ are distinct elements of S. We declare two formal sums equal if (i) they involve precisely the same elements of S with nonzero coefficients, and (ii) corresponding coefficients are equal. Order is irrelevant.

EXAMPLE Let $k = \mathbf{R}$ and let $S = \{s_0, s_1, s_2, \ldots\}$ where the s_n are simply symbols. Then the following relations hold among the formal sums in S over **R**.

$$2s_0 + 7s_1 + \pi \cdot s_2 = 7s_1 + \pi \cdot s_2 + 0s_4 + 2s_0$$

$$2s_0 + 7s_1 + \pi \cdot s_2 \neq 2s_0 + 6s_1 + \pi \cdot s_2$$

Furthermore, an expression such as $2s_0 + 7s_1 + \pi \cdot s_2 + 5s_0$ is not a formal sum since it involves two s_0-terms, and an expression such as

$$1s_0 + 1s_1 + 1s_2 + \cdots + 1s_n + \cdots$$

likewise fails as a formal sum since it involves infinitely many terms with nonzero coefficient.

Resuming the general discussion, the set of all formal sums in S over k is denoted k[S]. We shall write a typical element of k[S] in the form

$$(\#) \qquad \sum_{s \in S} \lambda_s \cdot s$$

where λ_s in k is the coefficient of s, and it is understood that $\lambda_s = 0$ for all but finitely many values of s in S.

One puts k[S] on solid ground set theoretically as follows. Identify the expression (#) with the function $g: S \to k$ defined by $g(s) = \lambda_s$. Then g is a function which is nonzero at only finitely many points of S. Conversely, with each such function g we may associate the formal sum

$$\sum_{s \in S} g(s) \cdot s \quad \text{in} \quad k[S]$$

and thus set theoretically k[S] is precisely the set of all maps $S \to k$ with finite support. Observe that the notion of equality in k[S] given above is, in this light, simply equality of functions.

Define addition in k[S] componentwise:

$$\sum_{s \in S} \lambda_s \cdot s \; + \; \sum_{s \in S} \mu_s \cdot s \; = \; \sum_{s \in S} (\lambda_s + \mu_s) \cdot s$$

that is, add the corresponding coefficients in k. Clearly if only finitely many λ_s and μ_s are nonzero, only finitely many of the sums $\lambda_s + \mu_s$ are nonzero. Thus addition is a bona fide operation on k[S]. In fact, $\langle k[S],+\rangle$ is an additive group:

(i) Associativity is inherited from $\langle k,+\rangle$.

(ii) The identity element is $\sum 0 \cdot s$.

(iii) The additive inverse of $\sum \lambda_s \cdot s$ is $\sum (-\lambda_s) \cdot s$, which is clearly also in k[S].

(iv) Commutativity is inherited from k.

Note that in our set-theoretic interpretation, $\langle k[S],+\rangle$ is indeed a subgroup of the set of all maps from S to k.

Now assume further that S is a monoid. (Recall that this is set together with an associative operation for which there is an identity element, say e.) Then one defines multiplication in k[S] as follows. First consider the case of two formal sums which involve at most one nonzero term: say, $\lambda \cdot s$ and $\mu \cdot t$ ($\lambda, \mu \in k$; $s, t \in S$). Since products are defined now both in k and S, it is natural to take

$$(\#) \quad (\lambda \cdot s) \cdot (\mu \cdot t) \; = \; (\lambda \cdot \mu) \cdot (s \cdot t)$$

Of course, arbitrary formal sums may have many nonzero terms, and our definition must accommodate these, too. The key is to extend (#) to all of k[S] by imposing a distributive law, thus defining the general product as a sum of formal sums:

$$(\sum \lambda_s \cdot s) \cdot (\sum \mu_t \cdot t) = \sum_{S \times S} (\lambda_s \cdot \mu_t) \cdot st$$

where the right-hand sum is taken over all pairs (s,t) in S × S. (Again, since only finitely many λ_s and μ_t are nonzero, only finitely many products $\lambda_s \cdot \mu_t$ are nonzero, so the product is a finite sum and a legitimate element of k[S].) Note that the right-hand expression may admit simplification through collection of like terms (as happens with polynomials), and thus we may express it as a single formal sum as shown:

$$(\sum \lambda_s \cdot s) \cdot (\sum \mu_t \cdot t) = \sum \nu_r \cdot r$$

where ν_r in k is defined by

$$\nu_r = \sum_{s \cdot t = r} \lambda_s \cdot \mu_t$$

This last sum is taken over all pairs (s,t) in S × S such that st=r.

EXAMPLE Let $S = \{s_0, s_1, s_2, \ldots\}$ as before. Define multiplication on S by

$$s_n \cdot s_m = s_{n+m}$$

Then S is a monoid with identity s_0. (In fact, S is isomorphic to **N**, the natural numbers.) Here are two products in k[S]:

$$(1 \cdot s_0 + \pi \cdot s_2) \cdot (2 \cdot s_0 + e \cdot s_1) = 2 \cdot s_0 + e \cdot s_1 + 2\pi \cdot s_2 + \pi \cdot e \cdot s_3$$

$$(1 \cdot s_0 + \pi \cdot s_2) \cdot (2 \cdot s_0 + e \cdot s_2) = 2 \cdot s_0 + (2\pi + e) \cdot s_2 + \pi \cdot e \cdot s_4$$

Note the familiar collection process in the second case.

(1.5) PROPOSITION Let k be a commutative ring and S a monoid. Then k[S] is a ring. Moreover, k[S] is commutative if and only if S is.

PROOF We have already shown that k[S] is an additive group, so we need only verify the axioms involving multiplication. It is convenient first to prove the distributive laws.

(i) Left Distributivity. Compute, applying the definitions and laws as noted. (Here sums are taken over S, S × S, or S × S × S as appropriate.)

$$(\sum_s \lambda_s \cdot s) \cdot (\sum_t \mu_t \cdot t + \sum_t \nu_t \cdot t)$$

$$= (\sum_s \lambda_s \cdot s) \cdot (\sum_t (\mu_t + \nu_t) \cdot t) \qquad \text{(addition in k[S])}$$

$$= \sum_{(s,t)} \lambda_s \cdot (\mu_t + \nu_t) \cdot st \qquad \text{(multiplication in k[S])}$$

$$= \sum_{(s,t)} (\lambda_s \cdot \mu_t + \lambda_s \cdot \nu_t) \cdot st \qquad \text{(distributivity in k)}$$

$$= \sum_{(s,t)} \lambda_s \cdot \mu_t \cdot st + \lambda_s \cdot \nu_t \cdot st \qquad \text{(addition in k[S])}$$

$$= \sum_{(s,t)} \lambda_s \cdot \mu_t \cdot st + \sum_{(s,t)} \lambda_s \cdot \nu_t \cdot st \qquad \text{(addition in k[S])}$$

$$= (\sum_s \lambda_s \cdot s) \cdot (\sum_t \mu_t \cdot t) + (\sum_s \lambda_s \cdot s) \cdot (\sum_t \nu_t \cdot t) \qquad \text{(multiplication in k[S])}$$

(ii) Right Distributivity is similar.

(iii) Associativity. Again compute.

$$(\sum_s \lambda_s \cdot s) \cdot \{(\sum_t \mu_t \cdot t)(\sum_r \nu_r \cdot r)\}$$

$$= (\sum_s \lambda_s \cdot s) \cdot (\sum_{(t,r)} \mu_t \nu_r \cdot tr) \qquad \text{(multiplication in k[S])}$$

$$= \sum_s (\lambda_s \cdot s \cdot \sum_{(t,r)} \mu_t \nu_r \cdot tr) \qquad \text{(right distributivity in k[S])}$$

$$= \sum_s \sum_{(t,r)} \lambda_s (\mu_t \nu_r) \cdot s(tr) \qquad \text{(left distributivity in k[S])}$$

$$= \sum_{(s,t,u)} \lambda_s (\mu_t \nu_r) \cdot s(tr) \qquad \text{(commutativity of addition)}$$

Similarly one may show that

$$\{(\sum_s \lambda_s \cdot s) \cdot (\sum_t \mu_t \cdot t)\} \cdot (\sum_r \nu_r \cdot r) = \sum_{(s,t,u)} (\lambda_s \mu_t) \nu_r \cdot (st)r$$

and one sees that the results of both calculations are identical by appeal to the associative laws in k and S.

Clearly $1 \cdot e$ in $k[S]$ is the multiplicative identity of $k[S]$ (where 1 and e are the identities of k and S, respectively), so $k[S]$ is indeed a ring. If S is commutative, it follows at once from the definition of multiplication and the commutativity of k that $k[S]$ is likewise commutative. Conversely, if there exist s,t in S such that $st \neq ts$ in S, then $(1s)\cdot(1t) = 1\cdot st \neq 1\cdot ts = (1t)\cdot(1s)$ in $k[S]$, whence $k[S]$ is not commutative.

(1.6) PROPOSITION Under the hypotheses of (1.5), there exist

(i) An injective homomorphism of rings

$$k \to k[S]$$
$$\lambda \to \lambda \cdot e$$

with image in the center of $k[S]$, and

(ii) An injective homomorphism of monoids

$$S \to k[S]$$
$$s \to 1 \cdot s$$

(Henceforth we regard both of these injections as identifications.) In particular, $k[S]$ is a k-algebra.

PROOF Exercise.

$k[S]$ is called the monoid algebra of S over k. Note how an element μ in k acts on an element of $k[S]$ under the identification (i):

$$\mu\cdot(\sum \lambda_s \cdot s) = (\mu\cdot e)\cdot(\sum \lambda_s \cdot s) = \sum (\mu\cdot\lambda_s)\cdot s$$

(1.7) PROPOSITION (Universal Property) Let k be a commutative ring and S a monoid. Let A be a k-algebra. Then every monoid homomorphism ϕ: $S \to \langle A, \cdot \rangle$ extends uniquely to a homomorphism of k-algebras ϕ_*: $k[S] \to A$. That is, the following diagram commutes.

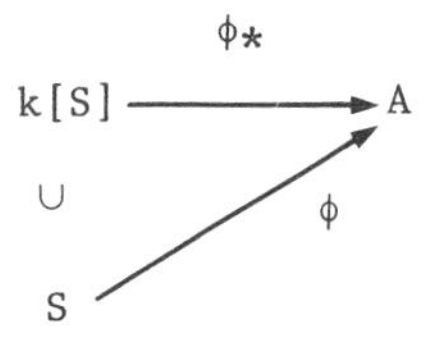

Here the inclusion of S into k[S] is via the identification of (1.6) (ii).

PROOF Define ϕ_* by

$$\phi_*(\sum_s \lambda_s \cdot s) = \sum_s \lambda_s \cdot \phi(s)$$

We will show that ϕ_* is multiplicative, leaving additivity to the reader. Let

$$a = \sum_s \lambda_s \cdot s \quad \text{and} \quad b = \sum_t \mu_t \cdot t$$

Then

$$a \cdot b = \sum_r \nu_r \cdot r \quad \text{where} \quad \nu_r = \sum_{st=r} \lambda_s \cdot \mu_t$$

and so by definition,

$$\phi_*(ab) = \sum_r \nu_r \cdot \phi(r) \quad \text{where} \quad \nu_r = \sum_{st=r} \lambda_s \cdot \mu_t$$

On the other hand,

$$\phi_*(a) \cdot \phi_*(b) = \{ \sum_s \lambda_s \cdot \phi(s) \} \cdot \{ \sum_t \mu_t \cdot \phi(t) \}$$

$$= \sum_{(s,t)} \lambda_s \mu_t \cdot \phi(s) \cdot \phi(t) \quad \text{(by the ring laws of A and centrality of } \lambda \text{ and } \mu)$$

$$= \sum_{(s,t)} \lambda_s \mu_t \cdot \phi(st) \quad (\phi \text{ is a monoid homomorphism})$$

$$= \sum_r \nu_r \cdot \phi(r)$$

where ν_r is defined as above. (This last step is merely collection of like terms.) Thus $\phi_*(ab) = \phi_*(a) \cdot \phi_*(b)$. Also $\phi_*(1 \cdot e) = 1 \cdot \phi(e) = \phi(e) = 1_A$, again since ϕ is a homomorphism of monoids, so ϕ_* is indeed a (unital) homomorphism of rings. Finally, to see that ϕ_* is moreover a homomorphism of k-algebras, we compute

$$\phi_*(\mu \cdot \sum_s \lambda_s \cdot s) = \phi_*(\sum_s (\mu \cdot \lambda_s) \cdot s)$$

$$= \sum_s (\mu \cdot \lambda_s) \cdot \phi(s)$$

$$= \mu \cdot \sum_s \lambda_s \cdot \phi(s)$$

$$= \mu \cdot \phi_*(\sum_s \lambda_s \cdot s)$$

This establishes the required k-linearity.

(1.8) PROPOSITION (Functoriality) Let k and k' be commutative rings and let S and S' be monoids.

(i) If ψ: k → k' is a homomorphism of rings, then

$$\begin{array}{ccc} k[S] & \xrightarrow{\psi_*} & k'[S] \\ \sum_s \lambda_s \cdot s & \longrightarrow & \sum_s \psi(\lambda_s) \cdot s \end{array}$$

together with ψ is a morphism of algebras from the k-algebra k[S] to the k'-algebra k'[S].

(ii) If ϕ: S → S' is a homomorphism of monoids, then

$$\begin{array}{ccc} k[S] & \xrightarrow{\phi_*} & k[S'] \\ \sum_s \lambda_s \cdot s & \longrightarrow & \sum_s \lambda_s \cdot \phi(s) \end{array}$$

is a homomorphism of k-algebras.

REMARK Part (i) shows that for fixed S, k → k[S] is a functor from the category of commutative rings to the category of algebras. Part (ii) shows that for fixed k, S → k[S] is a functor from the category of monoids to the category of k-algebras.

PROOF (i) The reader may show that ψ_* is a homomorphism of rings. We shall verify the central condition of Definition (1.3). Let μ be in k.

$$\psi_*(\mu \cdot \sum_s \lambda_s \cdot s) = \psi_*(\sum_s (\mu \cdot \lambda_s) \cdot s)$$

$$= \sum_s \psi(\mu \cdot \lambda_s) \cdot s$$

$$= \sum_s \psi(\mu) \cdot \psi(\lambda_s) \cdot s$$

$$= \psi(\mu) \cdot \sum_s \psi(\lambda_s)\cdot s$$

$$= \psi(\mu)\cdot\psi_*(\sum_s \lambda_s\cdot s)$$

Thus (ψ_*, ψ) is a morphism of algebras.

(ii) Composing the given map $\phi: S \to S'$ with the inclusion of S' into k[S'], we obtain a monoid homomorphism $S \to k[S']$. Applying the universal property (1.7) to this last map, we obtain the stated result.

EXERCISES

(1.1) Complete the proof of Proposition (1.2). Be sure to use explicitly the interpretation of $\lambda\cdot a$ as the product $f(\lambda)\cdot a$ where f is the ring homomorphism defining the algebra structure.

(1.2) Show that every ring with unity is an algebra over its center and an algebra over the integers.

(1.3) Let A and B be k-algebras via the homomorphisms $f: k \to A$, $g: k \to B$. Show that the direct product of rings $A \times B$ is likewise a k-algebra via the homomorphism

$$f \times g: k \longrightarrow A \times B$$
$$\lambda \longrightarrow (f(\lambda), g(\lambda))$$

Hence direct products are defined for algebras. Conclude that under our notational convention $\lambda\cdot(a,b) = (\lambda a, \lambda b)$.

(1.4) Show that the composition of a pair of morphisms of algebras is again a morphism of algebras.

(1.5) Prove Proposition (1.6).

(1.6) Let n be a positive integer. Recall that $C_n = \langle s\rangle$ is the cyclic group of order n. In particular, C_n is a monoid.

(a) Explicitly describe the **Q**-algebra $A = \mathbf{Q}[C_n]$ and the identification of **Q** with a subring of A.

(b) Is A an integral domain? If not, exhibit a pair of zero divisors.

(c) Is $A^{\times} = \mathbf{Q}^*$ under the identification of (a)? If not, exhibit a unit of A which is not in $\mathbf{Q}$.

(1.7) Write out a Cayley table for multiplication in the monoid algebra $\mathbf{F}_3[C_2]$, where $\mathbf{F}_3 = \mathbf{Z}/3\mathbf{Z}$ is the field of three elements. Explicitly identify all units and zero divisors.

(1.8) Complete the proof of Proposition (1.7).

(1.9) Suppose $\psi: k \to k'$ is a unital homomorphism of commutative rings and $\phi: S \to S'$ is a homomorphism of monoids. Define a function $g: k[S] \to k'[S']$ by

$$g\Big(\sum_s \lambda_s \cdot s \Big) = \sum_s \psi(\lambda_s) \cdot \phi(s)$$

Show that (g, ψ) is a morphism of algebras. [Hint: use Proposition (1.8) and Exercise (1.4).]

(1.10) Let S be a monoid and k a commutative ring. Suppose that s and t are distinct elements of S. Note that the expression s + t in k[S] is ambiguous in the sense that we may interpret it either as a single formal sum in the set k[S] or as the group-theoretic sum of the elements s and t in $\langle k[S], + \rangle$. (To be more precise, the plus sign is ambiguous, since it both occurs as a character in the string that defines a formal sum and denotes an operation on formal sums.) Appeal to the set-theoretic interpretation of k[S] to show that this ambiguity is harmless; i.e., both interpretations resolve to the same object.

GROUP RINGS

For the remaining exercises, k is a commutative ring with unity and G is a group. The monoid algebra k[G] is then called a <u>group ring</u>. [Exercises (1.6) and (1.7) above provide two excellent examples of this construction.] Group rings arise in connection with the theory of <u>group representations</u>, which is beyond the scope of this text. (See, for example, the celebrated <u>Representation Theory of Finite Groups and Associative Algebras</u> by Curtis and Reiner.)

(1.11) State and prove the universal property of group rings, analogous to Proposition (1.7). (Hint: start with group homomorphisms from G to $A^{\times}$.)

(1.12) State and prove functoriality for group rings, analogous to Proposition (1.8).

(1.13) Conclude from the previous exercise that there is a unique homomorphism of k-algebras α: $k[G] \to k$ which sends every element $s \in G$ to 1_k. This map α is called the augmentation homomorphism. (Here we have identified G with a subset of k[G] in the usual manner.)

(1.14) Show that the kernel of the augmentation homomorphism is the ideal in the ring k[G] generated by elements of the form (s - 1) for $s \in G$. This ideal is called the augmentation ideal and denoted $I_{k[G]}$. Conclude that $k[G]/I_{k[G]}$ is isomorphic to k.

(1.15) Suppose there is given a homomorphism of groups $G \to G'$, so that by functoriality there is an induced homomorphism of k-algebras $k[G] \to k[G']$. Show that this induced map sends $I_{k[G]}$ into $I_{k[G']}$.

(1.16) Generalize Exercise (1.14) as follows: Let N be a normal subgroup of G so that there is a short exact sequence of groups

(#) $1 \to N \to G \to G/N \to 1$

Show that the canonical map $G \to G/N$ induces a k-homomorphism of group rings $k[G] \to k[G/N]$ with kernel $I_{k[N]}k[G]$ (the ideal in k[G] generated by the augmentation ideal $I_{k[N]}$). Conclude that (#) induces a short exact sequence of k-algebras

$$0 \to I_{k[N]}k[G] \to k[G] \to k[G/N] \to 0$$

and that $k[G/N] \cong k[G]/I_{k[N]}k[G]$.

(1.17) Show that if $G \cong G'$, then $k[G] \cong k[G']$ as k-algebras.

The previous exercise shows that the group structure of G completely determines the ring structure of k[G]. The remaining exercises in this set address some naive observations concerning the converse relationship; i.e., the extent to which the ring structure of $k[G]$ determines the group structure of G. This problem is known as the Isomorphism Question.

(1.18) Let $\zeta \in k[G]$ satisfy $\zeta^n = 1$ for some positive integer n, and let $z = 1 + \zeta + \zeta^2 + \cdots + \zeta^{n-1}$. Show that $\zeta z = z$, and conclude that $z^2 = nz$.

(1.19) Continuing the exercise above, show that the subset $zk = \{ z\lambda : \lambda \in k \}$ of $k[G]$ is a commutative ring (not necessarily with unity) in its own right under the operations inherited from $k[G]$.

The reader may note at this point that the rings zk constructed above clearly admit multiplication by elements of k, although this action is not in general obtained via a homomorphism of rings $k \to zk$. In the terminology of the following chapter, zk is a k-module. Thus we have an essential extension of the notion of a k-algebra A: it is a ring which is also a k-module such that $\lambda(ab) = (\lambda a)b = a(\lambda b)$ for all $\lambda \in k$ and $a,b \in A$. (It is this last condition that generalizes the centrality of k in our original definition.)

(1.20) In the context of the previous exercise, describe multiplication in zk in the special case $n = p$, a prime, and $k = \mathbf{F}_p$, the field of p elements.

(1.21) In the context of Exercises (1.18) and (1.19), assume now that ζ takes the form ωs for $\omega \in k$, $s \in G$, with $(\omega s)^n = 1$. Assume further that $n = n \cdot 1_k \in k^\times$. Prove that there is an isomorphism of k-algebras $zk \to k$. What is the unity in zk?

(1.22) Consider the special case $k = \mathbf{C}$, the field of complex numbers, and $G = C_n = \langle s \rangle$, the cyclic group of order n with generator s. Define ω to be $e^{2\pi i/n}$, so that $o(\omega) = n \in k^\times$ (that is, ω is a primitive n-th root of unity). Let z_i $(i = 0,1,\ldots,n-1) \in k[G]$ be given by

$$z_i = \sum_{j=0}^{n-1} (\omega^i s)^j$$

(a) Show that for each i, $z_i k$ is a two-sided ideal of k[G] and in fact a ring with unity in its own right.

(b) Show that $z_0 + z_1 + \cdots + z_{n-1} = n$. [Hint: by choice of ω and arithmetic, $0 = \omega^n - 1 = (\omega - 1)(\omega^{n-1} + \omega^{n-2} + \cdots + \omega + 1)$.]

(c) Show that $z_i z_j = 0$ whenever $i \neq j$.

(d) Use Exercise (1.18) and (c) to conclude further that

$$z_i k \cap (\sum_{j \neq i} z_j k) = 0 \quad \text{for} \quad i \neq j$$

(e) Deduce from (b) and (d) that k[G] is isomorphic to the direct product of k-algebras $\Pi\ z_i k$, $i = 0,\ldots,n-1$.

(f) Conclude from (e) and Exercise (1.21) that k[G] is k-algebra isomorphic to k^n, the n-fold direct product of k with itself.

(1.23) Let $k = \mathbf{C}$ and $G = C_n \times C_m$, the direct product of cyclic groups. Show that k[G] is k-algebra isomorphic to k^{mn}. (Hint: as in the previous exercise, manufacture n elements z_i out of a generator for C_n and a primitive n-th root of unity and m elements y_j out of a generator for C_m and a primitive m-th root of unity. Now consider the ideals generated by the mn products $z_i y_j$.)

(1.24) Generalize the previous exercise, using the structure theorem for finite abelian groups to show that the complex group algebra $\mathbf{C}[G]$ of a finite abelian group G is $\mathbf{C}$-algebra isomorphic to $\mathbf{C}^{o(G)}$. Conclude that the ring structure of the complex group algebra of a finite abelian group in general determines the cardinality of the group but not its isomorphism class. (Identify the specific properties of the ring of complex numbers used in your proof.)

(1.25) Show that the rational group rings $\mathbf{Q}[C_4]$ and $\mathbf{Q}[C_2 \times C_2]$ are not isomorphic. (Hint: use the basic idea of Exercise (1.23) to achieve a direct product decomposition of $\mathbf{Q}[C_2 \times C_2]$ into copies of $\mathbf{Q}$ and use this decomposition to show that $\mathbf{Q}[C_2 \times C_2]$ has no units of order 4. Ah, but $\mathbf{Q}[C_4]$ does!)

(1.26) Show that if $\mathbf{Q}[C_4]$ is isomorphic to $\mathbf{Q}[G]$ for some group

G, then G is isomorphic to C_4. (Hint: show first that o(G) = 4 and then appeal to the paucity of isomorphism classes of such groups.)

A general theme in the isomorphism question might be this: the leaner the ring k, the more the ring structure of k[G] determines the group structure of G. Along these lines, we note that a good number of powerful positive results do hold for the special case $k = \mathbf{Z}$. For a brilliant, masterful exposition of these issues (and of group rings in general), see Passman's wonderful book The Algebraic Structure of Group Rings.

2. POLYNOMIALS

(2.1) MONOMIALS Let $x_1, x_2, \ldots, x_n$ be n symbols, to which we refer as indeterminates. Then a monomial in these indeterminates is an expression of the form

$$(\#) \quad x_1^{\nu_1} x_2^{\nu_2} \cdots x_n^{\nu_n}$$

where $\nu_1, \nu_2, \ldots, \nu_n$ are natural numbers. Set theoretically, we can identify (#) with the function from $\{x_1, \ldots, x_n\}$ to $\mathbf{N}$ which sends x_i to ν_i, $i = 1, 2, \ldots, n$. [See (1.4), the discussion of formal sums.]

We denote the set of all monomials in the indeterminates $x_1, x_2, \ldots, x_n$ by $\langle x_1, x_2, \ldots, x_n \rangle$.

(2.2) NOTATION We shall frequently omit indeterminates with exponent 0 from expressions such as (#) and furthermore omit those exponents which are 1. Thus, for instance, one has

$$x_1 x_3 x_5^2 = x_1^1 x_2^0 x_3^1 x_4^0 x_5^2$$

in $\langle x_1, x_2, x_3, x_4, x_5 \rangle$. Accordingly, one identifies the indeterminates $x_1, x_2, \ldots, x_n$ with elements of the larger collection $\langle x_1, x_2, \ldots, x_n \rangle$:

$$x_i = x_1^0 x_2^0 \cdots x_i^1 \cdots x_n^0$$

The monomial $x_1^0 x_2^0 \cdots x_n^0$ we usually write as 1. Finally, when convenient, (#) may be condensed to an indexed product

$$\prod_{i=1}^{n} x_i^{\nu_i}$$

(2.3) PRODUCTS OF MONOMIALS Define a product on $\langle x_1,\ldots,x_n\rangle$ by

$$\left(\prod_{i=1}^{n} x_i^{\nu_i}\right)\cdot\left(\prod_{i=1}^{n} x_i^{\mu_i}\right) = \prod_{i=1}^{n} x_i^{\nu_i+\mu_i}$$

One sees easily that $\langle x_1,\ldots,x_n\rangle$ is a commutative monoid with respect to this operation, with identity 1. If one interprets monomials as functions $\{x_1,\ldots,x_n\} \to \mathbf{N}$ as suggested above, this product formula is simply the usual addition of functions. Cosmetically, it is just what one would expect from high school algebra.

(2.4) PROPOSITION (Universal Property) Let M be a commutative monoid and let $f\colon \{x_1,\ldots,x_n\} \to M$ be an arbitrary function. Then there exists a unique homomorphism of monoids

$$g\colon \langle x_1,\ldots,x_n\rangle \to M$$

which extends f. That is, the following diagram commutes.

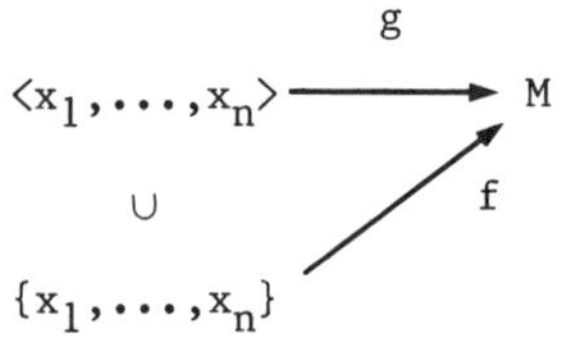

PROOF Define g by

$$g\left(\prod_{i=1}^{n} x_i^{\nu_i}\right) = \prod_{i=1}^{n} f(x_i)^{\nu_i}$$

Then

$$g\left(\prod_{i=1}^{n} x_i^{\nu_i} \cdot \prod_{i=1}^{n} x_i^{\mu_i}\right) = g\left(\prod_{i=1}^{n} x_i^{\nu_i+\mu_i}\right)$$

$$= \prod_{i=1}^{n} f(x_i)^{\nu_i+\mu_i}$$

$$= \prod_{i=1}^{n} f(x_i)^{\nu_i} \cdot \prod_{i=1}^{n} f(x_i)^{\mu_i}$$

$$= g\left(\prod_{i=1}^{n} x_i^{\nu_i}\right) \cdot g\left(\prod_{i=1}^{n} x_i^{\mu_i}\right)$$

(Note how the third line of this calculation depends on the commutativity of M.) Since clearly g sends 1 to the identity of M, it is indeed a homomorphism of monoids. It is immediate from the definition that g extends f. We leave uniqueness to the reader.

In consequence of this universal property, $\langle x_i,\ldots x_n\rangle$ is called the <u>free commutative monoid</u> on the set $\{x_1,\ldots,x_n\}$.

(2.5) POLYNOMIALS We now come to the central definition of this section. What follows depends heavily on the monoid algebra construction of Section 1.

DEFINITION Let k be a commutative ring. Then the monoid algebra $k[\langle x_1,\ldots,x_n\rangle]$ is called the <u>polynomial ring in the indeterminates $x_1,\ldots,x_n$ over k</u> and is denoted simply $k[x_1,\ldots,x_n]$. Elements of $k[x_1,\ldots,x_n]$ are called <u>polynomials</u> (in $x_1,\ldots,x_n$ with coefficients in k).

Let $\boldsymbol{\nu} = (\nu_1,\ldots,\nu_n)$ denote an n-tuple of natural numbers. Then an element of $k[x_1,\ldots,x_n]$ has the form

$$\sum_{\boldsymbol{\nu}} a_{\boldsymbol{\nu}} x_1^{\nu_1} x_2^{\nu_2} \cdots x_n^{\nu_n}$$

where $\boldsymbol{\nu}$ ranges over the n-fold product of $\mathbf{N}$ with itself and only finitely many $a_{\boldsymbol{\nu}}$ in k are nonzero. Recall that according to (1.6), we naturally identify k and $\langle x_1,\ldots,x_n\rangle$ with subsets of the k-algebra $k[x_1,\ldots,x_n]$. We speak of the elements of k as the <u>constant polynomials</u> in $k[x_1,\ldots,x_n]$. The zero of k is called the <u>zero polynomial</u> in $k[x_1,\ldots,x_n]$.

EXAMPLE Let $k = \mathbf{R}$, the real numbers. We consider the polynomial ring $\mathbf{R}[x,y]$. (When the number of indeterminates is small, we prefer symbols like x,y,z to subscripted variables.) A real number a is identified with the element $a \cdot x^0y^0$ in $\mathbf{R}[x,y]$, and the monomial x^ny^m in $\langle x,y \rangle$ is identified with $1 \cdot x^ny^m$ in $\mathbf{R}[x,y]$. Hence a typical element of $\mathbf{R}[x,y]$ might be written as

$$2 + \pi \cdot xy + \sqrt{3} \cdot x^2y^5 + x^4y^7$$

The following proposition formalizes the notion of evaluation of a polynomial via the familiar idea of substitution.

(2.6) PROPOSITION (Universal Property) Let A be a commutative k-algebra and let $\mathbf{b} = (b_1,..,b_n)$ be an n-tuple of elements of A (i.e., a point of A^n). Then there exists a unique homomorphism of k-algebras $\mathrm{val}[\mathbf{b}]: k[x_1,...,x_n] \to A$ such that $\mathrm{val}[\mathbf{b}](x_i) = b_i$, $i=1,...,n$. (The k-homomorphism $\mathrm{val}[\mathbf{b}]$ is called <u>evaluation at **b**.</u>)

PROOF Consider the set map

$$\begin{aligned} \{x_1,...,x_n\} &\to A \\ x_i &\to b_i \end{aligned}$$

which sends the indeterminate x_i to the algebra element b_i. By the universal property (2.4) of the free commutative monoid $\langle x_1,...,x_n \rangle$, there exists a unique extension of this map to a monoid homomorphism from $\langle x_1,...,x_n \rangle$ into the commutative monoid $\langle A,\cdot \rangle$. (Here the commutativity of A is paramount.) This latter homomorphism in turn extends uniquely to a homomorphism of k-algebras $k[x_1,...,x_n] \to A$ by the universal property of monoid algebras (1.7). (Remember that $k[x_1,...,x_n] = k[\langle x_1,...,x_n \rangle]$.) This second extension is the map $\mathrm{val}[\mathbf{b}]$. The reader may deduce its uniqueness according to Exercise (2.1).

Let p lie in $k[x_1,...,x_n]$. Henceforth we write $p(b_1,...,b_n)$ or $p(\mathbf{b})$ for $\mathrm{val}[\mathbf{b}](p)$. We call $p(\mathbf{b})$ <u>the value of p at **b**.</u> A fixed p in $k[x_1,...,x_n]$ thus defines a function

$$\begin{aligned} A^n &\to A \\ \mathbf{b} &\to p(\mathbf{b}) \end{aligned}$$

which is in general <u>not</u> a homomorphism. Of special importance here is the case A = k. We'll examine the connection between polynomials and functions in some detail below.

One final note on evaluation: If p lies in $k[x_1,\ldots,x_n]$, it makes sense to evaluate p at arguments in the commutative k-algebra $k[x_1,\ldots,x_n]$ itself. In particular, $p(x_1,\ldots,x_n)$ is defined, and it is immediate that in fact $p(x_1,\ldots,x_n) = p$. We sometimes use the longer expression to emphasize the indeterminates.

(2.7) DEGREE We define the degree of a monomial

$$\prod_{i=1}^{n} x_i^{\nu_i}$$

in $\langle x_1,\ldots,x_n\rangle$ to be $\nu_1 + \nu_2 + \ldots + \nu_n$; that is, we sum the exponents. Thus the degree of a monomial is always greater than or equal to zero. Now let p be a nonzero polynomial in $k[x_1,\ldots,x_n]$. Then

$$p = \sum_{\boldsymbol{\nu}} a_{\boldsymbol{\nu}} x_1^{\nu_1} \cdots x_n^{\nu_n}$$

where $\boldsymbol{\nu} = (\nu_1,\ldots,\nu_n)$ varies over $\mathbf{N}^n$ and at least one $a_{\boldsymbol{\nu}}$ is nonzero, but no more than finitely many such coefficients are nonzero. For each p define deg(p), the degree of p, to be the maximum of the degrees of all monomials occurring with nonzero coefficient. Note that the only polynomials of degree zero are then the nonzero constant polynomials (the nonzero elements of k). Finally, define the degree of the zero polynomial to be minus infinity ($-\infty$).

(2.8) PROPOSITION (i) Let p and q be elements of $k[x_1,\ldots,x_n]$. Then

$$\deg(p + q) \leqslant \max\{\deg(p),\deg(q)\}$$

(ii) Suppose further that k is an integral domain. Then

$$\deg(p\cdot q) = \deg(p) + \deg(q)$$

Moreover, this assertion fails in general when k is not an integral domain.

PROOF (i) Suppose $m = \max\{\deg(p),\deg(q)\}$. Then all monomials of degree greater than m have zero coefficient in both p and q and hence in the sum $p + q$ since addition of polynomials is just addition of corresponding coefficients. Therefore $\deg(p + q) \leq m$ as claimed.

(ii) First consider the case of polynomials p and q in one variable x, with coefficients in the integral domain k. We may assume that neither p nor q is the zero polynomial, else the assertion is trivial. So say $\deg(p) = r$ and $\deg(q) = s$ where r and s are nonnegative. Then

$$p(x) = ax^r + \text{terms of smaller degree}$$
$$q(x) = bx^s + \text{terms of smaller degree}$$

for some a,b in k; $a,b \neq 0$. Note that <u>in this one variable case there is a unique term of highest degree for each polynomial</u>. Thus

$$p(x)\cdot q(x) = (ab)x^{r+s} + \text{terms of smaller degree}$$

and $ab \neq 0$ since k is an integral domain. Hence $\deg(pq) = r + s = \deg(p) + \deg(q)$ as required.

The general multivariable case is somewhat more subtle since the underscored statement of the last paragraph fails in general for $k[x_1,\ldots x_n]$ when $n > 1$. The proof is given in Exercise (2.2) below.

Finally, if k is not an integral domain, there exist a,b in k, both nonzero, such that $ab = 0$. Identifying a,b, and 0 with constant polynomials in $k[x_1,\ldots,x_n]$, we at once contradict formula (ii):

$$-\infty = \deg(0) = \deg(ab) < \deg(a) + \deg(b) = 0$$

This completes the proof.

(2.9) COROLLARY Let k be an integral domain. Then

(i) $k[x_1,\ldots,x_n]$ is an integral domain;

(ii) the units of $k[x_1,\ldots,x_n]$ are precisely the units of k.

PROOF (i) Let p and q be polynomials such that $p\cdot q = 0$. Then by (2.8), $\deg(p) + \deg(q) = -\infty$, and thus either p or q is 0.

(ii) Say p is a unit, and $p\cdot q = 1$. Then by (2.8), $\deg(p) + \deg(q) = 0$. This forces p and q to be both of degree 0; that is, $p,q \in k$ and hence by assumption $p,q \in k^\times$. Conversely, the units of k are evidently units of $k[x_1,\dots,x_n]$.

REMARK Of course (2.9)(i) fails if k has zero divisors; in fact, even (2.9)(ii) may fail! Consider the one variable case k[x] where $k = \mathbf{Z}/4\mathbf{Z}$. We have

$$([2]x + 1)([2]x + 1) = 1$$

whence $(k[x])^\times \neq k^\times$. Nevertheless, there do exist commutative rings k which are not integral domains, such that polynomial algebras over k indeed satisfy (2.9)(ii). See Exercise (2.3) below.

(2.10) POLYNOMIALS AS FUNCTIONS We conclude this section with a discussion which relates the formal construction of polynomials in one variable, as given above, with the earlier notion familiar to us from calculus. Recall from (2.6) that for each $p \in k[x]$ there is an induced map

$$\begin{aligned} k &\to k \\ a &\to p(a) \end{aligned}$$

Let us denote this map f_p (the function induced by p). Then f_p lies in M(k), the set of all maps from k to k. Now M(k) is evidently a commutative ring under pointwise addition and multiplication of functions. Moreover, we may identify k with the constant functions in M(k), whence M(k) acquires the structure of a k-algebra. We have the following theorem.

THEOREM Let k be an infinite field. Then the map

$$\begin{aligned} k[x] &\to M(k) \\ p &\to f_p \end{aligned}$$

is an injective homomorphism of k-algebras. Hence we may identify polynomials in k[x] with functions $k \to k$.

PROOF One verifies by direct computation that $p \to f_p$ is indeed a homomorphism of k-algebras. We demonstrate injectivity by establishing the triviality of the kernel. Suppose $p \in k[x]$ and $f_p = 0$ (the zero map). Then $f_p(a) = p(a) = 0$ for all $a \in k$ and thus p has infinitely many roots in k. By Corollary (1.13) of Chapter V, this forces p to be the zero polynomial. Thus our map has trivial kernel.

This result is extended to the multivariable case in the exercises below. In contrast, for finite fields we have the following.

THEOREM Let k be a finite field of q elements. Then the map

$$k[x] \to M(k)$$
$$p \to f_p$$

has kernel generated by the polynomial $x^q - x$ in k[x]. In particular, this map is not injective.

PROOF Recall from the previous chapter that for k a field, k[x] is a euclidean domain with respect to degree. Hence if I is a nonzero ideal of k[x], and g(x) is a nonzero polynomial in I of minimal degree, then I = g(x)k[x], the principal ideal generated by g(x). We may apply this to the current situation, taking $I = \ker(p \to f_p)$ and $g(x) = x^q - x$, provided we can show that indeed g(x) both lies in I and is of minimal degree among the nonzero elements of I.

First, since the nonzero elements of k form a group of order q - 1 under multiplication, it follows by elementary group theory that every element of k (including 0) satisfies $x^q - x = 0$. Therefore, g(x) lies in I. Second, any element of I vanishes at all q points of k by definition and so must have degree at least q, again by Corollary (1.13) of Chapter V. Hence g(x) is of minimal degree, as required. This completes the proof.

EXAMPLES The first theorem shows in particular that the polynomial algebras **Q**[x], **R**[x], and **C**[x] may be identified with certain classes of functions on **Q**, **R**, and **C**, respectively, just as one would expect. As a special case of the second for q = 5, we have the nonzero polynomial $x^5 - x \in \mathbf{F}_5[x]$, which induces the zero function on $\mathbf{F}_5$, the field of five elements (**Z**/5**Z**).

EXERCISES

(2.1) Let f,g: $k[x_1,\ldots,x_n] \to A$ be homomorphisms of k-algebras. Show that f = g if and only if $f(x_i) = g(x_i)$, i = 1,...,n. Use this to establish uniqueness in Proposition (2.6).

(2.2) Let n be a positive integer and consider the set S of n-tuples of natural numbers:

$$S = \{ \boldsymbol{\nu} = (\nu_1,\ldots,\nu_n) \} = \mathbf{N}^n$$

For $\boldsymbol{\nu} \in S$, define $w(\boldsymbol{\nu})$, the weight of $\boldsymbol{\nu}$, to be the sum of its components, $\nu_1 + \nu_2 + \ldots + \nu_n$. Construct a relation $\leqslant$ on S as follows. For $\boldsymbol{\nu} = (\nu_1,\ldots,\nu_n)$ and $\boldsymbol{\mu} = (\mu_1,\ldots,\mu_n)$ in S, write $\boldsymbol{\nu} \leqslant \boldsymbol{\mu}$ if

(i) $w(\boldsymbol{\nu}) < w(\boldsymbol{\mu})$, or

(ii) $w(\boldsymbol{\nu}) = w(\boldsymbol{\mu})$ and there exists j, $1 \leqslant j \leqslant n$, such that $\nu_i = \mu_i$, $i < j$, and $\nu_j < \mu_j$, or

(iii) $\boldsymbol{\nu} = \boldsymbol{\mu}$.

(Thus S is ordered first by weight, and second lexicographically.) Write $\boldsymbol{\nu} < \boldsymbol{\mu}$ if $\boldsymbol{\nu} \leqslant \boldsymbol{\mu}$ as defined above and $\boldsymbol{\nu} \neq \boldsymbol{\mu}$.

(a) Show that $\leqslant$ is a total ordering on S. (See Appendix A.)

(b) Show that $\leqslant$ respects addition on S; that is, if $\boldsymbol{\nu} \leqslant \boldsymbol{\mu}$ and $\boldsymbol{\nu}' \leqslant \boldsymbol{\mu}'$, then $\boldsymbol{\nu} + \boldsymbol{\nu}' \leqslant \boldsymbol{\mu} + \boldsymbol{\mu}'$; moreover, the latter inequality is strict if either of the former inequalities is.

(c) Use (a) and (b) to complete the proof of Proposition (2.8).

(2.3) Let k_0 and k_1 be integral domains. Show that the direct product of rings $k = k_0 \times k_1$ is not an integral domain, but nevertheless $(k[x_1,\ldots,x_n])^\times = k^\times$. (Hint: consider the morphisms of algebras $k[x_1,\ldots,x_n] \to k_i[x_1,\ldots,x_n]$ induced by the projections $k_0 \times k_1 \to k_i$, $i = 0,1$.)

(2.4) Show that $(k[x])[y]$ is k-algebra isomorphic to $k[x,y]$. Extend this result to the general n-variable case. [Hint: make repeated and elegant use of Proposition (2.6).]

(2.5) Let k be an infinite field and $A = k[x_1,\ldots,x_n]$. Let B be the set of all maps $k^n \to k$, and identify k with the set of all constant functions in B.

(a) Show that B (as a ring under pointwise addition and multiplication of functions) is a k-algebra.

(b) If $p \in A$, define $f_p \in B$ by $f_p(\mathbf{b}) = p(\mathbf{b})$ (p evaluated at the point $\mathbf{b} \in k^n$). Show that the map

$$A \to B$$
$$p \to f_p$$

is an injective homomorphism of k-algebras. Conclude that in this case we may identify $k[x_1,\ldots,x_n]$ with a certain class of functions $k^n \to k$. (Hint: use induction on n.)

(2.6) Maintain the notation of the previous exercise, but now let k be a finite field of q elements.

(a) Show that for every polynomial $p \in A$ there is a polynomial $p^* \in A$ of degree less than q in each variable such that $f_p = f_{p^*}$.

(b) Suppose that $p \in A$ has degree less than q in each variable. Deduce that $f_p = 0$ if and only if $p = 0$.

For the remaining exercises, k is an infinite field and A is the polynomial algebra $k[x_1,\ldots,x_n]$. According to Exercise (2.5), we may identify elements of A with functions on k^n. If $p \in A$, we define Z(p), the <u>zero set of p</u>, to be the set of all points $\mathbf{b} \in k^n$ such that $p(\mathbf{b}) = 0$.

(2.7) Let S be a subset of A. Define the zero set of S, Z(S), to be the set of common zeros of all the elements of S; i.e.,

$$Z(S) = \bigcap_{p \in S} Z(p)$$

Let I be the ideal generated by S in A. Show that Z(S) = Z(I).

A subset X of k^n which arises as the zero set of some subset (hence ideal) I of A is called an <u>algebraic set</u>.

(2.8) Prove that the union of two algebraic sets is an algebraic set; the intersection of a family of algebraic sets is likewise an algebraic set; both the empty set and k^n itself are also algebraic sets.

As a consequence of Exercise (2.8) we find that the complements of the algebraic sets in k^n constitute the open subsets of a topology on k^n. This is the famous Zariski topology which arises in algebraic geometry.

(2.9) Let $k = \mathbf{C}$ and $n = 1$. Describe the algebraic sets for this case.

(2.10) Let X be any subset of k^n. Associate with X a subset J(X) of A defined by

$$J(X) = \{p \in A : p(\mathbf{b}) = 0 \text{ for all } \mathbf{b} \in X\}$$

Thus J(X) is the set of polynomials in A that vanish at all points of X.

(a) Show that J(X) is an ideal of A.

(b) In general, is $Z(J(X)) = X$ for all subsets X of k^n? If not, give a specific counterexample.

(c) In general, is $J(Z(I)) = I$ for all ideals I of A? If not, give a specific counterexample.

3. MATRICES AND DETERMINANTS

We now reconstruct the familiar notions of matrices and determinants over an arbitrary commutative ring k. Since the present theory is entirely similar to that of the well-known case $k = \mathbf{R}$, we leave most of the examples and a few of the arguments to the exercises. Be aware also that we make no attempt in this section to relate these notions to linear algebra. At this point we are only concerned with the basic properties of matrices as such; all connections with linear transformations on vector spaces are postponed until Chapter VII.

Throughout this section, k denotes a commutative ring (with unity).

(3.1) DEFINITION Let m and n be positive integers. Then an $m \times n$ matrix, with entries in k, is a doubly indexed family $A = (a_{ij})$, $1 \leqslant i \leqslant m$, $1 \leqslant j \leqslant n$, where each a_{ij} lies in k. Typically the matrix A is envisioned as a rectangular array, as shown.

$$A = \begin{pmatrix} a_{11} & a_{12} & \cdots & a_{1n} \\ a_{21} & a_{22} & \cdots & a_{2n} \\ \vdots & \vdots & & \vdots \\ a_{m1} & a_{m2} & \cdots & a_{mn} \end{pmatrix}$$

Thus the (i,j)-entry a_{ij} of A is the element found at the intersection of the i-th row and j-th column of the array. The set of all such m × n matrices is denoted $Mat_{m\times n}(k)$.

(3.2) NOTATION AND TERMINOLOGY The following elements of notation and terminology are standard in conjunction with matrices.

(i) $Mat_{n\times n}(k)$, the set of all n × n matrices with entries in k, is generally denoted $M_n(k)$. Elements of $M_n(k)$ are called <u>square matrices</u>.

(ii) A square matrix $A = (a_{ij}) \in M_n(k)$ is called a <u>diagonal matrix</u> if $a_{ij} = 0$ whenever $i \neq j$. Thus a diagonal matrix has the form shown.

$$A = \begin{pmatrix} \lambda_1 & & & 0 \\ & \lambda_2 & & \\ & & \ddots & \\ 0 & & & \lambda_n \end{pmatrix}$$

That is, each diagonal entry a_{ii} is equal to some ring element $\lambda_i \in k$, and each off-diagonal entry a_{ij} $(i \neq j)$ is zero. We sometimes write $A = \text{diag}(\lambda_1, \lambda_2, \ldots, \lambda_n)$ for the diagonal matrix whose successive diagonal entries are $\lambda_1, \lambda_2, \ldots, \lambda_n$.

(iii) The n × n diagonal matrix diag(1,1,...,1), all of whose diagonal entries are $1 \in k$, is called the <u>n × n identity matrix</u> and denoted I_n. That is,

$$I_n = \begin{pmatrix} 1 & & & 0 \\ & 1 & & \\ & & \ddots & \\ 0 & & & 1 \end{pmatrix}$$

In connection with this, we introduce a handy notational convention, the so-called Kronecker delta, δ_{ij}, defined by

$$\delta_{ij} = \begin{cases} 1 & \text{if } i = j \\ 0 & \text{if } i \neq j \end{cases}$$

(Of course, the meaning of '1' in this definition depends on the ambient ring k.) Then in fact $I_n = (\delta_{ij})$ $(1 \leqslant i,j \leqslant n)$.

(iv) Let $A = (a_{ij}) \in \mathrm{Mat}_{m\times n}(k)$ be an arbitrary $m \times n$ matrix. Then we define an $n \times m$ matrix $B = (b_{ij}) \in \mathrm{Mat}_{n\times m}(k)$, called the transpose of A, by $b_{ij} = a_{ji}$, $1 \leqslant i \leqslant n$, $1 \leqslant j \leqslant m$. (Note the reversal of dimensions.) We write $B = {}^tA$. Thus we have the pair of arrays

$$A = \begin{pmatrix} a_{11} & a_{12} & \cdots & a_{1n} \\ a_{21} & a_{22} & \cdots & a_{2n} \\ \cdot & \cdot & & \cdot \\ \cdot & \cdot & & \cdot \\ \cdot & \cdot & & \cdot \\ a_{m1} & a_{m2} & \cdots & a_{mn} \end{pmatrix} \qquad {}^tA = \begin{pmatrix} a_{11} & a_{21} & \cdots & a_{m1} \\ a_{12} & a_{22} & \cdots & a_{m2} \\ \cdot & \cdot & & \cdot \\ \cdot & \cdot & & \cdot \\ \cdot & \cdot & & \cdot \\ a_{1n} & a_{2n} & \cdots & a_{mn} \end{pmatrix}$$

and we see that geometrically ${}^tA$ is obtained from A by reflection through a line parallel to the diagonal. A square matrix is called symmetric if it is equal to its own transpose.

(v) The zero matrix of $\mathrm{Mat}_{m\times n}$ is the $m \times n$ matrix all of whose entries are zero. It is in this context that we interpret the statement '$0 \in \mathrm{Mat}_{m\times n}(k)$'.

(3.3) ADDITION AND MULTIPLICATION OF MATRICES Let $A = (a_{ij})$ lie in $\mathrm{Mat}_{m\times n}(k)$, and let $B = (b_{ij})$ lie in $\mathrm{Mat}_{p\times r}(k)$.

(i) The matrix sum $A + B$ is defined if and only if $m = p$ and $n = r$, in which case it is the $m \times n$ matrix $C = (c_{ij})$ defined by

$$c_{ij} = a_{ij} + b_{ij} \quad (1 \leq i \leq m,\ 1 \leq j \leq n)$$

Thus only matrices of the same dimension may be added, in which case the addition is componentwise.

(ii) The <u>matrix product</u> AB is defined if and only if $n = p$, in which case it is the $m \times r$ matrix $C = (c_{ij})$ defined by

$$c_{ij} = \sum_{k=1}^{n} a_{ik}b_{kj} \quad (1 \leq i \leq m,\ 1 \leq j \leq r)$$

Thus the product AB is defined only if A has as many columns as B has rows, in which case the (i,j)-th entry of the product is given by the 'dot product' of the i-th row of A with the j-th column of B. [The reader may recall from linear algebra that the dot product of vectors $(x_1,x_2,\ldots,x_n)$ and $(y_1,y_2,\ldots,y_n)$ is the sum $x_1y_1 + x_2y_2 + \cdots + x_ny_n$.]

The following proposition summarizes the elementary arithmetic properties of matrices. For reasons cited in the introduction, the proofs of all but one of the assertions are left to the exercises.

(3.4) PROPOSITION (i) For all positive integers m and n, $\langle Mat_{m\times n},+\rangle$ is an additive group.

(ii) Let $A \in Mat_{m\times n}(k)$, $B \in Mat_{n\times p}(k)$, $C \in Mat_{p\times r}(k)$. Then $(AB)C = A(BC)$.

(iii) Let $A \in Mat_{m\times n}(k)$. Then $I_mA = A = AI_n$.

(iv) Let $A \in Mat_{m\times n}(k)$, and let $B,C \in Mat_{n\times p}(k)$. Then $A(B + C) = AB + AC$. Similarly, matrix arithmetic is right distributive whenever the appropriate sums and products are defined.

PROOF OF ASSOCIATIVITY Let $A = (a_{ij})$, $B = (b_{ij})$, $C = (c_{ij})$ be as in statement (ii). We compute the (i,j)-entry of both A(BC) and (AB)C according to Definition (3.3)(ii), obtaining identical results:

$$\text{the } (i,j)\text{-entry of } A(BC) = \sum_{k=1}^{n} a_{ik}\cdot(\text{the } (k,j)\text{-entry of } BC)$$

$$= \sum_{k=1}^{n} a_{ik}\cdot\left(\sum_{m=1}^{p} b_{km}c_{mj}\right)$$

$$= \sum_{k=1}^{n}\sum_{m=1}^{p} a_{ik}b_{km}c_{mj}$$

$$\text{the } (i,j)\text{-entry of } (AB)C = \sum_{m=1}^{p} (\text{the } (i,m)\text{-entry of } AB)\cdot c_{mj}$$

$$= \sum_{m=1}^{p}\left(\sum_{k=1}^{n} a_{ik}b_{km}\right)\cdot c_{mj}$$

$$= \sum_{m=1}^{p}\sum_{k=1}^{n} a_{ik}b_{km}c_{mj}$$

(Note that we have used the ring properties of k freely.) It follows that the products A(BC) and (AB)C agree in each entry and hence are equal.

Our next result is an immediate consequence of Proposition (3.4) as applied to the case of square matrices of equal size.

COROLLARY For all positive n, the square matrices $M_n(k)$ constitute a ring with identity I_n.

The matrix rings $M_n(k)$ are the focus of the remainder of this section.

(3.5) SCALAR MATRICES Let $n > 0$ be fixed. An element of $M_n(k)$ of the form $\text{diag}(\lambda,\lambda,\ldots,\lambda)$ is called a scalar matrix. Thus a scalar matrix is a diagonal matrix with identical diagonal entries, as shown.

$$\text{diag}(\lambda,\lambda,\ldots,\lambda) = \begin{pmatrix} \lambda & & & O \\ & \lambda & & \\ & & \ddots & \\ O & & & \lambda \end{pmatrix}$$

LEMMA The mapping $k \to M_n(k)$ sending $\lambda \in k$ to the scalar matrix

$\mathrm{diag}(\lambda,\lambda,\ldots,\lambda)$ is an injective homomorphism of rings whose image is the subring of $M_n(k)$ consisting of all scalar matrices. (Henceforth, we regard this embedding as an identification.)

PROOF One readily verifies the equalities

$$\mathrm{diag}(\lambda + \mu,\lambda + \mu,\ldots,\lambda + \mu) = \mathrm{diag}(\lambda,\lambda,\ldots,\lambda) + \mathrm{diag}(\mu,\mu,\ldots,\mu)$$

$$\mathrm{diag}(\lambda\mu,\lambda\mu,\ldots,\lambda\mu) = \mathrm{diag}(\lambda,\lambda,\ldots,\lambda)\cdot\mathrm{diag}(\mu,\mu,\ldots,\mu)$$

Moreover, $1 \to \mathrm{diag}(1,1,\ldots,1) = I_n$, whence the given map is transparently an injective homomorphism of rings. In particular its image is a subring of $M_n(k)$.

This leads us to the main point of the section.

(3.6) PROPOSITION For each positive integer n, $M_n(k)$ is a k-algebra via the mapping $k \to M_n(k)$ described in the previous lemma.

PROOF The only remaining assertion to be verified is that the scalar matrices -- the image of k under the diagonal map -- lie in the center of $M_n(k)$. Recalling our identification, we have for all $\lambda \in k$, $A \in M_n(k)$,

$$\lambda A = \begin{pmatrix} \lambda & & & & & O \\ & \lambda & & & & \\ & & \cdot & & & \\ & & & \cdot & & \\ & & & & \cdot & \\ O & & & & & \lambda \end{pmatrix} \begin{pmatrix} a_{11} & a_{12} & \cdots & a_{1n} \\ a_{21} & a_{22} & \cdots & a_{2n} \\ \cdot & \cdot & & \cdot \\ \cdot & \cdot & & \cdot \\ \cdot & \cdot & & \cdot \\ a_{n1} & a_{n2} & \cdots & a_{nn} \end{pmatrix}$$

$$= \begin{pmatrix} \lambda a_{11} & \lambda a_{12} & \cdots & \lambda a_{1n} \\ \lambda a_{21} & \lambda a_{22} & \cdots & \lambda a_{2n} \\ \cdot & \cdot & & \cdot \\ \cdot & \cdot & & \cdot \\ \cdot & \cdot & & \cdot \\ \lambda a_{n1} & \lambda a_{n2} & \cdots & \lambda a_{nn} \end{pmatrix}$$

By similar calculation, we also find that

$$A\lambda = \begin{pmatrix} a_{11}\lambda & a_{12}\lambda & \cdots & a_{1n}\lambda \\ a_{21}\lambda & a_{22}\lambda & \cdots & a_{2n}\lambda \\ \vdots & \vdots & & \vdots \\ a_{n1}\lambda & a_{n2}\lambda & \cdots & a_{nn}\lambda \end{pmatrix}$$

and hence $\lambda A = A\lambda$ since k is commutative. Thus the scalar matrices are central, and $M_n(k)$ is therefore a bona fide k-algebra.

REMARK Applying our identification of k with the scalar matrices of $M_n(k)$ in its most grandiose form, we may write $k \subseteq M_n(k)$. Then according to (3.6), we have moreover $k \subseteq Z(M_n(k))$; that is, k lies in the center of $M_n(k)$. In fact, equality holds:

THEOREM For all positive integers n, $k = Z(M_n(k))$.

PROOF We shall sketch a proof of this remarkable fact in the exercises.

(3.7) PROPOSITION (Functoriality) Let $\psi: k \to k'$ be a (unital) homomorphism of commutative rings. For any given $n > 0$, define the map $\psi_*: M_n(k) \to M_n(k')$ by

$$\begin{pmatrix} a_{11} & a_{12} & \cdots & a_{1n} \\ a_{21} & a_{22} & \cdots & a_{2n} \\ \vdots & \vdots & & \vdots \\ a_{n1} & a_{n2} & \cdots & a_{nn} \end{pmatrix} \xrightarrow{\psi_*} \begin{pmatrix} \psi(a_{11}) & \psi(a_{12}) & \cdots & \psi(a_{1n}) \\ \psi(a_{21}) & \psi(a_{22}) & \cdots & \psi(a_{2n}) \\ \vdots & \vdots & & \vdots \\ \psi(a_{n1}) & \psi(a_{n2}) & \cdots & \psi(a_{nn}) \end{pmatrix}$$

(That is, if $A = (a_{ij}) \in M_n(k)$, then $\psi_*(A)$ is the element of $M_n(k')$ whose (i,j)-entry is $\psi(a_{ij})$.) Then ψ_* together with ψ is a morphism of algebras.

PROOF We must first show that $\psi_*: M_n(k) \to M_n(k')$ is a homomorphism. Let $A = (a_{ij})$, $B = (b_{ij}) \in M_n(k)$.

(i) Additivity. Use the additivity of $\psi: k \to k'$ and compute:

$$\begin{aligned} \text{the } (i,j)\text{-entry of } \psi_*(A + B) &= \psi(\text{the } (i,j)\text{-entry of } (A + B)) \\ &= \psi(a_{ij} + b_{ij}) \\ &= \psi(a_{ij}) + \psi(b_{ij}) \\ &= \text{the } (i,j)\text{-entry of } \psi_*(A) + \\ &\quad\ \text{the } (i,j)\text{-entry of } \psi_*(B) \end{aligned}$$

Thus $\psi_*(A + B) = \psi_*(A) + \psi_*(B)$ as required.

(ii) Multiplicativity. Again compute:

$$\begin{aligned} \text{the } (i,j)\text{-entry of } \psi_*(AB) &= \psi(\text{the } (i,j)\text{-entry of } AB) \\ &= \psi\left(\sum_{k=1}^{n} a_{ik}b_{kj}\right) \\ &= \sum_{k=1}^{n} \psi(a_{ik})\psi(b_{kj}) \\ &= \text{the } (i,j)\text{-entry of } \psi_*(A)\psi_*(B) \end{aligned}$$

Thus $\psi_*(AB) = \psi_*(A)\psi_*(B)$.

(iii) ψ_* is unital. Since ψ is a (unital) homomorphism, it maps 0 to 0 and 1 to 1. Hence $\psi(\delta_{ij}) = \delta_{ij}$, and it follows that ψ_* maps I_n in $M_n(k)$ to I_n in $M_n(k')$.

To conclude the proof, we must lastly check that ψ and ψ_* are compatible in the sense of (1.3). In preparation for this, note that $\psi_*(\mathrm{diag}(\lambda_1,\lambda_2,\ldots,\lambda_n)) = \mathrm{diag}(\psi(\lambda_1),\psi(\lambda_2),\ldots,\psi(\lambda_n))$ for any family of $\lambda_i \in k$. We now have a straightforward calculation before us, the only delicate points being the identifications.

$$\begin{aligned} \psi_*(\lambda A) &= \psi_*(\mathrm{diag}(\lambda,\lambda,\ldots,\lambda)A) \\ &= \psi_*(\mathrm{diag}(\lambda,\lambda,\ldots,\lambda)) \cdot \psi_*(A) && \text{[by (ii)]} \\ &= \mathrm{diag}(\psi(\lambda),\psi(\lambda),\ldots,\psi(\lambda)) \cdot \psi_*(A) \\ &= \psi(\lambda)\psi_*(A) \end{aligned}$$

This is the required compatibility condition.

REMARK It follows at once from the proposition that for fixed n, $k \to M_n(k)$ is a functor from the category of commutative rings to the category of algebras.

DETERMINANTS

We continue to assume that k is a commutative ring with unity, and now construct the determinant map from $M_n(k)$ to k. We shall see that most of the key properties encountered in the special case $k = \mathbf{R}$ persist in this more abstract setting.

(3.8) MORE NOTATION AND TERMINOLOGY

(i) Column Matrices. An element of $\mathrm{Mat}_{n\times 1}(k)$ is called a column matrix and has the general appearance

$$C = \begin{pmatrix} c_1 \\ c_2 \\ \cdot \\ \cdot \\ \cdot \\ c_n \end{pmatrix} \qquad (c_j \in k,\ 1 \leqslant j \leqslant n)$$

Clearly an $n \times 1$ column matrix is susceptible to multiplication by elements of $M_n(k)$, and hence in particular susceptible to multiplication by $\lambda \in k$ via the identification of (3.5). In fact, for C as above,

$$\lambda C = \mathrm{diag}(\lambda,\lambda,\ldots,\lambda)C = \begin{pmatrix} \lambda c_1 \\ \lambda c_2 \\ \cdot \\ \cdot \\ \cdot \\ \lambda c_n \end{pmatrix}$$

An especially important family of column matrices, denoted $E_1, E_2, \ldots, E_n$, is defined as follows: E_j is the $n \times 1$ column matrix whose j-th entry is 1, all other entries 0. Thus

$$E_1 = \begin{pmatrix} 1 \\ 0 \\ \cdot \\ \cdot \\ \cdot \\ 0 \end{pmatrix} \quad E_2 = \begin{pmatrix} 0 \\ 1 \\ \cdot \\ \cdot \\ \cdot \\ 0 \end{pmatrix} \quad , \ldots\ldots , \quad E_n = \begin{pmatrix} 0 \\ 0 \\ \cdot \\ \cdot \\ \cdot \\ 1 \end{pmatrix}$$

(ii) Amalgamation; Column Decomposition. Let $C_1, C_2, \ldots, C_n$ be a family of $n \times 1$ column matrices. Then $C = (C_1, C_2, \ldots, C_n)$ is the $n \times n$ matrix whose j-th column is C_j. For instance,

$$I_n = (E_1, E_2, \ldots, E_n)$$

where the columns E_j are as defined above.

Now let $A = (a_{ij}) \in M_n(k)$. Then we let $A^{(j)}$, $1 \leqslant j \leqslant n$, denote the column matrix which is the j-th column of A. That is,

$$A^{(j)} = \begin{pmatrix} a_{1j} \\ a_{2j} \\ \cdot \\ \cdot \\ \cdot \\ a_{nj} \end{pmatrix}$$

Thus $A = (A^{(1)}, A^{(2)}, \ldots, A^{(n)})$ and we can view A as the amalgamation of its associated column matrices.

(iii) Deletion. If $A \in M_n(k)$, $n > 1$, then henceforth $\partial_{ij}A$, $1 \leqslant i,j \leqslant n$, denotes the $(n - 1) \times (n - 1)$ matrix which results upon the deletion of the i-th row and j-th column from A. For example, if $A = (a_{ij}) \in M_4(k)$,

$$\partial_{11}A = \begin{pmatrix} a_{22} & a_{23} & a_{24} \\ a_{32} & a_{33} & a_{34} \\ a_{42} & a_{43} & a_{44} \end{pmatrix}, \qquad \partial_{23}A = \begin{pmatrix} a_{11} & a_{12} & a_{14} \\ a_{31} & a_{32} & a_{34} \\ a_{41} & a_{42} & a_{44} \end{pmatrix}$$

(3.9) THEOREM (The Fundamental Theorem of Determinants) For each $n \geqslant 1$, there exists a map $\det: M_n(k) \to k$, called the determinant, satisfying the following rules.

(i) Multilinearity. Suppose $A = (A^{(1)}, \ldots, A^{(n)}) \in M_n(k)$ and for some j, $A^{(j)} = \lambda C + \lambda' C'$ where $\lambda, \lambda' \in k$ and C and C' are column matrices. Then

$$\det(A^{(1)}, \ldots, \lambda C + \lambda' C', \ldots, A^{(n)}) = \lambda \cdot \det(A^{(1)}, \ldots, C, \ldots, A^{(n)}) + \lambda' \cdot \det(A^{(1)}, \ldots, C', \ldots, A^{(n)})$$

(ii) Alternation of Sign. Suppose $A = (A^{(1)},\dots,A^{(n)}) \in M_n(k)$ and $A^{(j)} = A^{(j+1)}$ for some j. Then $\det(A) = 0$.

(iii) Normalization. $\det(I_n) = 1_k$.

Moreover, the rules above uniquely characterize the determinant.

PROOF OF EXISTENCE [Uniqueness will be shown in (3.12).] We begin by giving an inductive definition for the determinant.

For n = 1, define det: $M_n(k) \to k$ to be the identity map on k. Clearly properties (i), (ii), and (iii) hold -- (ii) vacuously. For $n > 1$, define det: $M_n(k) \to k$ inductively by the formula

$$(\#) \quad \det(A) = \sum_{j=1}^{n} (-1)^{j+1} a_{1j} \det(\partial_{1j} A)$$

Thus the evaluation of det(A) for a given matrix proceeds in terms of matrices of smaller sizes until we finally reach the scalar case, for which the map is given explicitly. (The linear algebra student may recognize (#) as expansion by the first row.) Properties (i), (ii), and (iii) are, of course, verified by induction.

(i) For notational simplicity, we establish linearity in the first column; clearly the argument applies in the general case. So suppose $A^{(1)} = \lambda C + \lambda' C'$, and in particular $a_{11} = \lambda c + \lambda' c'$ where c and c' are the first entries of C and C' respectively. Then by (#)

$$\det(A) = (\lambda c + \lambda' c')\det(\partial_{11} A) + \sum_{j=2}^{n} (-1)^{j+1} a_{1j} \det(\partial_{1j} A)$$

and so by induction and the arithmetic of matrices,

$$\begin{aligned} \det(A) &= \lambda c \cdot \det(\partial_{11} A) + \lambda' c' \cdot \det(\partial_{11} A) \\ &\quad + \sum_{j=2}^{n} (-1)^{j+1} a_{1j} \cdot \lambda \cdot \det(\partial_{1j}(C, A^{(2)},\dots,A^{(n)})) \\ &\quad + \sum_{j=2}^{n} (-1)^{j+1} a_{1j} \cdot \lambda' \cdot \det(\partial_{1j}(C', A^{(2)},\dots,A^{(n)})) \end{aligned}$$

But $\partial_{11}(A) = \partial_{11}(C,\dots,A^{(n)}) = \partial_{11}(C',\dots,A^{(n)})$ since none of these involves the disputed first column of A. Hence combining the first and third terms and the second and fourth terms, we have with appeal to (#)

$$\det(A) = \lambda\det(C, A^{(2)}, \ldots, A^{(n)}) + \lambda'\det(C', A^{(2)}, \ldots, A^{(n)})$$

as claimed. (Note: the student who feels that the calculation above is overly terse has no compassion for the typist.)

(ii) We now demonstrate the second property in the special case of the first two columns. Once more the argument carries over to the general case without issue.

Suppose that $A^{(1)} = A^{(2)}$. Write (#) as

$$\det(A) = a_{11}\det(\partial_{11}A) - a_{12}\det(\partial_{12}A) + \sum_{j=3}^{n} (-1)^{j+1} a_{1j}\det(\partial_{1j}A)$$

Now for j=3,...,n we see that $\partial_{1j}A$ has first and second columns identical since A does. (For each, the deletion ∂_{1j} only removes the top row and some column to the right of the second.) Hence the indexed sum in our expansion of det(A) vanishes by induction. Now again since the first two columns of A are identical, $a_{11} = a_{12}$ and $\partial_{11}A = \partial_{12}A$. Thus the two surviving terms cancel, and the assertion follows.

(iii) Normalization is immediate: Since the (1,1)-entry of I_n is the only nonzero element of the top row, (#) reduces to

$$\det(I_n) = 1 \cdot \det(\partial_{11}I_n)$$

But $\partial_{11}I_n = I_{n-1}$, and by induction $\det(I_{n-1}) = 1$. Thus $\det(I_n) = 1$.

(3.10) COROLLARY (i) Suppose $A' \in M_n(k)$ is obtained from $A \in M_n(k)$ by transposition of two adjacent columns. Then det(A') = -det(A).

(ii) Suppose A' is obtained from A by transposition of <u>any</u> two columns. Then again det(A') = -det(A).

(iii) Suppose <u>any</u> two columns of A are identical. Then det(A)= 0.

PROOF (i) Suppose A = (...,C,C',....) and A' = (...,C',C,...) where C and C' are two adjacent columns. Consider the matrix A'' obtained from A by replacing both columns in question by their sum, C + C', while leaving the rest of A intact:

$$A'' = (\ldots, C + C', C + C', \ldots)$$

By alternation of sign, clearly $\det(A'') = 0$. But by multilinearity

$$\begin{aligned}\det(A'') &= \det(\ldots,C,C + C',\ldots) + \det(\ldots,C',C + C',\ldots)\\ &= \det(\ldots,C,C,\ldots) + \det(\ldots,C,C',\ldots)\\ &+ \det(\ldots,C',C,\ldots) + \det(\ldots,C',C',\ldots)\end{aligned}$$

and again by alternation the first and fourth of these terms are 0. Thus

$$0 = \det(A'') = \det(A) + \det(A')$$

and the claim follows.

(ii) Any transposition can be obtained as the composition of an odd number of adjacent transpositions (set theory!), whence by (i) the net effect on the determinant is a factor $(-1)^{\text{odd}} = -1$.

(iii) Exercise: use the idea of (i) and the result of (ii).

(3.11) COROLLARY (Expansion by Rows and Columns) (i) For <u>any</u> fixed row index i, $1 \leqslant i \leqslant n$,

$$\det(A) = \sum_{j=1}^{n} (-1)^{i+j} a_{ij} \det(\partial_{ij} A)$$

(ii) For <u>any</u> fixed column index j, $1 \leqslant j \leqslant n$,

$$\det(A) = \sum_{i=1}^{n} (-1)^{i+j} a_{ij} \det(\partial_{ij} A)$$

PROOF Following the existence proof in the Fundamental Theorem, one shows that both formulas above also satisfy multilinearity, alternation of sign, and normality. The result then follows by uniqueness, which is established below.

We now give a closed formula for the determinant; this is the key to demonstrating that there can be only one such map.

(3.12) THEOREM For all $A \in M_n(k)$,

$$\det(A) = \sum_{\pi \in S_n} \sigma(\pi) a_{\pi(1)1} \cdots a_{\pi(n)n}$$

where π varies over the symmetric group S_n and $\sigma: S_n \to \{\pm 1\}$ is the sign homomorphism.

PROOF Write A as

$$(a_{11}E_1 + a_{21}E_2 + \cdots + a_{n1}E_n, A^{(2)}, \ldots, A^{(n)})$$

where $E_1, \ldots, E_n$ are the special column matrices introduced in (3.8). Then by multilinearity,

$$\det(A) = \sum_{i=1}^{n} a_{i1}\det(E_i, A^{(2)}, \ldots, A^{(n)})$$

Repeating this expansion in the second column, we find further that

$$\det(A) = \sum_{i=1}^{n} \sum_{j=1}^{n} a_{i1}a_{j2}\det(E_i, E_j, A^{(3)}, \ldots, A^{(n)})$$

Continuing over all columns,

$$\det(A) = \sum_{\phi} a_{\phi(1)1}a_{\phi(2)2}\cdots a_{\phi(n)n}\det(E_{\phi(1)}, \ldots, E_{\phi(n)})$$

where ϕ varies over all maps $\{1,\ldots,n\} \to \{1,\ldots,n\}$. But consider: if any given ϕ is not injective, then the matrix $(E_{\phi(1)}, \ldots, E_{\phi(n)})$ has two identical columns and therefore vanishing determinant by (3.10)(iii). Thus we can restrict the ϕ to injective, hence bijective, maps; that is, to elements of S_n. Summarizing,

$$\det(A) = \sum_{\pi \in S_n} a_{\pi(1)1}\cdots a_{\pi(n)n}\det(E_{\pi(1)}, \ldots, E_{\pi(n)})$$

Now $(E_{\pi(1)}, \ldots, E_{\pi(n)})$ can be obtained from $I_n = (E_1, \ldots, E_n)$ by successive transpositions, the number of such being even or odd according to $\sigma(\pi)$. Therefore by (3.10)(ii),

$$\begin{aligned}\det(E_{\pi(1)}, \ldots, E_{\pi(n)}) &= \sigma(\pi)\det(I_n) \\ &= \sigma(\pi)\end{aligned}$$

This establishes the formula.

COROLLARY For all $A \in M_n(k)$, $\det(A) = \det({}^tA)$.

PROOF In order to maintain typographical sanity, for $\pi \in S_n$ we write π^* for the inverse of π. We now sketch the calculation, leaving verification of each step to the reader as an exercise. The main points to bear in mind are (i) that as π varies over S_n, so does π^*, and (ii) that π and π^* always have the same sign. Thus:

$$\begin{aligned}\det(A) &= \sum_{\pi \in S_n} \sigma(\pi) a_{\pi(1)1} \cdots a_{\pi(n)n} \\ &= \sum_{\pi \in S_n} \sigma(\pi^*) a_{1\pi^*(1)} \cdots a_{n\pi^*(n)} \\ &= \sum_{\pi \in S_n} \sigma(\pi) a_{1\pi(1)} \cdots a_{n\pi(n)} \\ &= \det({}^tA)\end{aligned}$$

Uniqueness of the Determinant. We are now in a position to observe the uniqueness of the determinant map. For the proof of (3.12) depends only upon multilinearity, alternation, and normality. Thus any map $M_n(k) \to k$ satisfying these same three properties satisfies the formula (3.12), and consequently there can only be one such map.

We conclude this section by developing some remarkable multiplicative properties of the determinant. First some notation and terminology.

The group of units of the ring $M_n(k)$ is denoted $GL_n(k)$. (GL for general linear group.) Elements of $GL_n(k)$ are called invertible or, if k is a field, nonsingular matrices.

(3.13) THEOREM Let $A, B \in M_n(k)$. Then

$$\det(AB) = \det(A) \cdot \det(B)$$

Thus the determinant is a homomorphism of monoids $\langle M_n(k), \cdot \rangle \to \langle k, \cdot \rangle$.

PROOF Let $A = (a_{ij})$, $B = (b_{ij})$. Then by the definition of the matrix product and formula (3.12) we have

$$\begin{aligned}\det(AB) &= \det\Big(\sum_k a_{ik} b_{kj} \Big) \\ &= \sum_{\pi \in S_n} \sigma(\pi) \prod_j \Big(\sum_k a_{\pi(j)k} b_{kj} \Big)\end{aligned}$$

By the arithmetic properties of commutative rings, this may be rewritten as

$$\sum_{\pi} \sigma(\pi) \sum_{\phi} \prod_{j} a_{\pi(j)\phi(j)} b_{\phi(j)j}$$

where ϕ varies over the n^n function $\{1,2,\ldots,n\} \to \{1,2,\ldots,n\}$. Again by arithmetic, this last expression admits rearrangement, giving

$$\det(AB) = \sum_{\phi} \{\sum_{\pi} \sigma(\pi) \prod_{j} a_{\pi(j)\phi(j)}\} \prod_{j} b_{\phi(j)j}$$

Now consider the inner sum, contained within the braces. Let $\bar{a}_{ij}$ be defined as $a_{i\phi(j)}$. Then this inner sum is precisely $\det(\bar{a}_{ij})$. Hence if ϕ is not injective, $(\bar{a}_{ij})$ has repeated columns and the sum in question is zero. Thus it suffices to sum over injective, and therefore bijective, maps ϕ; that is, over elements of the symmetric group S_n. Summarizing,

$$\det(AB) = \sum_{\tau \varepsilon S_n} \{ \sum_{\pi \varepsilon S_n} \sigma(\pi) \prod_{j} a_{\pi(j)\tau(j)} \} \prod_{j} b_{\tau(j)j}$$

It is easily verified that the sequence of ordered pairs $(\pi(j),\tau(j))$ $(j = 1,2,\ldots,n)$ is just a permutation of the sequence $(\pi\tau^*(j),j)$, where again we are using '*' for 'inverse'. Therefore

$$\det(AB) = \sum_{\tau} \{\sum_{\pi} \sigma(\pi) \prod_{j} a_{\pi\tau^*(j),j}\} \prod_{j} b_{\tau(j)j}$$

$$= \sum_{\tau} \{\sum_{\pi} \sigma(\pi\tau^*) \prod_{j} a_{\pi\tau^*(j),j}\} \sigma(\tau) \prod_{j} b_{\tau(j)j}$$

because σ: $S_n \to \{\pm 1\}$ is a homomorphism. Continuing,

$$\det(AB) = \sum_{\tau} \{\sum_{\pi} \sigma(\pi) \prod_{j} a_{\pi(j)j}\} \cdot \sigma(\tau) \prod_{j} b_{\tau(j)j}$$

since $\pi\tau^*$ varies over S_n as π does. But now the inner sum coalesces:

$$\det(AB) = \sum_{\tau} \det(A)\, \sigma(\tau) \prod_{j} b_{\tau(j)j}$$

$$= \det(A) \sum_{\tau} \sigma(\tau) \prod_{j} b_{\tau(j)j}$$

$$= \det(A) \cdot \det(B)$$

This completes the calculation.

Our next result is a special case of Exercise (3.8).

(3.14) THEOREM (Cramer's Rule for Matrix Inversion) Suppose that $A \in M_n(k)$ and $\det(A) \in k^\times$. Then $A \in GL_n(k)$ with A^{-1} given by $B = (b_{ij})$ where

$$b_{ij} = \det(A)^{-1} \cdot \det(A^{(1)},\ldots,\underset{\substack{\uparrow \\ \text{i-th column}}}{E_j},\ldots,A^{(n)})$$

(Here the matrix indicated in the second factor is obtained from A by replacing the i-th column of A by E_j.)

PROOF By direct calculation, the (i,j)-entry of BA is

$$\sum_k b_{ik}a_{kj} = \det(A)^{-1} \cdot \sum_k a_{kj} \det(A^{(1)},\ldots,\underset{\substack{\uparrow \\ \text{i-th column}}}{E_k},\ldots,A^{(n)})$$

$$= \det(A)^{-1} \cdot \det(A^{(1)},\ldots,\overset{\substack{\downarrow}}{\sum_k a_{kj}E_k},\ldots,A^{(n)})$$

where this last equality follows by multilinearity. But the interior summation is precisely $A^{(j)}$. Thus if $i = j$, the matrix indicated in the second factor is just A again, and the product is 1. Otherwise, $i \neq j$, and the indicated matrix has identical, distinct columns (the i-th and the j-th), whence by alternation of sign the determinant is 0. Therefore $BA = (\delta_{ij}) = I_n$, and A has at least a left inverse.

Now as we shall see shortly in the sequel, the multiplicativity of the determinant map implies at once that $\det(B) \in k^\times$ also, whence B, too, has a left inverse -- call it C. Then $CB = I_n$, and accordingly one computes $C = CI_n = C(BA) = (CB)A = I_nA = A$. Thus $AB = I_n$ and B is a certified (two-sided) inverse for A, as required.

The next and final theorem of this section applies both the multiplicativity of the determinant and Cramer's Rule. This elegant result is one of the cornerstones of linear algebra.

(3.15) THEOREM Let $A \in M_n(k)$. Then $A \in GL_n(k)$ if and only if $\det(A) \in k^\times$.

PROOF If $\det(A) \in k^\times$, by Cramer's rule we can invert it. Conversely, if A is invertible with inverse B, then

$$\det(A)\det(B) = \det(AB) = \det(I_n) = 1.$$

Thus $\det(A) \in k^\times$.

COROLLARY Let $SL_n(k)$ be the subset of $M_n(k)$ consisting of all matrices of determinant 1_k. ($SL_n(k)$ is called the special linear group of $n \times n$ matrices over k.) Then we have a short exact sequence of groups

$$1 \to SL_n(k) \xrightarrow{\text{inc}} GL_n(k) \xrightarrow{\det} k^\times \to 1$$

PROOF It follows from the theorem that det: $M_n(k) \to k$ induces a map $GL_n(k) \to k^\times$ by restriction. This map is moreover a homomorphism of groups by (3.13), with kernel $SL_n(k)$ by definition. The surjectivity follows from Exercise (3.10).

EXERCISES

(3.1) Complete the proof of Proposition (3.4).

(3.2) Let $A,B \in M_n(k)$ have the form

$$A = \begin{pmatrix} A_1 & 0 \\ 0 & A_2 \end{pmatrix} \qquad B = \begin{pmatrix} B_1 & 0 \\ 0 & B_2 \end{pmatrix}$$

where $A_1,B_1 \in M_p(k)$, $A_2,B_2 \in M_r(k)$, and $p + r = n$. Show that

$$AB = \begin{pmatrix} A_1B_1 & 0 \\ 0 & A_2B_2 \end{pmatrix}$$

(Here the zeros indicate blocks of zeros of the appropriate sizes.)

(3.3) We call $A = (a_{ij}) \in M_n(k)$ an upper triangular matrix if $a_{ij} = 0$ whenever $j < i$. Show that the upper triangular matrices constitute a subring of $M_n(k)$.

(3.4) Let A and B be matrices such that the product AB is defined. Show that ${}^t(AB) = {}^tB\,{}^tA$.

(3.5) Let k be a field. Show that the ring $M_n(k)$ is simple. (That is, it has no two-sided ideals other than (0) and itself.) Show, moreover, that this assertion fails if k is not a field.

(3.6) We compute the center of the matrix ring $M_n(k)$:

(a) Show by direct calculation that $Z(M_2(k)) = k$. [Hint: consider the multiplicative action of the matrices

$$\begin{pmatrix} 0 & 1 \\ 0 & 0 \end{pmatrix} \quad \text{and} \quad \begin{pmatrix} 0 & 0 \\ 1 & 0 \end{pmatrix}$$

on a typical element of $M_2(k)$.]

(b) Let $A \in M_n(k)$, $n > 1$, and suppose A commutes with all elements $B \in M_n(k)$ such that both the last row and last column of B consist entirely of zeros. Show that $\partial_{nn}(A) \in Z(M_{n-1}(k))$.

(c) Let $A \in M_n(k)$, $n > 1$, and suppose A commutes with all elements $B \in M_n(k)$ such that both the first row and first column of B consist entirely of zeros. Show that $\partial_{11}(A) \in Z(M_{n-1}(k))$.

(d) Conclude by induction that $Z(M_n(k)) = k$ under the identification of Lemma (3.5). [Hint: for $n \geqslant 3$, use (b) and (c) to reduce the center to matrices of the form $\lambda I_n + B$, where $\lambda \in k$ and the only nonzero entries of B are the $(n,1)$ and $(1,n)$ entries. Next apply the hint from part (a), suitably modified, to show that B is in fact the zero matrix.]

(3.7) Show that the determinant of an upper triangular matrix is the product of its diagonal entries.

(3.8) Let $A \in M_n(k)$ and define a map

$$L_A: k^n \to k^n$$
$$\mathbf{b} \to A\mathbf{b}$$

(Here we regard k^n as a set of column matrices, so that the matrix multiplication indicated is defined.)

(a) Show that $L_{AB} = L_A \circ L_B$ for all $A,B \in M_n(k)$.

(b) Use (a) and Proposition (3.14) to show that the following four statements are equivalent:

(i) L_A is surjective
(ii) $A \in GL_n(k)$

(iii) $\det(A) \in k^{\times}$

(iv) L_A is bijective

(c) Show by example that (b) fails in general if we replace (i) by the statement that L_A is injective. (But as all good linear algebraists know, this modification succeeds if k is a field.)

(d) Prove the full statement of Cramer's Rule: If $A \in GL_n(k)$ and $\mathbf{b} \in k^n$, then the matrix equation $A\mathbf{x} = \mathbf{b}$ has solution

$$(\#) \quad x_j = \det(A^{(1)},\ldots,\mathbf{b},\ldots,A^{(n)}) \cdot \det(A)^{-1}$$

where x_j is the j-th component of $\mathbf{x}$ and the matrix indicated in the first factor is obtained from A by replacing its j-th column by the column matrix $\mathbf{b}$. [Hint: According to (b), such an $\mathbf{x}$ exists, and hence by the definition of matrix multiplication,

$$\mathbf{b} = x_1A^{(1)} + x_2A^{(2)} + \cdots + x_nA^{(n)}$$

Substitute this equation into the right-hand side of (#) and make liberal use of the Fundamental Theorem of Determinants.]

(3.9) Show that the map $\phi\colon \langle \mathbf{R},+\rangle \to SL_2(\mathbf{R})$ defined by

$$\phi(\alpha) = \begin{pmatrix} \cos(\alpha) & -\sin(\alpha) \\ \sin(\alpha) & \cos(\alpha) \end{pmatrix}$$

is a homomorphism of groups. Be sure to verify explicitly that $\mathrm{Im}(\phi)$ in fact lies in $SL_2(\mathbf{R})$.

(3.10) Use Exercise (3.4) to show that $\det\colon M_n(k) \to k$ is indeed surjective, as claimed in Corollary (3.15).

(3.11) Let $A \in M_n(k)$ have the form

$$A = \begin{pmatrix} A_1 & 0 \\ 0 & A_2 \end{pmatrix}$$

where $A_1 \in M_p(k)$, $A_2 \in M_r(k)$, $p + r = n$, and again the zeros indicate blocks of zeros. Deduce that $\det(A) = \det(A_1)\cdot\det(A_2)$.

7
Modules and Vector Spaces

In this chapter we will be studying the interaction between a ring R, with 1, and an additive group M. This interaction will be effected with a unital ring homomorphism from R to End(M), the ring of endomorphisms of M. The elements of the ring R will be called scalars and will usually be denoted with the letters r, s, and t. The elements of the additive group M will usually be denoted with the letters x, y, and z. This convention is particularly helpful since we will be using the symbol + to denote two different operations: addition in R and addition in M. However, the ambiguity is removed by observing that the relevant operation is determined by the location of the elements involved. All rings in this chapter are rings with 1 and all ring homomorphisms are unital.

1. LEFT R-MODULES

In this section we will define and discuss left R-modules, or for brevity, R-modules. Since the proofs follow well worn lines, many will be omitted. As always, the student is urged to supply all missing details. We begin with the definition of a left R-module.

(1.1) DEFINITION Let R be a ring with 1 and let M be an additive group. Then a left R-module is a unital ring homomorphism

$$\theta: R \to \text{End}(M)$$

Note that if θ is a left R-module, then θ induces a function from $R \times M \to M$ whose action is defined by

$$(r, x) \longmapsto \theta(r)(x) \text{ (where } \theta(r)(x) = \theta(r) \text{ evaluated at } x)$$

This function will be called the <u>scalar multiplication</u> corresponding to θ. For convenience, we will denote

$$rx = \theta(r)(x)$$

Note that we are using juxtaposition ambiguously: rx denotes the action of scalar multiplication on x via θ and rs denotes the product of r and s in the ring R. Again, the relevant usage is determined by the location of the elements involved. Also note that, in our notation, the scalars are always written on the left. This is the reason for the term left R-module.

We will first explore the connection between the homomorphism θ and the scalar multiplication mapping.

(1.2) PROPOSITION Let $\theta: R \to \text{End}(M)$ be a left R-module. Then for each $r,s \in R$ and $x,y \in M$,

(M1) $r(x + y) = rx + ry$

(M2) $(r + s)x = rx + sx$

(M3) $r(sx) = (rs)x$

(M4) $1x = x$

Conversely, let a mapping f be defined from $R \times M$ to M whose action is specified by

$$f(r, x) = rx$$

and which satisfies properties (M1) to (M4). Then there is an R-module θ such that the scalar multiplication corresponding to θ is f.

PROOF We will establish only properties (M1) and (M3). The other proofs are similar and will be left as exercises. To establish (M1), let $r \in R$ and $x,y \in M$. Then

$$\begin{aligned} r(x + y) &= \theta(r)(x + y) && \text{(definition)} \\ &= \theta(r)(x) + \theta(r)(y) && [\theta(r) \text{ is a homomorphism}] \\ &= rx + ry && \text{(definition)} \end{aligned}$$

To establish (M3), let $r,s \in R$ and let $x \in M$. Then

$$\begin{aligned} r(sx) &= \theta(r)(sx) \\ &= \theta(r)(\theta(s)(x)) \\ &= [\theta(r) \circ \theta(s)](x) \\ &= \theta(rs)(x) \\ &= (rs)x \end{aligned}$$

The next to the last equality follows since θ is a ring homomorphism and multiplication in the ring End(M) is composition of functions.

We will now let $f\colon R \times M \to M$ be a function with action specified above and which satisfies (M1) to (M4). Let $r \in R$ and define a function $\theta(r)\colon M \to M$ by

$$\theta(r)(x) = rx \text{ for each } x \in M$$

Then θ defines a function from R to End(M). We will leave the proof that θ is a unital ring homomorphism to the student.

As a result of (1.2), we can specify a left R-module by either the unital ring homomorphism or the scalar multiplication. If either θ has been clearly specified, or a scalar multiplication has been defined which satisfies (M1) to (M4), then we will say that M is a <u>left R-module</u>. In the case that R is a field, we will call M an <u>R-vector space</u>. In case that the ring of scalars is clear from context, we will sometimes say that M is a module. We will also usually refer to R-modules instead of the more bulky left R-modules. In the sequel, we will usually give an additive group M the structure of a left R-module by defining a scalar multiplication on M.

(1.3) PROPOSITION Let M be a left R-module and let $r \in R$, $x \in M$ and $n \in \mathbf{Z}$. Then

(a) $0x = 0$

(b) $r0 = 0$

(c) $-(rx) = (-r)x = r(-x)$

(d) $n(rx) = (nr)x = r(nx)$

PROOF The proof of this proposition is computational and will be left as an exercise.

(1.4) NOTATION AND TERMINOLOGY Let M be an R-module and let N be an additive subgroup of M. Then N is said to be an R-submodule, or a submodule, of M if the restriction of scalar multiplication to $R \times N$ has its image in N. More precisely, for every $r \in R$ and $x \in N$, we have $rx \in N$. In case that M is an R-vector space, we will refer to R-submodules as R-subspaces.

If M and N are R-modules and if $f: M \to N$ is a homomorphism of additive groups, then we will say that f is an R-homomorphism if for every $r \in R$ and $x \in M$, we have

(*) $f(rx) = r(f(x))$

If f satisfies (*), then we will sometimes say that f is compatible with scalar multiplication. In case that M and N are R-vector spaces, we will sometimes say that f is R-linear, or f is linear. Evidently, the composition of R-homomorphisms is an R-homomorphism whenever it is defined.

If M and N are R-modules, then we will denote

$\mathrm{Hom}_R(M, N) = \{f: f \text{ is an R-homomorphism from M to N}\}$

As before, we will say that f is an isomorphism of R-modules, or f is an isomorphism, if f is a bijective R-homomorphism. If there is an R-isomorphism from M to N, then we will say that M is isomorphic to N and write $M \cong N$. If f is an isomorphism, we will leave as an exercise to the reader to show that f^{-1} is an isomorphism.

As we will see, many of the facts about subgroups of groups and group homomorphisms translate to facts about R-submodules of an R-module and R-homomorphisms. First we will note that we can define quotient structures.

(1.5) PROPOSITION Let M be an R-module and let N be an R-submodule of M. Define scalar multiplication on the additive group M/N by

$$r(x + N) = rx + N \qquad (r \in R;\ x \in M)$$

Then M/N is an R-module and the canonical mapping k is an R-homomorphism.

PROOF We must first show that scalar multiplication is well defined. The proofs of properties M1 through M4 are then easily established and are left to the student.

Suppose that $x + N = y + N$ $(x,y \in M)$ and $r \in R$. We must show that $r(x + N) = r(y + N)$ or, equivalently, that $rx + N = ry + N$. Since $x + N = y + N$, $x - y \in N$. Hence, since N is an R-submodule of M, $r(x - y) \in N$. Thus $rx - ry \in N$ so that $rx + N = ry + N$..

To see that k is an R-homomorphism we need only show that k is compatible with scalar multiplication. Let $r \in R$ and $x \in M$. Then

$$\begin{aligned} k(rx) &= (rx) + N \\ &= r(x + N) \end{aligned}$$

Therefore, k is an R-homomorphism.

The concepts of ker(f) and im(f) remain the same as well as many of the facts connecting them.

(1.6) PROPOSITION Let M and N be R-modules and let $f\colon M \to N$ be an R-homomorphism. Then

(i) ker(f) is an R-submodule of M.

(ii) im(f) is an R-submodule of N.

(iii) $\mathrm{im}(f) \cong M/\ker(f)$.

(iv) f is injective if and only if $\ker(f) = \{0\}$.

PROOF We will establish (i). Proofs of the other facts are similar and will be left to the exercises. We have seen previously that ker(f) is an additive subgroup of M. Hence we need only show that f is compatible with respect to scalar multiplication. Let $r \in R$ and $x \in \ker(f)$. Then

$$\begin{aligned} f(rx) &= rf(x) \\ &= r0 \\ &= 0 \end{aligned}$$

Hence, $rx \in \ker(f)$ and $\ker(f)$ is an R-submodule of M.

(1.7) PROPOSITION Let M, N, K, and L be R-modules and let $f, g \in \mathrm{Hom}_R(M, N)$, $h \in \mathrm{Hom}_R(K, M)$ and $h' \in \mathrm{Hom}_R(N, L)$. Define $(f + g)\colon M \to N$ by

$$(f + g)(x) = f(x) + g(x) \qquad (x \in M)$$

Then

(i) $f + g$ is an R-homomorphism.

(ii) $\mathrm{Hom}_R(M, N)$ is an additive group under +.

(iii) $(f + g) \circ h = f \circ h + g \circ h$.

(iv) $h' \circ (f + g) = h' \circ f + h' \circ g$.

PROOF We will prove (i). The remaining facts will be left to the reader. Let $r \in R$ and $x \in M$. Then

$$\begin{aligned} (f + g)(rx) &= f(rx) + g(rx) \\ &= rf(x) + rg(x) \\ &= r(f(x) + g(x)) \\ &= (r(f + g))(x) \end{aligned}$$

Therefore $f + g$ is compatible with scalar multiplication.

To show that $f + g$ is an additive homomorphism, let $x, y \in M$, then

$$\begin{aligned} (f + g)(x + y) &= f(x + y) + g(x + y) \\ &= f(x) + f(y) + g(x) + g(y) \\ &= f(x) + g(x) + f(y) + g(y) \\ &= (f + g)(x) + (f + g)(y) \end{aligned}$$

Hence, $f + g$ is an additive homomorphism and (i) is established.

(1.8) EXAMPLES We will now give some examples of R-modules. As before, the student is urged to fill in all missing details.

(1.8.1) Additive Groups. Let M be an additive group and let $x \in M$ and $n \in \mathbf{Z}$. If we define nx as before, then M is a $\mathbf{Z}$-module. Note that, in this case, our previous convention, using m, n as integers, takes precedence over the convention of using the letters r, s, and t for elements of the ring.

If N is an additive subgroup of M, then N is a $\mathbf{Z}$-submodule of M and conversely. Moreover, additive homomorphisms are $\mathbf{Z}$-homomorphisms and conversely. Thus, the category of $\mathbf{Z}$-modules is the same as the category of additive groups.

(1.8.2) $\mathbf{R}^n$ as an $\mathbf{R}$-module. As in chapter I we define addition on $\mathbf{R}^n$ and observe that with this operation, $\mathbf{R}^n$ is an additive group. Define

$$\mathbf{R} \times \mathbf{R}^n \to \mathbf{R}^n$$
$$(r, (r_1,\ldots,r_n)) \mapsto (rr_1,\ldots,rr_n).$$

With scalar multiplication so defined, $\mathbf{R}^n$ is an $\mathbf{R}$-module, or an $\mathbf{R}$-vector space.

Recall that, if B is a matrix, then B^T denotes the transpose of B. Let A be a real $n \times m$ matrix and define a function f_A by

$$f_A\colon \mathbf{R}^m \to \mathbf{R}^n$$
$$(r_1,\ldots,r_m) \mapsto [A(r_1,\ldots,r_m)^T]^T$$

Then f_A is an $\mathbf{R}$-homomorphism, or is $\mathbf{R}$ linear. We will see later [cf. Exercise (1.23)] that all $\mathbf{R}$ linear functions from $\mathbf{R}^m$ to $\mathbf{R}^n$ arise in this manner.

(1.8.3) The cyclic submodule generated by x. Let M be an R-module and let $x \in M$. Then

$$Rx = \{rx\colon r \in R\}$$

is an R-submodule of M which is called the cyclic submodule generated by x. Note that in the case that $R = \mathbf{Z}$, our terminology agrees with the terminology used for abelian groups.

(1.8.4) Let R be a ring with 1 and let S be a subring of R such that $1 \in S$. Then R admits the structure of an S-module if we take scalar multiplication to be multiplication in R.

In the special case S = R, R is itself an R-module. Whenever we refer to R as an R-module, we will be refering to R with this scalar multiplication. Note that R-submodules are just left ideals in R. Also, the cyclic submodule generated by x is just the principle left ideal, Rx, generated by x.

(1.8.5) The R-vector space $C^n(\mathbf{R})$. As in II (1.20.5), we define $C^n(\mathbf{R})$ to be the set of all n times continuously differentiable real valued functions of one real variable. If addition is defined as usual, then $C^n(\mathbf{R})$ is an additive group. Define scalar multiplication on $C^n(\mathbf{R})$ as follows

$$\mathbf{R} \times C^n(\mathbf{R}) \to C^n(\mathbf{R})$$
$$(r, f) \mapsto rf \qquad (r \in \mathbf{R},\ f \in C^n(\mathbf{R}))$$

We define the function rf: $\mathbf{R} \to \mathbf{R}$ by the rule $(rf)(x) = r(f(x))$ for each $x \in \mathbf{R}$. With scalar multiplication so defined, $C^n(\mathbf{R})$ is an **R**-vector space.

Let **P** be the set of all real polynomials and observe that $\mathbf{P} \subseteq C^n(\mathbf{R})$ for each $n \in \mathbf{N}$. As we have previously seen, $C^n(\mathbf{R})$ is a ring and **P** is a subring of $C^n(\mathbf{R})$. Thus, $C^n(\mathbf{R})$ will admit the structure of a **P**-module. [cf. Example (1.8.4)]

Define D: $C^{n+1} \to C^n(\mathbf{R})$ to be the differentiation operator. Then D is an **R**-linear function but D is not a **P**-homomorphism.

(1.8.6) Sums and intersections of submodules. Let $\{M_i: i \in I\}$ be a family of submodules of an R-module M. Then $\bigcap_{i \in I} M_i$ is a submodule of M.

Let N be the set of all elements of M which can be formed by taking finite sums of elements in $\bigcup_{i \in I} M_i$. Then N is a submodule of M which we will call the sum of the family $\{M_i: i \in I\}$. In this case, we will write

$$N = \sum_{i \in I} M_i$$

(1.8.7) The pullback module with respect to f. Let R, S be rings with 1 and let $f: R \to S$ be a unital ring homomorphism. Let M be an S-module. Define a function by

$$R \times M \to M$$
$$(r, x) \mapsto f(r)x.$$

With scalar multiplication so defined, M is an R-module. This module is called the pullback module with respect to f.

(1.8.8) $\mathrm{End}_K(K^n)$ as a K[x]-module with respect to T. Let K be a field. If K^n is defined in a way similar to that of $\mathbf{R}^n$, then K^n is a K vector space. Let $\mathrm{End}_K(K^n)$ denote the set of all K-linear mappings from K^n to K^n. Let T be a K-linear function from K^n to K^n. Then rT^n is also a K-linear function. It follows that polynomials in T are K-linear functions. (In a polynomial in T, we interpret the constant term r_0 as the function which multiplies by r_0.) Let f be a polynomial in K[x] and define

$$fg = f(T) \circ g. \qquad (g \in \mathrm{End}_K(K^n))$$

With scalar multiplication so defined, $\mathrm{End}_K(K^n)$ is a K[x]-module. The structure of this module is very important in the study of certain canonical forms of K-matrices.

(1.8.9) $\mathrm{Hom}_R(M, N)$ as an $\mathrm{Hom}_R(N, N)$-module. Let $S = \mathrm{Hom}_R(N, N)$. Then S is a ring with respect to composition and addition of functions. Define a scalar multiplication from $S \times \mathrm{Hom}_R(M, N)$ to $\mathrm{Hom}_R(M, N)$ by

$$gf = g \circ f \qquad (g \in S;\ f \in \mathrm{Hom}_R(M, N))$$

With scalar multiplication so defined, $\mathrm{Hom}_R(M, N)$ is a left S-module.

(1.8.10) Let M, N, and K be left R-modules and let $f: M \to N$ be an R-homomorphism. Define mappings $f_*: \mathrm{Hom}_R(K, M) \to \mathrm{Hom}_R(K, N)$ and $f^*: \mathrm{Hom}_R(N, K) \to \mathrm{Hom}_R(M, K)$ by

$$f^*(h) = h \circ f$$

$$f_*(g) = f \circ g$$

for every $g \in \mathrm{Hom}_R(K, M)$ and $h \in \mathrm{Hom}_R(N, K)$. Then f_* and f^* are homomorphisms of additive groups. [cf. (1.7)]

As a final remark, we note that exact sequences of R-modules can be defined as in II, (4.12). All results given in Chapter II dealing with exact sequences of groups, or additive groups, obtain if the word 'group' is replaced by 'R-module' and group homomorphisms are replaced by R-homomorphisms.

EXERCISES

(1.1) Let M, N, and K be R-modules and let $f: M \to N$ and $g: N \to K$ be R-homomorphisms. Show that $g \circ f$ is an R-homomorphism.

(1.2) Prove (M2) and (M4) in (1.2).

(1.3) Show that the function θ defined in the proof of (1.2) is a unital ring homomorphism.

(1.4) Prove (1.3).

(1.5) Let M and N be R-modules and let $f: M \to N$ be an isomorphism. Show that f^{-1} is an isomorphism.

(1.6) Prove (M1), (M3) and (M4) in (1.5).

(1.7) Prove (1.6) (ii), (iii) and (iv).

(1.8) Prove (1.7) (ii), (iii) and (iv).

(1.9) Let $\theta: R \to \mathrm{End}(M)$ be an R-module and let $I \subseteq \ker(\theta)$. Then there is a unique R/I module structure on M making the following diagram commutative.

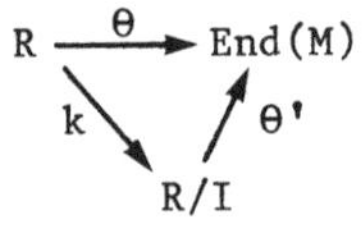

(1.10) In Example (1.8.7), what is the relationship between the unital ring homomorphism for the S-module M and the unital ring homomorphism for the R-module M?

(1.11) Let θ: R $\to$ End(M) be an R-module. Show that an additive subgroup N of M is an R-submodule of M if and only if for every r $\in$ R,

$\theta(r)(N) \subseteq N$

(1.12) Let θ: R $\to$ End(M) be an R module and let N be an R-submodule of M. Show that for every r $\in$ R, there is a unique $\theta'(r)$ which makes the following diagram commutative.

$$\begin{array}{ccc} M & \xrightarrow{\theta(r)} & M \\ {\scriptstyle k}\downarrow & & \downarrow{\scriptstyle k} \\ M/N & \xrightarrow{\theta'(r)} & M/N \end{array}$$

Show that θ' defines a ring homomorphism from R into End(M/N). Show that this is the ring homomorphism corresponding to the R-module structure defined on M/N in (1.5).

(1.13) Let θ: R $\to$ End(M) and θ': R $\to$ End(N) be R-modules. Show that an additive homomorphism f: M $\to$ N is an R-homomorphism if and only if for every r $\in$ R, the following diagram is commutative.

$$\begin{array}{ccc} M & \xrightarrow{\theta(r)} & M \\ {\scriptstyle f}\downarrow & & \downarrow{\scriptstyle f} \\ N & \xrightarrow{\theta'(r)} & N \end{array}$$

(1.14) Define the mapping D as in (1.8.5). Show that D is not a **P**-homomorphism.

(1.15) Let $\mathbf{R}^2$ be a **R** vector space as in (1.8.2). Let

x = (1, 3)

y = (2, 5)

z = (1, -1)

Describe

(a) $\mathbf{R}x$

(b) $\mathbf{R}x + \mathbf{R}y$

Show

(c) $\mathbf{R}z \cap \mathbf{R}x = \mathbf{R}z \cap \mathbf{R}y = \mathbf{R}x \cap \mathbf{R}y = \{0\}$

(d) $\mathbf{R}z \cap (\mathbf{R}x + \mathbf{R}y) \neq (\mathbf{R}z \cap \mathbf{R}x) + (\mathbf{R}z \cap \mathbf{R}y)$

(1.16) Let $\mathbf{R}^3$ be a $\mathbf{R}$ vector space as in (1.8.2). Let

$x = (1, 0, 1)$

$y = (2, 1, -1)$

Describe

(a) $\mathbf{R}x$

(b) $\mathbf{R}x + \mathbf{R}y$

(1.17) Let M be an R-module and N be an R-submodule of M. Define

$(N: M) = \{r: r \in R, rM \subseteq N\}$

Show that (N: M) is an ideal in R.

(1.18) Let M be an R-module and $\{M_i: i \in I\}$ be a family of R-submodules of M. Show that

(a) $\sum_{i \in I} M_i$ is a submodule of M.

(b) $\bigcap_{i \in I} M_i$ is a submodule of M.

(1.19) Let M be an R-module and let S be a subset of M. By the previous exercise, there is a smallest R-submodule of M which contains S. This R-submodule of M will be denoted $\langle S \rangle$. Describe the R-submodule $\langle S \rangle$.

(1.20) An R-module is said to be faithful if the unital ring homomorphism θ is injective. Show that an R-module M is faithful if and only if for every $r \in R$, $rx \neq 0$ for some $x \in M$.

(1.21) Let R be a ring and regard R as a left R-module as in (1.8.4). Let $r \in R$ and define $f: R \to R$ by

$f(x) = xr \quad (x \in R)$

Show that f is an R-homomorphism.

(1.22) Let M and N be R-modules. Define scalar multiplication on the additive group $M \times N$ by

$$r(x, y) = (rx, ry)$$

where $r \in R$, $x \in M$ and $y \in N$. Show that $M \times N$ is an R-module.

(1.23) Let M be an R-module and $x \in M$. Define

$$(0: x) = \{r: r \in R \text{ and } rx = 0\}$$

Show that:

(a) $(0: x)$ is a left ideal of R.

(b) If R is regarded as a left R-module then $Rx \cong R/(0: x)$.

(1.24) Let $f: \mathbf{R}^m \to \mathbf{R}^n$ be $\mathbf{R}$-linear. Show that there is an $n \times m$ matrix A such that $f = f_A$. See (1.8.2).

(1.25) Let A be an $n \times m$ matrix and let B be an $m \times k$ matrix. Show that

$$f_A f_B = f_{AB}$$

(1.26) Let R be an integral domain and Q(R) be the field of fractions of R. [IV, (4.10)] Let M be an R-module and define a relation $\sim$ on $M \times R^*$ by

$$(x, r) \sim (y, s) \text{ if } ry = sx$$

(a) Show that $\sim$ is an equivalence relation on $M \times R^*$.

(b) Let $x/r = [(x, r)]$ and let $Q(M) = \{x/r: x \in M, r \in R^*\}$. Show that Q(M) is an additive group under

$$x/r + y/s = (sx + ry)/rs$$

(c) Define a scalar multiplication from $Q(R) \times Q(M)$ to Q(M) by

$$(r/s)(x/s') = rx/ss' \qquad (r \in R;\ s, s' \in R^*;\ x \in M)$$

Show that Q(M) is a Q(R)-vector space.

(d) Let M, N, and K be R-modules and let $f\colon M \to N$ and $g\colon N \to K$ be R-homomorphisms. Define $Q(f)\colon Q(M) \to Q(N)$ by

$$Q(f)(x/r) = f(x)/r \quad (x \in M;\ r \in R^*)$$

Show that Q(f) is a Q(R)-homomorphism and $Q(g \circ f) = Q(g) \circ Q(f)$.

(e) Let

$$0 \to M \to N \to K \to 0$$

be an exact sequence of R-modules. Show that

$$0 \to Q(M) \to Q(N) \to Q(K) \to 0$$

is an exact sequence of Q(R)-vector spaces.

(1.27) Let R be a ring and let R^o be the opposite ring. That is the ring with the same elements and addition as R, and with the multiplication reversed. Define a right R-module as a unital ring homomorphism

$$\theta\colon R^o \to \mathrm{End}(M)$$

where M is an additive group. State and prove the analog to (1.2).

(1.28) Let

$$0 \to M \xrightarrow{f} N \xrightarrow{g} K \to 0$$

be an exact sequence of R-modules and let L be an R-module. Show that

$$0 \to \mathrm{Hom}_R(K, L) \xrightarrow{g^*} \mathrm{Hom}_R(N, L) \xrightarrow{f^*} \mathrm{Hom}_R(M, L)$$

$$0 \to \mathrm{Hom}_R(L, M) \xrightarrow{f_*} \mathrm{Hom}_R(L, N) \xrightarrow{g_*} \mathrm{Hom}_R(L, K)$$

are exact sequences of abelian groups. [cf. (1.8.10)]

2. DIRECT PRODUCTS AND DIRECT SUMS

The study of complex mathematical structures is often facilitated by somehow decomposing the complicated structure into simpler structures and then studying the simpler structures. In this section, we will study one technique, called direct decomposition, for decomposing modules. The reader is reminded that all results dealing with R-modules obtain for abelian groups as a special case. Thus, more familiar structures are available for constructing examples.

(2.1) DEFINITION Let $\{M_i : i \in I\}$ be a family of R-modules such that $i, j \in I$ and $i \neq j$ implies $M_i \cap M_j = \emptyset$. Then $\prod_{i \in I} M_i$ is the set of all functions of the following form

$$\begin{aligned} I &\xrightarrow{x} \bigcup_{i \in I} M_i \\ i &\mapsto x_i \end{aligned}$$

such that $x_i \in M_i$. We will call x_i the i-th component of the function x.

We will define addition and scalar multiplication on $\prod_{i \in I} M_i$ as follows

$$(x + y)_i = x_i + y_i \qquad (x, y \in \prod_{i \in I} M_i,\ i \in I);$$

$$(rx)_i = rx_i \qquad (x \in \prod_{i \in I} M_i,\ r \in R,\ i \in I).$$

With this definition of addition and scalar multiplication, it is easily verified that $\prod_{i \in I} M_i$ is an R-module which we will call the direct product of the family $\{M_i : i \in I\}$.

We should note that the restriction $M_i \cap M_j = \emptyset$ for $i \neq j$ is not really a restriction. If distinct modules of the family are not disjoint, then we can replace them by isomorphic copies which are disjoint.

For each $i \in I$ we define a mapping $p_i : \prod_{i \in I} M_i \to M_i$, called the i-th projection mapping, by the formula

$$p_i(x) = x_i \qquad (x \in \prod_{i \in I} M_i,\ i \in I)$$

The i-th projection mappings are easily seen to be R-homomorphisms.

Note that in the case $I = \{1, 2\}$, $\prod_{i \in I} M_i \cong M_1 \times M_2$. [cf. Exercise (1.21)] We will sometimes prefer to use the notation $M_1 \times M_2$.

We will now extend a result which was previously seen to be true for finite families of groups. [II, Exercise (2.24)]

(2.2) PROPOSITION (Universality of the direct product). Let $\{M_i: i \in I\}$ be a family of R-modules and let p_i be the i-th projection mapping, for each $i \in I$. Let M be any R-module and let $\{f_i: i \in I\}$ be a family of R-homomorphisms such that $f_i: M \to M_i$ for each $i \in I$. Then there is a unique R-homomorphism $f: M \to \prod_{i\in I} M_i$ rendering commutative the following family of diagrams.

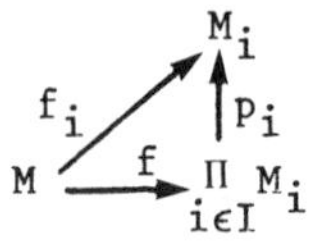

(i ∈ I)

PROOF Define $f: M \to \prod_{i\in I} M_i$ by $(f(x))_i = f_i(x)$ for each $x \in M$, $i \in I$. We will show that f preserves scalar multiplication. Let $x \in M$, $r \in R$, and $i \in I$. Then

$$\begin{aligned} (f(rx))_i &= f_i(rx) && \text{(definition of } f\text{)} \\ &= rf_i(x) && (f_i \text{ is an R-homomorphism}) \\ &= (rf(x))_i && \text{(definition of scalar multiplication)} \end{aligned}$$

Hence, $f(rx) = rf(x)$. Similarly, we can show that f preserves addition. Thus, f is an R-homomorphism.

We will now show that (1) is commutative for each $i \in I$. Let $x \in M$. Then

$$p_i(f(x)) = (f(x))_i = f_i(x)$$

Hence, $p_i \circ f = f_i$ and (1) is commutative for each $i \in I$.

To establish the uniqueness of f, suppose that $g: M \to \prod_{i\in I} M_i$ is an R-homomorphism such that for each $i \in I$, $p_i \circ g = f_i$. Then for each $x \in M$, $p_i(f(x)) = f_i(x) = p_i(g(x))$. But, since the i-th component of $f(x)$ and $g(x)$ agree for each $i \in I$, we have $f(x) = g(x)$ for each $x \in M$. Thus, $f = g$, and uniqueness is established.

(2.3) COROLLARY Let $\{M_i: i \in I\}$ be a family of R-modules and let

$f, g: M \to \prod_{i \in I} M_i$ be R-homomorphisms such that $p_i \circ f = p_i \circ g$ for each $i \in I$. Then $f = g$.

(2.4) DEFINITION Let $\{M_i: i \in I\}$ be a family of R-modules. Then $\bigoplus_{i \in I} M_i$ is defined to be the subset of $\prod_{i \in I} M_i$ of all elements $x \in \prod_{i \in I} M_i$ such that $x_i = 0$ for all but finitely many i. It is easily verified that $\bigoplus_{i \in I} M_i$ is an R-submodule of $\prod_{i \in I} M_i$. We will call $\bigoplus_{i \in I} M_i$ the direct sum of the family $\{M_i: i \in I\}$. In case that $I = \{1, 2\}$, we will sometimes write $M_1 \oplus M_2$ for the direct sum.

For each $i \in I$ define a mapping $u_i: M_i \to \bigoplus_{i \in I} M_i$, which we will call the i-th injection mapping, by

$$(u_i(x))_j = \begin{cases} 0 & i \neq j \\ x & i = j \end{cases} \qquad (i, j \in I,\ x \in M_i)$$

The injection mappings are easily seen to be R-homomorphisms which satisfy the following universal property.

(2.5) PROPOSITION (Universality of the direct sum) Let $\{M_i: i \in I\}$ be a family of R-modules and let $\{u_i: i \in I\}$ be the family of injection mappings. If M is any R-module and $\{f_i: i \in I\}$ is a family of R-homomorphisms such that $f_i: M_i \to M$ for each $i \in I$, then there is a unique R-homomorphism $f: \bigoplus_{i \in I} M_i \to M$ rendering commutative the following family of diagrams.

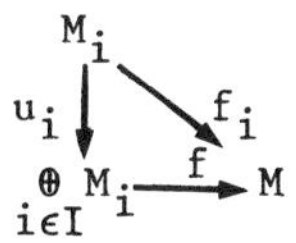

$(i \in I)$

PROOF Define $f: \bigoplus_{i \in I} M_i \to M$ by the rule $f(x) = \sum_{i \in I} f_i(x_i)$. Since $x_i = 0$ for all but finitely many i, this summation is well defined. It is also easy to show that f is an R-homomorphism.

To see that (2) is commutative, let $i \in I$ and $x \in M_i$. Then $f(u_i(x)) = f_i(x)$, since there is only one nonzero term in the summation corresponding to $f(u_i(x))$. Therefore, (2) is commutative. The proof of uniqueness is similar to that in (2.2).

(2.6) COROLLARY Let $\{M_i: i \in I\}$ be a family of R-modules and let N be an R-module. Suppose that $f, g: \bigoplus_{i\in I} M_i \to N$ are R-homomorphisms such that

$$f \circ u_i = g \circ u_i \qquad (i \in I)$$

then $f = g$.

Note that if I is finite, then the direct product of $\{M_i: i \in I\}$ and the direct sum of $\{M_i: i \in I\}$ are the same. We will see in an exercise when I is infinite, they are not necessarily even isomorphic.

(2.7) PROPOSITION Let $\{M_i: i \in I\}$, $\{N_j: j \in J\}$ be families of R-modules and let

$$p_j: \prod_{i\in I} N_j \to N_j \qquad (j \in J)$$

$$u_i: M_i \to \bigoplus_{i\in I} M_i \qquad (i \in I)$$

be the projection and injection mappings. Define

$$F: \mathrm{Hom}_R(\bigoplus_{i\in I} M_i, \prod_{j\in J} N_j) \to \prod_{(i,j)\in I\times J} \mathrm{Hom}_R(M_i, N_j)$$

by $F(f)_{ji} = (p_j \circ f \circ u_i)$, $i \in I$, $j \in J$. Then F is an isomorphism of additive groups.

PROOF We will first establish that F is injective. Suppose that $F(f) = F(g)$; then for every $i \in I$ and $j \in J$, we have $p_j \circ f \circ u_i = p_j \circ g \circ u_i$. By Corollary (2.3) we have $f \circ u_i = g \circ u_i$ for every $i \in I$. But then, by Corollary (2.6), we have $f = g$. Hence, F is injective.

To establish that F is surjective, let

$$(f_{ji}) \in \prod_{(i,j)\in I\times J} \mathrm{Hom}_R(M_i, N_j)$$

Then by (2.2), there is a unique f_i rendering commutative the following class of diagrams.

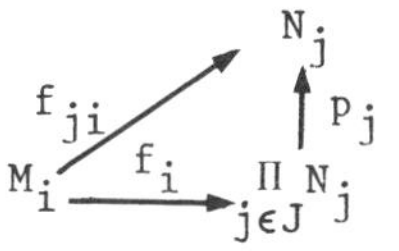

$(i \in I;\ j \in J)$

By (2.5), there is a unique f rendering commutative the following class of diagrams.

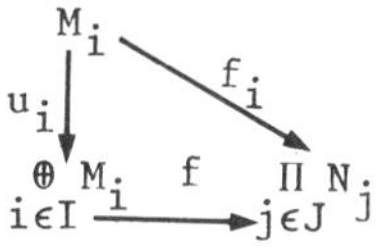

By definition of F, $F(f) = (f_{ji})$.

The fact that F preserves addition will be left as an exercise to the reader.

In the special case I and J are finite sets, we will write

$$F(f) = \begin{pmatrix} f_{11} & f_{12} & \cdots & f_{1n} \\ f_{21} & f_{22} & \cdots & f_{2n} \\ \\ f_{m1} & f_{m2} & \cdots & f_{mn} \end{pmatrix}$$

where $f_{ji} = p_j \circ f \circ u_i$; $i = 1, 2, \ldots, n$; $j = 1, 2, \ldots, m$.

In Exercise (2.14), we will see that in case that $I = J$ are finite, then

$$F(f \circ g) = F(f)F(g)$$

where the operation on the right is multiplication of matrices.

(2.8) DEFINITION Let M be an R-module and let $\{M_i : i \in I\}$ be a family of nonzero R-submodules of M. Then $\{M_i : i \in I\}$ is said to be an independent family of submodules of M, or more simply independent, if for any set $F \subseteq I$ and for any $i \in I - F$, we have

$$M_i \cap \left(\sum_{j \in F} M_j \right) = \{0\}$$

(2.9) REMARK Let $\{M_i: i \in I\}$ be a family of R-submodules of an R-module M. Note that $\{M_i: i \in I\}$ is an independent family of R-submodules if and only if for any family $\{x_i: i \in I, x_i \in M_i\}$ such that

$$\sum_{i\in I} x_i = 0$$

we have, $x_i = 0$ for each $i \in I$. The proof of this fact will be left as an exercise.

(2.10) PROPOSITION Let M be an R-module and let $\{M_i: i \in I\}$ be an independent family of R-submodules of M such that $\sum_{i\in I} M_i = M$. Then $M \cong \bigoplus_{i\in I} M_i$. Conversely, if $M \cong \bigoplus_{i\in I} M_i$, then there is an independent family $\{N_i: i \in I\}$ of submodules of M such that

$$\sum_{i\in I} N_i = M$$

$$N_i \cong M_i \qquad (i \in I)$$

PROOF Define $f_i: M_i \to M$ to be the inclusion mapping. Then by (2.5), there is an $f: \bigoplus_{i\in I} M_i \to M$ rendering commutative the following family of diagrams.

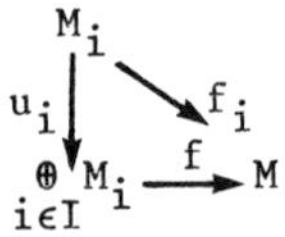

$(i \in I)$

Surjectivity of f is immediate since im f contains M_i for each $i \in I$ and $\sum_{i\in I} M_i = M$.

From the proof of (2.5) f is defined by

$$f(x) = \sum_{i\in I} x_i \qquad (x \in \bigoplus_{i\in I} M_i)$$

Let $x \in \bigoplus_{i\in I} M_i$ be such that $f(x) = 0$. Then $\sum_{i\in I} x_i = 0$, and hence $x = 0$, by (2.9). Therefore, f is injective.

Conversely, if $M \cong \bigoplus_{i\in I} M_i$, then suppose that $f: \bigoplus_{i\in I} M_i \to M$ is an isomorphism. Set $N_i = (f \circ u_i)(M_i)$; then $\{N_i: i \in ;I\}$ is an independent family of submodules of M, $\sum_{i \in I} N_i = M$ and $N_i \cong M_i$ for each $i \in I$.

(2.11) COROLLARY Let $\{M_i: i \in I\}$ be an independent family of R-submodules of an R-module M and let $x_i, y_i \in M_i$ for each $i \in I$ be such that $\sum_{i \in I} x_i = \sum_{i \in I} y_i$. Then $x_i = y_i$ for each $i \in I$.

PROOF If $\sum_{i \in I} x_i = \sum_{i \in I} y_i$, then $\sum_{i \in I}(x_i - y_i) = 0$. We can now apply Remark (2.9) to conclude that $x_i - y_i = 0$ for each $i \in I$. Therefore, $x_i = y_i$ for each $i \in I$.

(2.12) REMARK If $\{M_i: i \in I\}$ is an independent family of submodules of M such that $\sum_{i \in I} M_i = M$, then it follows from the previous Corollary that every element $x \in M$ admits a unique representation in the form

$$x = \sum_{i \in I} x_i$$

where $x_i \in M_i$ for each $i \in I$.

(2.13) DEFINITION A subset S of an R-module M is said to be an independent subset of M if the family $\{Rs: s \in S\}$ is an independent family of submodules of M. In the case that M is a vector space we will sometimes say that S is a linearly independent subset of M. A subset of M which is not independent will be called dependent.

The following proposition is an immediate consequence of (2.9) and characterizes independent subsets of an R-module M.

(2.14) PROPOSITION Let M be an R-module and let $\{x_i: i \in I\}$ be a family of nonzero elements of M. Then $\{x_i: i \in I\}$ is an independent subset of M if and only if

$$\sum_{i \in I} r_i x_i = 0 \qquad (r_i \in R)$$

implies $r_i x_i = 0$ for each $i \in I$.

The following corollaries are immediate and will be stated without proof.

(2.15) COROLLARY Let M be an R-module and let $\{x_i: i \in I\}$ be a subset of nonzero elements of M. Suppose that

(i) $\sum_{i\in I} r_i x_i = 0$ implies $r_i x_i = 0$ for all $i \in I$
(ii) every $x \in M$ admits a representation of the form

$$x = \sum_{i\in I} r_i x_i$$

where all but finitely many of the r_i's are 0.

Then $M \cong \bigoplus_{i\in I} Rx_i$.

A subset of nonzero elements $\{x_i: i \in I\}$ which satisfies (i) and (ii) above is said to be a basis for the R-module M.

(2.16) COROLLARY Let M_1 and M_2 be R-modules and let $f: M_1 \to M_2$ be an R-homomorphism. Let $\{x_i: I \in I\}$ be an independent subset of M_1 such that

$$(\sum_{i\in I} Rx_i) \cap \ker f = \{0\}.$$

Then $\{f(x_i): i \in I\}$ is an independent subset of M_2.

Note that one consequence of corollary (2.16) is that injective R-homomorphisms preserve independent subsets. Note further that bijective R-homomorphisms preserve basis.

We will now establish what has come to be called The Fundamental Theorem for Finitely Generated Modules over a Euclidean Domain.

(2.17) THEOREM Let R be a Euclidean domain and let M be a finitely generated R-module. Then M is a finite direct sum of cyclic R-modules.

PROOF Since M is finitely generated, there is a finite set which generates M. We will be searching for a finite independent set which generates M. If F is a generating set and F is not independent, then there is a nontrivial linear combination of elements of F which is zero. We will be concerned with what we will call minimal generating relations.

An equation of the form

$$\sum_{i \in I} r_i x_i = 0$$

will be called a minimal generating relation if $\{x_i: i \in I\}$ generates M and the number of terms in the equation is minimal. Since M is a finitely generated R-module, minimal generating relations exist. A minimal generating relation will be called trivial if all the terms are zero. If for some generating set F, all minimal generating relations with elements from F are trivial, then that set is clearly a basis for M. Assuming that no such F exists, among all nontrivial relations, we will choose a relation

$$r_1x_1 + \cdots + r_nx_n = 0$$

such that $r_1 \neq 0$ and $d(r_1)$ is minimal.

We claim that $r_1 | r_i$ for $i = 2, \ldots, n$. We show only $r_1 | r_2$, let

$$r_2 = qr_1 + r \quad \text{(either } r = 0 \text{ or } d(r) < d(r_1))$$

Note that,

$$\{x_1 + qx_2, x_2, \ldots, x_n\}$$

generates M and

$$r_1(x_1 + qx_2) + rx_2 + r_3x_3 + \cdots + r_nx_n = 0$$

is a minimal generating relation. But then, $d(r) < d(r_1)$ would contradict the minimality of $d(r_1)$. Thus $r = 0$, and $r_1 | r_2$. Similarly, $r_1 | r_i$ for $i = 3, \ldots, n$.

Let

$$s_1x_1 + s_2x_2 + \cdots + s_nx_n = 0$$

be another minimal generating relation. Then, we claim that $r_1 | s_1$. To show this, let

$$s_1 = qr_1 + r \quad \text{(either } d(r) < d(r_1) \text{ or } r = 0)$$

Then note that

$$(s_1 - qr_1)x_1 + \cdots + (s_n - qr_n)x_n = 0$$

But $s_1 - qr_1 = r$, and $d(r) < d(r_1)$ would contradict the minimality of $d(r_1)$. Thus $r = 0$, and our second claim is established.

Let

$$r_i = s_i r_1 \quad i = 2, \ldots, n$$

and set

$$(1) \quad y = x_1 + s_2x_2 + \cdots + s_nx_n$$

Then, $\{y, x_2, \ldots, x_n\}$ generates M. Let $N = \langle\{x_2, \ldots, x_n\}\rangle$. We will show that $\{Ry, N\}$ is an independent family of submodules of M whose sum is necessarily M. Suppose that

$$(2) \quad t_1y + t_2x_2 + \cdots + t_nx_n = 0$$

Then, combine (1) and (2) to get

$$t_1x_1 + (t_2 + s_2)x_2 + \cdots + (t_n + s_n)x_n = 0$$

which implies that $r_1 | t_1$. Let $t_1 = qr_1$, then, from (1)

$$\begin{aligned} qr_1y &= q(r_1x_1 + r_1s_2x_2 + \cdots + r_1s_nx_n) \\ &= q(r_1x_1 + \cdots + r_nx_n) \qquad (\text{since } r_i = r_1s_i) \\ &= 0 \end{aligned}$$

Therefore, relation (2) implies $t_1y = 0$. It follows that $\{Ry, N\}$ is an independent family of R-submodules of M whose sum is M. Therefore

$$M \cong Ry \oplus N$$

The theorem now follows by induction.

The previous theorem has many consequences. Since any field is a Euclidean domain, any finitely generated vector space is a direct sum of cyclic subspaces. Since this result obtains in a more general form, we will postpone a formal statement at this time. Since **Z** is also a Euclidean domain, the previous theorem also applies to finitely generated abelian groups. We will now state that fact.

(2.18) COROLLARY Let G be a finitely generated abelian group. Then G is a direct sum of cyclic groups. In particular, any finite abelian group is a direct sum of cyclic groups.

(2.19) DEFINITION Let M be an R-module and let $\{x_i: i \in I\}$ be a subset of M. Then M is said to be freely generated by $\{x_i: i \in I\}$, or $\{x_i: i \in I\}$ is said to be a free basis for M, if $M = \langle\{x_i: i \in I\}\rangle$ and whenever

$$\sum_{i \in I} r_i x_i = 0$$

where $r_i \in R$, then $r_i = 0$ for every $i \in I$.

An R-module M is said to be free if M has a free basis. The following proposition characterizes free R-modules.

(2.20) PROPOSITION Let M be a free R-module with free basis $\{x_i: i \in I\}$. Let N be an R-module and let $\{y_i: i \in I\}$ be a subset of N. Then there is a unique R-homomorphism $f: M \to N$ such that

$$f(x_i) = y_i \qquad (i \in I)$$

PROOF Let $x \in M$. Then x admits a unique representation in the form

$$x = \sum_{i \in I} r_i x_i$$

Define $f: M \to N$ by

$$f(x) = \sum_{i \in I} r_i y_i$$

Then f is the required R-homomorphism.

(2.21) PROPOSITION Let F be an R-module. Then F is a free module if and only if

$$F \cong \bigoplus_{i \in I} R$$

for some indexing set I.

PROOF We will first show that $\bigoplus_{i \in I} R$ is free for any indexing set I. Define $d_i \in \bigoplus_{i \in I} R$ as follows

$$d_{ij} = \begin{cases} 1 & i = j \\ 0 & i \neq j \end{cases} \qquad (i,j \in I)$$

Let $S = \{d_i: i \in I\}$. Then it will be left to the student to show that $\bigoplus_{i \in I} R$ is freely generated by S.

Conversely, if F is a free module with free basis $\{x_i: i \in I\}$, then it is easily shown, Exercise (2.17), that

$$F \cong \bigoplus_{i \in I} R$$

EXERCISES

(2.1) Let $\{M_i: i \in I\}$ be a family of R-modules. Show that $\prod_{i \in I} M_i$ is an R-module.

(2.2) Show that the projection mappings $p_i: \prod_{i \in I} M_i \to M_i$ are R-homomorphisms.

(2.3) Show that the function $f: M \to \prod_{i \in I} M_i$ defined in the proof of (2.2) is an additive homomorphism.

(2.4) Write a complete proof of (2.3).

(2.5) Show that $\bigoplus_{i \in I} M_i$ is an R-submodule of $\prod_{i \in I} M_i$.

(2.6) Show that the injection mappings $u_i: M_i \to \bigoplus_{i \in I} M_i$ are R-homomorphisms.

(2.7) Show that the function $f: \bigoplus_{i \in I} M_i \to M$ as defined in the proof of (2.5) is an R-homomorphism.

(2.8) Prove the uniqueness assertion in (2.5).

(2.9) Prove (2.6).

(2.10) Let x, y, and z be defined as elements of $\mathbf{R}^2$ as

$x = (1, 3)$

$y = (2, 5)$

$z = (1, -1)$

Show that {x, y}, {x, z}, and {y, z} are all bases for $\mathbf{R}^2$. Show that {x, y, z} is not an independent subset of $\mathbf{R}^2$.

(2.11) Let x, y be defined as elements of $\mathbf{R}^3$ by

$$x = (1, 0, 1)$$
$$y = (2, 1, -1)$$

Show that {x, y} is an independent subset of $\mathbf{R}^3$.

(2.12) Let F be a field and let M be an F-vector space. Show that a subset {x, y} of M is independent if and only if for every $r \in F$, we have $x \neq ry$.

(2.13) Prove the statement in (2.9).

(2.14) Let $f: \mathbf{R}^n \to \mathbf{R}^m$ be an $\mathbf{R}$-linear mapping and let A be an $m \times n$ matrix with real entries such that $f = f_A$. Referring to (2.7), show that $F(f) = A$. Thus deduce the remark at the end of the proof of (2.7). See exercises (1.20) and (1.21).

(2.15) Let $\{N_i: i \in I\}$ be defined as in the proof of (2.10). Prove that $\{N_i: i \in I\}$ is an independent family of R-submodules of M such that

(a) $\sum_{i \in I} N_i = M$

(b) $N_i \cong M_i$

(2.16) Let $\{d_i: i \in I\}$ be defined as in the proof of (2.21). Show that $\{d_i: i \in I\}$ is a free basis for $\bigoplus_{i \in I} R$.

(2.17) Let $\{x_i: i \in I\}$ be a free basis for an R-module M. Show that for every $x \in M$, x admits a unique representation of the form

$$x = \sum_{i \in I} r_i x_i$$

(2.18) Prove that the function f defined in the proof of (2.20) is an R-homomorphism.

(2.19) Let $\mathbf{P}_n$ be the set of all polynomials in $\mathbf{P}$ with degree less than or equal to n. Find a basis for $\mathbf{P}_n$. For $\mathbf{P}$.

(2.20) Let M be an R-module. Show that there is a free R-module F and a surjective R-homomorphism $f: F \to M$. If M is finitely generated with n generators, show that F can be taken to have a free basis with n generators.

(2.21) Show that the only non-isomorphic abelian groups of order 12 are

$$\mathbf{Z}_4 \times \mathbf{Z}_3$$
$$\mathbf{Z}_2 \times \mathbf{Z}_2 \times \mathbf{Z}_3$$

(2.22) List the non-isomorphic abelian groups of order 18, 120, 180.

(2.23) Show that there is no surjective function

$$f: \bigoplus_{i\in I} \mathbf{Z} \to \prod_{i\in I} \mathbf{Z}$$

where $I = \mathbf{N}$. Hence, $\bigoplus_{i\in I} \mathbf{Z}$ is not isomorphic to $\prod_{i\in I} \mathbf{Z}$.

(2.24) Let G be a finite abelian group. Show that there are positive integers $m_1 | m_2 | \ldots | m_n$ such that

$$G \cong \mathbf{Z}_{m_1} \times \mathbf{Z}_{m_2} \times \cdots \times \mathbf{Z}_{m_n}$$

(2.25) An abelian group is said to be torsion if every element has finite order. Let G be a torsion abelian group. Show that:

(a) If p is a prime, then the set of all elements of G whose order is a power of p, which we will call G_p, is a subgroup of G.

(b) If $I = \{p\colon p \text{ is a prime and } G_p \neq \{e\}\}$, then show that the family

$$\{G_p\colon p \in I\}$$

is an independent family of subgroups of G.

(c) $G \cong \bigoplus_{p\in I} G_p$

(2.26) Let $\{M_i\colon I \in I\}$ be a family of R-modules. Show that the projection mapping $p_i\colon \prod_{i\in I} M_i \to M_i$ can be restricted to the module $\bigoplus_{i\in I} M_i$. We will denote the restricted map, somewhat ambiguously, by p_i. Show that the restricted projection mappings satisfy the following

(a) $p_i u_j = 0$ if $i \neq j$

(b) $p_i u_i = 1$

(c) $\sum_{i \in I} u_i p_i = 1$

Where 1 denotes the appropriate identity map. This property is characteristic of the direct sum; more precisely, if M is an R-module and p_i: M → M_i, u_i: M_i → M are R-homomorphisms which satisfy conditions (a), (b), and (c), then $M \cong \bigoplus_{i \in I} M_i$.

(2.27) Regard $M_n(K)$ as a K[x] module as in (1.8.8). Show that $M_n(K)$ is a finitely generated K[x]-module and as such is a direct sum of cyclic K[x]-modules.

3. VECTOR SPACES

In this section, we will restrict our attention to vector spaces. Necessarily, many of the results will be special cases of more general results which were previously stated. Throughout this section we will use the letter F to denote a field and U, V, and W to denote vector spaces. For the sake of clarity, we will continue with our convention of denoting scalars with the letters r, s, and t. Elements of vector spaces, which will be called vectors, will be denoted with the letters x, y, and z.

(3.1) PROPOSITION Let V be a vector space over a field F and let $x \in V$ and $r \in F$. If $rx = 0$, then either $r = 0$ or $x = 0$.

PROOF If $rx = 0$ and $r \neq 0$, then $r^{-1}(rx) = 1x = x$. Thus $x = 0$.

(3.2) PROPOSITION Let V be a vector space over a field F and let $\{x_i: i \in I\}$ be a subset of V. Then, $\{x_i: i \in I\}$ is a basis for V if and only if $\{x_i: i \in I\}$ is a free basis for V.

PROOF Let $\{x_i: i \in I\}$ be a basis for V and $r_1, \ldots, r_n \in F$ be such that

$$r_1x_1 + \cdots + r_nx_n = 0$$

Then, $r_ix_i = 0$ for every $i = 1, \ldots, n$. Since $\{x_i: i \in I\}$ is a basis,

$x_i \neq 0$. Therefore, $r_i = 0$ for each $i = 1, \ldots, n$.

The converse is clear.

(3.3) COROLLARY Let V be a vector space over a field F and let $S \subseteq V$. Then, S is linearly independent if and only if, whenever

$$r_1x_1 + \cdots + r_nx_n = 0$$

where $r_1,\ldots,r_n \in F$ and $x_1,\ldots,x_n \in S$, then

$$r_1 = r_2 = \cdots = r_n = 0$$

(3.4) COROLLARY Let V be a finitely generated vector space over a field F. Then V is free as an F-vector space.

(3.5) COROLLARY Let V be a vector space over a field F and let $\{x_i: i \in I\}$ be a basis for V. Let U be a vector space over F and let $\{y_i: i \in I\}$ be a subset of U. Then there is a unique F-linear function $f: V \to U$ such that $f(x_i) = y_i$ for each $i \in I$.

PROOF See (2.20).

(3.6) COROLLARY Let V be a vector space over a field F and let $\{x_1, \ldots, x_n\}$ be a basis for V. Then $V \cong F^n$.

PROOF See (2.21).

(3.7) COROLLARY Let V be a vector space over a field F and let $\{x_i: i \in I\}$ be a basis for V. If $x \in V$, then x admits a unique representation

$$(*) \quad x = \sum_{i\in I} r_ix_i$$

where $r_i \in R$.

PROOF The existence of a representation (*) is guaranteed by definition of basis. Let $r_i, s_i \in R$ for each $i \in I$ be such that

$$\sum_{i\in I} r_ix_i = \sum_{i\in I} s_ix_i$$

Then

$$\sum_{i \in I} (r_i - s_i)x_i = 0$$

Therefore, since $\{x_i: i \in I\}$ is independent, $r_i = s_i$ for each $i \in I$ and uniqueness is established.

Corollary (3.7) is sometimes phrased: Every element $x \in V$ is uniquely expressible as a linear combination of basis elements.

We should note that since any field is necessarily a Euclidean domain, then any finitely generated vector space over a field has a basis. [cf. (2.17)] We will see later that any vector space has a basis.

The following result, known as the exchange theorem, is essential to establish the uniqueness of dimension for finite dimensional vector spaces.

(3.8) THEOREM The Exchange Theorem. Let V be a vector space over a field F and let B be a basis for V. Let x be any nonzero element of V. Then there is a $y \in B$ such that

$$B' = (B - \{y\}) \cup \{x\}$$

is a basis for V.

PROOF Let x be a nonzero element of V and let

$$x = r_1x_1 + \cdots + r_nx_n$$

where $r_i \in R$ and $x_i \in B$. Without loss of generality, we will assume that $r_1 \neq 0$. Set $y = x_1$; then B', as defined above, is a basis for V. The proof of this fact will be left as an exercise.

(3.9) PROPOSITION Let V be a vector space over a field F and let $\{x_1, \ldots, x_n\}$, $\{y_1, \ldots, y_m\}$ be bases for V. Then $m = n$.

PROOF Certainly, $y_1 \neq 0$; hence, by the exchange theorem, with a possible reordering of the x_i's, we have

$$\{y_1, x_2, \ldots, x_n\}$$

is a basis for V. Continue with y_2, noting that, from the proof of the exchange theorem, the element replaced by y_2 is not y_1. By induction, we have $m \le n$. Similarly, $n \le m$.

(3.10) COROLLARY Let V be a finitely generated vector space over a field F and let $X = \{x_1, \ldots, x_n\}$ be a linearly independent subset of V. Then there is a basis B for V such that $X \subseteq B$.

PROOF Let $B = \{y_1, \ldots, y_m\}$ be a basis for V. Use the exchange theorem to replace elements of B with elements of X.

(3.11) COROLLARY Let V be a vector space over a field F and let $\{x_1, \ldots, x_n\}$ be a basis for V. Let $S \subseteq V$ be a set with more than n elements. Then S is dependent.

(3.12) COROLLARY Let V be a vector space over a field F and let $\{x_1, \ldots, x_n\}$ be a basis for V. Let S be a linearly independent subset of V with n elements. Then S is a basis for V.

(3.13) DEFINITION Let V be a vector space over a field F and let $\{x_1, \ldots, x_n\}$ be a basis for V. Then we will define the dimension of V to be n and we write

$$\dim V = n \quad (\dim V = 0 \text{ for the 0 vector space})$$

Note that the previous proposition guarantees us that the dimension of V is unique. In case V has a finite basis, we will refer to V as a finite dimensional vector space.

(3.14) PROPOSITION Let V and W be vector spaces over a field F such that $\dim V = \dim W$. Then $V \cong W$.

PROOF See (3.6).

(3.15) PROPOSITION Let V be a finite dimensional vector space over a field F and let W be a subspace of V. Then

$$\dim V/W = \dim V - \dim W$$

PROOF Let $\{x_1, \ldots, x_m\}$ be a basis for W. By (3.10), there is a basis

$$B = \{x_1, \ldots, x_m, \ldots, x_n\}$$

for V. Let $k: V \to V/W$ be the canonical map. Then, by (2.16),

$$B' = \{k(x_{m+1}), \ldots, k(x_n)\}$$

is a linearly independent subset of V/W. Since k(B) generates V/W, then B' generates V/W, and B' is a basis for V/W. Therefore,

$$\dim V/W = n - m = \dim V - \dim W$$

(3.16) COROLLARY Let V and W be finite dimensional vector spaces over a field F and let $f: V \to W$ be a surjective linear function. Then

$$\dim W = \dim V - \dim \ker(f)$$

(3.17) COROLLARY Let V be a finite dimensional vector space over a field F and let $U < W < V$ be a chain of subspaces of V. Then

$$\dim V/U = \dim V/W + \dim W/U$$

(3.18) PROPOSITION Let V be a finite dimensional vector space over a field F and let $f: V \to V$ be linear. Then the following conditions are equivalent:

(i) $\ker(f) = \{0\}$.

(ii) f is injective.

(iii) f is surjective.

PROOF We have already seen that (i) is equivalent to (ii). If f is injective, then $\dim f(V) = n$. But $\dim V = n$ implies $f(V) = V$. Hence (ii) implies (iii).

If f is surjective, let $W = \ker(f)$. Then,

$$\dim W = \dim V - \dim V/W = n - n = 0$$

Hence, $W = \{0\}$, and (iii) implies (i).

Note that if F is a field, then F is a vector space with dimension 1 over F. Note further that any linear function f: F → F must have the form

$$f(x) = rx$$

for some fixed $r \in F$. Thus, by (2.7), $\mathrm{Hom}_F(F^n, F^m)$ is isomorphic as an abelian group to the set of all $n \times m$ matrices over F. By Exercise (2.14), we have the following proposition.

(3.19) PROPOSITION Let F be a field. Then $\mathrm{Hom}_F(F^n, F^n)$ and $\mathrm{Mat}_n(F)$ are isomorphic as rings.

We close this section with a result about infinite dimensional vector spaces. The use of Zorn's lemma is essential here. The student is referred to Appendix A.

(3.20) PROPOSITION Let S be a linearly independent subset of a vector space V over a field F. Then there is a basis B for V such that $S \subseteq B$ (in particular, every nonzero vector space has a basis).

PROOF Let X be the set of all linearly independent subsets of V which contain S. Then $X \neq \emptyset$. Furthermore, the union of an upward directed chain of linearly independent subsets is a linearly independent subset of V. Thus, if X is ordered by set inclusion, X is inductive. Let B be a maximal element of X. We claim that B is a basis for V. For if not, then there is an

$$x \in V - \langle B \rangle$$

Then $B \cup \{x\}$ is a linearly independent subset of V which strictly contains B. But this is a contradiction, since B was maximal.

The definition of dimension for arbitrary vector spaces involves infinite cardinals and is beyond the scope of this book. However, surprisingly enough, any two vector spaces with the same dimension are isomorphic.

EXERCISES

(3.1) Prove Corollary (3.3).

(3.2) Prove that B', as defined in the proof of (3.8), is a basis for V.

(3.3) Prove (3.14).

(3.4) Prove (3.16).

(3.5) Let F be a field and regard F as a vector space over itself. Show that the only linear mappings from F to F have the form $f(x) = rx$

(3.6) Show that {cos x, sin x} is a linearly independent subset of $C^n(\mathbf{R})$.

(3.7) Let I be a left ideal of R. Then, I is an R-submodule of R; therefore R/I is an R-module. Suppose that R is the ring of 2 × 2 matrices over **R** and I is the set of all matrices of the form

$$\begin{pmatrix} 0 & a \\ 0 & b \end{pmatrix}$$

Show that, besides being an R-module, R/I is also a **R**-vector space. Find a basis for R/I both as an R-module and as a **R**-vector space.

(3.8) Let $\mathbf{P}_n$ be the set of all polynomials in **P** with degree less than or equal to n. What is the dimension of $\mathbf{P}_n$?

(3.9) Find an isomorphism from $\mathbf{R}^3$ to $\mathbf{P}_2$.

(3.10) Show that the set of all n × m matrices over **R** is a **R**-vector space. What is the dimension of this vector space?

(3.11) Find a linear function $f: \mathbf{R}^3 \to \mathbf{R}^3$ such that

$f(1, 2, 1) = (1, 3, 2)$

$f(1, 0, 1) = (1, 1, 0)$

$f(1, -1, 0) = (3, 1, 4)$

(3.12) Let V be the subspace of $\mathbf{R}^4$ generated by the following vectors

(1, 2, 1, -1)
(2, 1, 3, 1)
(3, 3, 4, 0)
(2, -1, 4, 1)

Find dim V.

(3.13) Define $L: C^2(\mathbf{R}) \to C^0(\mathbf{R})$ by

$$L(f) = D^2(f) + f$$

Show that L is linear. Show that the kernel of L is generated by {cos x, sin x}.

(3.14) A transcendental number is a real number which is not a root of a polynomial equation with rational coefficients. Given that e is transcendental, regard **R** as a **Q** vector space, and show that

$$\{e, e^2, \ldots, e^n, \ldots\}$$

is a linearly independent subset of **R**.

(3.15) Let F be a field and $m,n \in \mathbf{N}$ with $m > n$. Show that any system of n homogeneous linear equations with m unknowns has a nonzero solution.

(3.16) Let R be the ring of all two by two matrices over **R** and let S be the set of all matrices of the form

$$\begin{pmatrix} a & 0 \\ b & 0 \end{pmatrix}$$

Show that S is a cyclic R-module, but S is 2 dimensional when regarded as an **R**-vector space.

(3.17) Show that the following family of functions is a linearly independent family of functions in $C^n(\mathbf{R})$.

$$\{\sin x, \sin 2x, \ldots, \sin mx, \ldots, \cos x, \cos 2x, \ldots, \cos mx, \ldots\}$$

(3.18) An abelian group G is said to be divisible if for every $x \in G$ and for every $n \in \mathbf{N}$, there is a $y \in G$ such that $ny = x$. Let G be a divisible abelian group in which every nonzero element has infinite order. Show that:

(a) If $x \in G$ and $n \in \mathbf{N}$, then there is a unique $y \in G$ such that $ny = x$.

(b) If $m/n \in \mathbf{Q}$, $x \in G$, and $y \in G$ is the unique element such that $ny = x$, then

$$(m/n)x = my$$

gives G the structure of a $\mathbf{Q}$ vector space.

(c) G is a direct sum of copies of $\mathbf{Q}$.

(3.19) Let G be a divisible p-group, where p is a prime. Show that for each $x \in G$ and $n \in \mathbf{N}$ such that $(p, n) = 1$, there is a unique $y \in G$ such that $ny = x$. For each $x \in G$ such that $o(x) = p$, define

$$H_x = \{y\colon y \in G,\ p^m y = kx \text{ for some } m \in \mathbf{N} \text{ and } 0 < k < p\}$$

Show that:

(a) H_x is a subgroup of G.

(b) If $x \notin \langle y\rangle$, then $H_x \cap H_y = \{0\}$.

(3.20) Let $\mathbf{Q}_p$ be the set of all rational numbers of the form m/p^n, where p is a prime and $n \in \mathbf{N}$. Then $\mathbf{Q}_p$ is a subgroup of $\mathbf{Q}$ under addition. Define $\mathbf{Z}_p(\infty) = \mathbf{Q}_p/\mathbf{Z}$. If G is a divisible p-group and $x \in G$ is such that $o(x) = p$, then show that $H_x \cong \mathbf{Z}_p(\infty)$.

(3.21) Let G be a divisible p-group. Show that the set of all elements of order p is a $\mathbf{Z}_p$-vector space. Thus deduce that this subgroup of G is a direct sum of copies of $\mathbf{Z}_p$ under addition.

(3.22) Let G be a divisible p-group and let $\{x_i\colon i \in I\}$ be a family of elements of G of order p such that $i \neq j$ implies $x_i \notin \langle x_j\rangle$, and for every $x \in G$ such that $o(x) = p$, there is an $i \in I$ such that $x \in \langle x_i\rangle$. Show that

$$\{H_{x_i}\colon i \in I\}$$

is an independent family of subgroups of G. Thus deduce that G is isomorphic to a direct sum of copies of $\mathbf{Z}_p(\infty)$.

8
Field Extensions

<u>In this chapter 'ring' means ring with one and 'ring homomorphism' means unital ring homomorphism.</u>

If F is a field then a subring K of F containing 1_F is said to be a subfield of F provided it is itself a field under the operations of F. If $u,v \in F$ with $v \neq 0$ we use the notation u/v in place of uv^{-1}. Then a nonempty subset K of F is a subfield of F if and only if for all $u,v \in K$, $u - v \in K$ and, provided $v \neq 0$, $u/v \in K$.

1. FINITELY GENERATED EXTENSIONS

(1.1) DEFINITION Let K be a field. An extension field of K is a field F together with an injective homomorphism of rings $\alpha: K \to F$.

If F is an extension field of K with respect to $\alpha: K \to F$ then, since α is an injective homomorphism of rings, K is embedded as a subfield of F via the identification of k with $\alpha(k)$ for $k \in K$. Unless the homomorphism α is specifically needed for purposes of clarity [as in (2.6)], we will suppress it entirely and hence K will be considered to be a subfield of F.

Before proceeding we make the following general observation. Let T be a set, C a nonempty collection of subsets of T and suppose that the set $W = \bigcap_{A \in C} A$ is an element of C. Then W is the minimal element of C; that is, $W \in C$ and $W \subseteq A$ for all $A \in C$. It is readily verified that the intersection of a nonempty collection of subrings of a given ring R is again a subring of R and the intersection of a nonempty collection of subfields of a given field F is again a subfield of F.

NOTATION Let F be an extension field of K and S a subset of F.

$$K[S] = \bigcap_{A \in C} A$$

where C is the collection of all subrings of F which contain $K \cup S$. Then K[S] is the minimal <u>subring</u> of F containing $K \cup S$. If $S = \{u_1, \ldots, u_n\}$ we write $K[u_1, \ldots, u_n]$ instead of $K[\{u_1, \ldots, u_n\}]$.

$$K(S) = \bigcap_{E \in D} E$$

where D is the collection of all subfields of F which contain $K \cup S$. Then K(S) is the minimal <u>subfield</u> of F containing $K \cup S$ and is said to be the subfield of F obtained from K by adjoining S. As above, if $S = \{u_1, \ldots, u_n\}$ we write $K(u_1, \ldots, u_n)$ in place of $K(\{u_1, \ldots, u_n\})$.

We observe that $K[S] \subseteq K(S) \subseteq F$ and K(S) is in fact the minimal subfield of F containing K[S].

(1.2) PROPOSITION Let F be an extension field of K and S a subset of F. Then K[S] is an integral domain with fraction field K(S) [cf. IV, (4.10)]. It follows that

$$K(S) = \{u/v: u,v \in K[S], v \neq 0\}$$

PROOF Since K[S] is a subring of the field F containing 1, K[S] is an integral domain. The result now follows from the preceding comment and IV, (4.12).

An extension field F of a field K is said to be a <u>finitely generated</u> extension of K if $F = K(S)$ for some finite subset S of F. If $F = K(u)$ for some $u \in F$, then F is said to be a <u>simple</u> extension of K.

REMARK A simple set theoretic argument [see exercise (1.1)] shows that if F is an extension field of K and $u,v \in F$, then

$$(K(u))(v) = K(u, v) = (K(v))(u)$$

Hence if $u_1, \ldots, u_n \in F$, $K(u_1, \ldots, u_n)$ is the subfield of F obtained from K by adjoining the elements $u_1, \ldots, u_n$ one at a time and in any order.

(1.3) THEOREM Let F be an extension field of K and $u \in F$. Let $K[x]$ be the ring of polynomials in the indeterminate x over the field K. Define a mapping

$$\tau_u: K[x] \to F \text{ by } \tau_u(f) = f(u) \quad (f \in K[x])$$

Then the following assertions hold.

(i) τ_u is a homomorphism of rings.

(ii) Im $\tau_u = K[u]$ and hence

$$K[u] = \{f(u): f \in K[x]\} \text{ and}$$
$$K(u) = \{f(u)/g(u): f,g \in K[x],\ g(u) \neq 0\}$$

PROOF (i) Since $\tau_u(1) = 1$ (the constant polynomial 1 evaluated at u), τ_u is unital. For $f,g \in K[x]$,

$$\tau_u(f + g) = (f + g)(u) = f(u) + g(u) = \tau_u(f) + \tau_u(g)$$
$$\tau_u(fg) = (fg)(u) = f(u)g(u) = \tau_u(f)\tau_u(g)$$

Hence τ_u is a homomorphism of rings.

(ii) Since $k = \tau_u(k)$ for all $k \in K$ and $u = \tau_u(x)$, $K \cup \{u\} \subseteq$ Im τ_u. Hence Im τ_u is a subring of F (since τ_u is a homomorphism of rings) containing K and u. Since $K[u]$ is the minimal such subring, $K[u] \subseteq$ Im τ_u.

Now let A be any subring of F containing K and u. We wish to show that A necessarily contains Im τ_u. Let $v \in$ Im τ_u. Then $v = \tau_u(f)$ for some $f = a_0 + a_1x + \cdots + a_nx^n$ ($a_i \in K$). Since A contains K and u, and A is closed under products and sums,

$$v = \tau_u(f) = f(u) = a_0 + a_1u + \cdots + a_nu^n$$

is an element of A as desired. It follows that Im τ_u is contained in the intersection of all such subrings A; that is, Im $\tau_u \subseteq K[u]$. Hence Im $\tau_u = K[u]$ so that

$$K[u] = \{f(u)\colon f \in K[x]\}$$

and, since by (1.2) $K(u)$ is the fraction field of $K[u]$,

$$K(u) = \{f(u)/g(u)\colon f,g \in K[x],\ g(u) \neq 0\}$$

(1.4) DEFINITION Let F be an extension field of K and $u \in F$. The element u is said to be <u>algebraic</u> over K if there is a nonzero element $f \in K[x]$ such that $f(u) = 0$.

We say in this case that u <u>satisfies</u> the nonzero polynomial f. Note that u is algebraic over K if and only if the map τ_u described in (1.3) has nonzero kernel.

If u is not algebraic over K then u is said to be <u>transcendental</u> over K. Thus u is transcendental over K provided that u does not satisfy a nonzero polynomial with coefficients in K. We observe that u is transcendental over K if and only if $\ker \tau_u = (0)$; that is, if and only if τ_u is an injective map.

An extension F of K is said to be an algebraic extension of K if u is algebraic over K for all $u \in F$. Otherwise F is said to be a transcendental extension of K.

(1.5) DEFINITION Let E be an extension field of K, F an extension field of L, and $\alpha\colon K \to L$, $\tau\colon E \to F$ homomorphisms of rings such that $\tau|_K = \alpha$. We say in this case that the map τ is an <u>extension</u> of the map α and illustrate this fact as follows.

$$\begin{array}{ccc} E & \xrightarrow{\tau} & F \\ | & & | \\ K & \xrightarrow{\alpha} & L \end{array}$$

(1.6) THEOREM Let F be an extension field of K and $u \in F$ be such that u is transcendental over K. Let $K(x)$ be the fraction field of the polynomial ring $K[x]$; that is,

$$K(x) = \{f/g\colon f,g \in K[x],\ g \neq 0\}$$

Then the following assertions hold.

(i) The map

$$\tau_u\colon K[x] \to K[u] \text{ defined by } f \mapsto f(u)$$

is an isomorphism of rings.

(ii) τ_u extends to an isomorphism of fields

$$\tau_u'\colon K(x) \to K(u) \text{ defined by } f/g \mapsto f(u)/g(u)$$

PROOF (i) By (1.3) τ_u is a surjective homomorphism of rings. Since u is transcendental, τ_u is also injective.

(ii) This statement follows from IV, (4.13).

Note that if u is transcendental over K, u behaves exactly like an indeterminate over K: elements of K[u] are polynomials in u and two such polynomials are equal if and only if coefficients of like terms are equal. Then the simple extension field K(u) of K looks exactly like the field of rational functions in the 'indeterminant u'. The preceding proposition may be used to show that if u and v are elements of an extension field F over K with u and v both transcendental over K then there is an isomorphism τ of rings from K(u) to K(v) such that $\tau(u) = v$ and $\tau|_K = 1_K$ [cf. exercise (1.2)].

EXAMPLES The elements π and e of **R** are both transcendental over **Q**. The proof of these facts is beyond the scope of this text.

We begin our study of algebraic extensions by considering simple extensions K(u) of K with u algebraic over K.

(1.7) THEOREM Let F be an extension field of K and $u \in F$ with u algebraic over K. Let $\tau_u\colon K[x] \to K[u]$ be defined by $g \mapsto g(u)$ as in (1.3) and let $I = \ker \tau_u$. Then there is a unique <u>monic</u> polynomial f $\in K[x]$ (i.e, f has leading coefficient 1) such that $I = (f)$ (the principal ideal generated by f). In this case the following hold.

(i) τ_u induces an isomorphism of rings

$$\tau_u'\colon K[x]/I \to K[u] \quad \text{by} \quad [g] \mapsto g(u)$$

(ii) If $g \in K[x]$ then $g(u) = 0$ if and only if f divides g.

(iii) f is the unique monic irreducible polynomial in $K[x]$ such that $f(u) = 0$.

(iv) $K(u) = K[u] = \{g(u): g \in K[x]\}$.

PROOF Since u is algebraic, $I = \ker \tau_u \neq (0)$. Since I is an ideal of $K[x]$, and $K[x]$ is a principal ideal domain, $I = (f)$ for some unique monic polynomial $f \in K[x]$.

(i) This statement follows from the fundamental theorem of ring homomorphisms, IV, (2.16) since $I = \ker \tau_u$ and $\operatorname{Im} \tau_u = K[u]$.

(ii) Let $g \in K[x]$. Then $g(u) = 0$ if and only if $g \in I$. Since $I = (f)$ this occurs if and only if f divides g [cf. V, (2.1)].

(iii) Since by (i) $K[x]/I$ is isomorphic to the integral domain $K[u]$, it is itself an integral domain. Hence by IV, (4.6), I is a prime ideal. It now follows by V, (3.8) that f is an irreducible polynomial. If $g \in K[x]$ is also monic and irreducible and $g(u) = 0$, then by (i) f divides g. Since f and g are both monic and g is irreducible, this implies that $g = f$.

(iv) Since f is an irreducible polynomial by (iii), I is a maximal ideal of $K[x]$ by V, (3.8). It follows by IV, (3.3) that $K[x]/I$, and hence also $K[u]$, is a field. Since $K(u)$ is the minimal subfield of F containing $K[u]$, $K(u) = K[u]$. Hence, by (1.3),

$$K(u) = K[u] = \{g(u): g \in K[x]\}$$

If the element $u \in F$ is algebraic over K, the unique polynomial f given in the preceding theorem is called the <u>minimal polynomial</u> of u. Thus, by (iii) above, the minimal polynomial of u is the unique monic irreducible polynomial $f \in K[x]$ such that $f(u) = 0$. We will see that the degree of the minimal polynomial f of u over K plays an important role in the description of the field $K(u)$.

If F is an extension field of K, then F is a vector space over K [cf. VII, (1.8.4)]. The dimension of F as a vector space over K will be denoted by $[F: K]$. F is then said to be a finite dimensional extension of K (or merely a finite extension of K) if $[F: K]$ is finite and an infinite dimensional extension otherwise. If $[F: K] =$

n, we say that F has dimension n over K or that the extension of F over K has dimension n.

We observe that [F: K] = 1 if and only if F = K. The relation between finite dimensional extensions of K and finitely generated extensions of K is discussed in (2.2). For future reference we also observe that if F is an extension field of K, E an extension field of L, and $\alpha: K \to L$, $\tau: F \to E$ isomorphisms with τ an extension of α, then [F: K] = [E: L] [cf. exercise (1.4)].

(1.8) THEOREM Let F be an extension field of K and let $u \in F$ be algebraic over K with minimal polynomial f of degree n. Then $\{1, u, u^2, \ldots, u^{n-1}\}$ is a basis of the vector space K(u) over K. Hence [K(u): K] = n and every element of K(u) can be written <u>uniquely</u> as

$$\sum_{i=0}^{n-1} a_i u^i \quad (a_i \in K)$$

PROOF Let $B = \{1, u, u^2, \ldots, u^{n-1}\}$. We must prove that (1), B spans the vector space K(u) over K and (2), B is a linearly independent set over K.

(1) By (1.7) and (1.3), $K(u) = K[u] = \text{Im } \tau_u$. Let $w \in K(u)$. We wish to show that w is a K-linear combination of elements of B. Since $w \in \text{Im } \tau_u$, there is an element $g \in K[x]$ such that $\tau_u(g) = w$. By the division algorithm there are elements $q, h \in K[x]$ such that $g = qf + h$ and either $h = 0$ or $\deg h < \deg f = n$. Then

$$\begin{aligned} w &= \tau_u(g) \\ &= \tau_u(qf + h) \\ &= \tau_u(q)\tau_u(f) + \tau_u(h) \\ &= \tau_u(h) \qquad\qquad [\text{since } \tau_u(f) = 0] \\ &= h(u) \end{aligned}$$

Since either $h = 0$ or $\deg h < n$, $h = a_0 + a_1x + \cdots + a_mx^m$ with $a_i \in K$ and $0 \le m < n$. Hence $w = h(u) = a_0 + a_1u + \cdots + a_mu^m$ so that, since $m \le n - 1$, w is a K-linear combination of elements from B. It follows that B spans the vector space K(u) over K.

(2) We now show that B is a linearly independent set over K. Let

$a_i \in K$ $(0 \le i \le n - 1)$ be such that $a_0 + a_1u + \cdots + a_{n-1}u^{n-1} = 0$. We wish to show that each a_i is zero. Let $g \in K[x]$ be the polynomial defined by $g = a_0 + a_1x + \cdots + a_{n-1}x^{n-1}$. By hypothesis $g(u) = 0$. Hence, by (1.7), (ii), f divides g. Since either $g = 0$ or $\deg g \le n - 1 < n = \deg f$, it must be the case that $g = 0$. Thus $a_i = 0$ for all i as required.

We have now shown that B is a basis for K(u) over K. Consequently $[K(u): K] = n$ and each element in K(u) can be written uniquely as a K-linear combination of elements of B.

If $u \in F$ is algebraic over K with minimal polynomial f of degree n, then $n = \deg f = [K(u): K]$ is called the degree of u over K.

(1.9) EXAMPLES

(1.9.1) We first consider the simple extension $\mathbf{R}(i)$ of the field $\mathbf{R}$ of real numbers where $i^2 = -1$. Since i satisfies the monic irreducible polynomial $f = x^2 + 1$ in $\mathbf{R}[x]$, i is algebraic over $\mathbf{R}$ with minimal polynomial $f = x^2 + 1$. Then by (1.8)

$$[\mathbf{R}(i): \mathbf{R}] = \deg f = 2$$

and $\{1, i\}$ is a basis for the vector space $\mathbf{R}(i)$ over $\mathbf{R}$. Hence every element w of $\mathbf{R}(i)$ can be written uniquely as

$$w = a + bi \qquad (a,b \in \mathbf{R})$$

We see then that $\mathbf{R}(i) = \mathbf{C}$, the field of complex numbers.

(1.9.2) Let w be the positive real fourth root of 2 and consider the simple extension $\mathbf{Q}(w)$ of $\mathbf{Q}$. Since w satisfies the monic irreducible polynomial $f = x^4 - 2 \in \mathbf{Q}[x]$, [observe that f is irreducible over $\mathbf{Q}$ by Eisenstein's Theorem, V, (3.18)], w is algebraic over Q with minimal polynomial f. Then by (1.7)

$$[\mathbf{Q}(w): \mathbf{Q}] = \deg f = 4$$

and the set $B = \{1, w, w^2, w^3\}$ is a basis for the vector space Q(w) over $\mathbf{Q}$. Hence elements of $\mathbf{Q}(w)$ can be written uniquely in the form

(*) $a + bw + cw^2 + dw^3$ $(a,b,c,d \in \mathbf{Q})$

For example, since $\mathbf{Q}(w)$ is a field containing $\mathbf{Q}$ and w, $v = 2 - w^4 + 3w^5$ must also be an element of $\mathbf{Q}(w)$. We wish to use the technique of the preceding theorem to write v in the form (*). We calculate

$$\begin{aligned}
v &= 2 - w^4 + 3w^5 \\
&= T_w(2 - x^4 + 3x^5) \\
&= T_w((x^4 - 2)(3x - 1) + 6x) \quad \text{(by the division algorithm)} \\
&= T_w(x^4 - 2)T_w(3x - 1) + T_w(6x) \\
&= T_w(6x) \quad [\text{since } T_w(x^4 - 2) = w^4 - 2 = 0] \\
&= 6w
\end{aligned}$$

We could also have used the fact that $w^4 = 2$ to reduce larger powers of w as follows:

$$\begin{aligned}
v &= 2 - w^4 + 3w^5 \\
&= 2 - 2 + 3w(2) \quad (\text{since } w^4 = 2) \\
&= 6w
\end{aligned}$$

(1.9.3) Let ρ be a primitive fifth root of unity (a <u>primitive n-th root of unity</u> is a ρ such that $\rho^n = 1$ and $\rho^m \neq 1$ if $m < n$). Recalling DeMoivre's Theorem we observe that in fact we may take

$$\rho = e^{2\pi i/5} = \cos(2\pi/5) + i \sin(2\pi/5)$$

We wish to consider the simple extension $\mathbf{Q}(\rho)$ of $\mathbf{Q}$. We observe that although ρ satisfies the monic polynomial $g = x^5 - 1 \in \mathbf{Q}[x]$, g is not the minimal polynomial of ρ since it is not irreducible over $\mathbf{Q}$. In fact $g = (x - 1)(x^4 + x^3 + x^2 + x + 1)$ with both factors elements of $\mathbf{Q}[x]$. Let $f = x^4 + x^3 + x^2 + x + 1$. Then f is monic and irreducible over $\mathbf{Q}$ [by Eisenstein's Criterion applied to $f(x + 1)$ - cf. V, (3.19)] and $f(\rho) = 0$. Hence ρ is algebraic over $\mathbf{Q}$ of degree 4 and $\{1, \rho, \rho^2, \rho^3\}$ is a basis for $\mathbf{Q}(\rho)$ over $\mathbf{Q}$.

Let F be an extension field of K. We recall that if u and v are transcendental over K ($u,v \in F$) then there is an isomorphism of fields from K(u) to K(v) which sends u to v and fixes K. We will see that if $u,v \in F$ are algebraic over K, then such an isomorphism exists if and only if u and v have the same minimal polynomial over K.

(1.10) PROPOSITION Let E be an extension field of K, F an extension field of L, $\alpha: K \to L$, and $\tau: E \to F$ isomorphisms with τ an extension of α. If $f = \sum_{i=0}^{n} a_i x^i \in K[x]$, define $\alpha f \in L[x]$ by

$$\alpha f = \sum_{i=0}^{n} \alpha(a_i)x^i$$

Then if $u \in E$ is a root of the polynomial $f \in K[x]$, $v = \tau(u) \in F$ is a root of the corresponding polynomial $\alpha f \in L[x]$. If u is algebraic over K with minimal polynomial f, then v is algebraic over K with minimal polynomial αf.

PROOF Let $f = \sum_{i=0}^{n} a_i x^i$ $(a_i \in K, n \in \mathbf{N})$, $u \in E$ and $v = \tau(u)$. Then

$$\begin{aligned} \tau(f(u)) &= \tau(\sum_{i=0}^{n} a_i u^i) \\ &= \sum_{i=0}^{n} \tau(a_i)\ [\tau(u)]^i \quad \text{(since } \tau \text{ is a homomorphism)} \\ &= \sum_{i=0}^{n} \alpha(a_i)\ v^i \quad [\text{since } \tau|_K = \alpha \text{ and } \tau(u) = v] \\ &= \alpha f(v) \end{aligned}$$

Since τ is an isomorphism of rings $\tau(f(u)) = 0$ if and only if $f(u) = 0$. Hence u is a root of f if and only if $v = \tau(u)$ is a root of αf.

Now suppose that the element $u \in E$ is algebraic over K with minimal polynomial f. Since $v = \tau(u)$ satisfies the nonzero polynomial $\alpha f \in L[x]$, v is algebraic over L. Let g be the minimal polynomial of v over L. Then by (1.7), (ii), $g | \alpha f$. By the previous paragraph (using the maps α^{-1} and τ^{-1}), since $v = \tau(u)$ is a root of g, $u = \tau^{-1}(v)$ is a root of $\alpha^{-1}g \in K[x]$. Thus, since f is the minimal polynomial of u over K, f divides $\alpha^{-1}g$; hence αf divides g. Since g and f are both monic, we conclude that $g = \alpha f$.

(1.11) COROLLARY Let F be an extension field of K and $\tau: F \to F$ an isomorphism of rings such that $\tau|_K = 1_K$. Then if the element $u \in F$ is a root of the polynomial $f \in K[x]$, the element $v = \tau(u)$ is also a root of f. If u is algebraic over K with minimal polynomial f, then $v = \tau(u)$ is also algebraic over K with minimal polynomial f.

(1.12) THEOREM (Extension Property For Simple Extensions) Let K and L be fields and $\alpha: K \to L$ an isomorphism of rings. Suppose that E is an extension field of K, F is an extension field of L, the element $u \in E$ is algebraic over K with minimal polynomial $f \in K[x]$ and the element $v \in F$ is algebraic over L with minimal polynomial $\alpha f \in L[x]$. Then α can be extended uniquely to an isomorphism $\tau: K(u) \to L(v)$ such that $\tau(u) = v$.

PROOF Suppose that the minimal polynomial f of u over K has degree n (so that the minimal polynomial αf of v over L also has degree n). By (1.8) every nonzero element w of K(u) can be written as $w = g(u)$ for some unique polynomial $g \in K[x]$ of degree smaller than n. Similarly every nonzero element of L(v) can be written as h(v) for some unique polynomial $h \in L[x]$ of degree smaller than n. We may therefore define a map

$$\tau: K(u) \to L(v) \text{ by } g(u) \mapsto \alpha g(v) \quad (g \in K[x],\ \deg g < n \text{ or } g = 0)$$

Then $\tau(u) = v$. If $g,h \in K[x]$ then $\alpha(g + h) = \alpha g + \alpha h$ and $\alpha(gh) = \alpha g \alpha h$ so that τ is indeed a homomorphism of rings. If $w = g(u) \in K(u)$ ($g \in K[x]$, $\deg g < n$) then $w = 0$ if and only if $\tau(w) = \alpha g(v) = 0$ by (1.10). Hence τ is both well-defined and injective.

We next verify that τ is a surjective map. Let $w \in L(v)$. Then $w = h(v)$ for some $h \in L[x]$ with $\deg h < n$. Let $g = \alpha^{-1}h$. Then $g \in K[x]$, $\deg g < n$ and if $w_1 = g(u)$, $\tau(w_1) = \alpha g(v) = \alpha\alpha^{-1}h(v) = h(v) = w$. Hence τ is surjective.

Finally, we establish the uniqueness of τ. Suppose that $\rho: K(u) \to L(v)$ is also an isomorphism of rings such that $\rho|_K = \alpha$ and $\rho(u) = v$. Let $w \in K(u)$. Then, since $w = g(u)$ for some $g \in K[x]$, $\rho|_K = \alpha$ and $\rho(u) = v$, $\rho(w) = \alpha g(v) = \tau(w)$.

(1.13) COROLLARY Let F be an extension field of K and $u,v \in F$ algebraic over K with the same minimal polynomial $f \in K[x]$. Then if f has degree n the map

$$\tau: K(u) \to K(v) \text{ by } g(u) \mapsto g(v) \quad (g \in K[x],\ \deg g < n \text{ or } g = 0)$$

is a unique isomorphism of rings such that $\tau|_K = 1_K$ and $\tau(u) = v$.

In fact Corollary (1.13) may be used to show that if F is an extension field of K and u,v $\in$ F are algebraic over K, then u and v have the same minimal polynomial over K if and only if there is an isomorphism of rings from K(u) to K(v) which sends u to v and fixes K [cf. exercise (1.3)].

(1.14) EXAMPLES

(1.14.1) Let w be the positive real fourth root of 2. Then w is algebraic over **Q** with minimal polynomial $f = x^4 - 2$ [see Example (1.9.2)]. If v is one of the four roots of f (namely w, -w, wi, -wi), there is an isomorphism of fields

$$\tau: \mathbf{Q}(w) \to \mathbf{Q}(v) \text{ defined by}$$
$$a + bw + cw^2 + dw^3 \mapsto a + bv + cv^2 + dv^3 \quad (a,b,c,d \in \mathbf{Q})$$

The isomorphism τ fixes **Q** and sends w to v.

(1.14.2) The elements $\sqrt{2}$ and $\sqrt{3}$ of **R** are both algebraic over **Q** of degree 2. Hence $\mathbf{Q}(\sqrt{2})$ and $\mathbf{Q}(\sqrt{3})$ are both vector spaces over **Q** of degree 2. The map

$$\tau: \mathbf{Q}(\sqrt{2}) \to \mathbf{Q}(\sqrt{3}) \text{ defined by } a + b\sqrt{2} \mapsto a + b\sqrt{3}$$

is a **Q**-vector space homomorphism but it is not a homomorphism of rings since $\sqrt{2}$ and $\sqrt{3}$ have different minimal polynomials over **Q**. In fact we observe

$$\tau(\sqrt{2}\sqrt{2}) = \tau(2) = 2, \text{ while } \tau(\sqrt{2})\tau(\sqrt{2}) = \sqrt{3}\sqrt{3} = 3$$

Before giving more examples of finitely generated extensions we present a theorem on dimensions. Although this theorem could have been proved in the chapter on vector spaces, we preferred to wait until we could present some examples of its application.

If F is an extension field of K, a field E is said to be an intermediate field of F over K if $K \subseteq E \subseteq F$.

(1.15) THEOREM Let F be an extension field of K and E an intermediate field. Then [F: K] is finite if and only if both [F: E] and [E: K] are finite and in this case

$$[F: K] = [F: E]\,[E: K]$$

In fact, if $\{u_1, \ldots, u_n\}$ is a basis of F over E and $\{v_1, \ldots, v_m\}$ is a basis of E over K then the set $\{u_iv_j: 1 \le i \le n, 1 \le j \le m\}$ is a basis of F over K.

PROOF ($\Longrightarrow$) Suppose that [F: K] is finite. We show first that [E: K] is finite. Since E is an extension field of K, E is a vector space over K. Hence, since $E \subseteq F$, E is a subspace of the vector space F over K. Since [F: K] is finite, it follows that [E: K] is also finite.

If [F: E] is not finite then F has an infinite subset which is linearly independent over E. Such a subset is also linearly independent over K contradicting VII, (3.11).

($\Longleftarrow$) Now suppose that [F: E] and [E: K] are both finite with [F: E] = n and [E: K] = m. Let $\{u_1, \ldots, u_n\}$ be a basis of F over E and $\{v_1, \ldots, v_m\}$ a basis of E over K. We claim that the set B = $\{u_iv_j: 1 \le i \le n, 1 \le j \le m\}$ of mn elements is a basis of F over K and hence [F: K] = mn as desired. We need to prove that (1), B spans F over K and (2), B is linearly independent over K. For ease of notation, let I = $\{1, \ldots, n\}$ and J = $\{1, \ldots, m\}$. Then

$$\sum_i, \quad \sum_j, \quad \sum_{(i,j)}$$

will indicate sums taken over I, J, and I × J respectively.

(1) Let $w \in F$. We wish to show that w can be written as a K-linear combination of elements of B. Since $\{u_i: i \in I\}$ spans F over E, there are elements $b_i \in E$ $(i \in I)$ such that $w = \sum_i b_iu_i$. Then, since $\{v_j: j \in J\}$ spans E over K, for each $i \in I$ there are elements $c_{ij} \in K$ $(j \in J)$ such that $b_i = \sum_j c_{ij}v_j$. Then

$$\begin{aligned} w &= \sum_i b_iu_i \\ &= \sum_i \left(\sum_j c_{ij}v_j\right)u_i \end{aligned}$$

$$= \sum_{(i,j)} c_{ij}(u_i v_j)$$

Hence B spans F over K.

(2) We now show that B is linearly independent over K. Suppose that

$$\sum_{(i,j)} c_{ij}u_i v_j = 0 \text{ with } c_{ij} \in K$$

We need to show that each c_{ij} is 0. Let $b_i = \sum_j c_{ij}v_j$ $(i \in I)$. Then for each $i \in I$, $b_i \in E$. We compute

$$\begin{aligned} 0 &= \sum_{(i,j)} c_{ij}u_i v_j \\ &= \sum_i \left(\sum_j c_{ij}v_j \right) u_i \\ &= \sum_i b_i u_i \end{aligned}$$

Since $\{u_i: i \in I\}$ is linearly independent over E, each of the coefficients b_i $(i \in I)$ must be 0; that is, for each $i \in I$, $\sum_j c_{ij}v_j = 0$. Now, since $\{v_j: j \in J\}$ is linearly independent over K, for each $i \in I$ and $j \in J$, $c_{ij} = 0$. The result now follows.

(1.16) COROLLARY Let F be an algebraic extension field of K and $u_1,\ldots,u_k, v_1,\ldots,v_n \in F$. Then $K(u_1, \ldots, u_k) \subseteq K(v_1, \ldots, v_n)$ if and only if each u_i is a K-linear combination of finite products of the v_i.

PROOF Let

$S = \{w \in F: w \text{ is a finite product of the } u_i \text{ for } 1 \le i \le k\}$

$T = \{w \in F: w \text{ is a finite product of the } v_i \text{ for } 1 \le i \le n\}$

By (1.15) S generates $K(u_1, \ldots, u_k)$ over K and T generates $K(v_1, \ldots, v_n)$ over K. Hence $K(u_1, \ldots, u_k) \subseteq K(v_1, \ldots, v_n)$ if and only if each element of S is a K-linear combination of elements of T. But this occurs if and only if each u_i is a K-linear combination of elements of T.

(1.17) EXAMPLES

(1.17.1) Let $F = \mathbf{Q}(\sqrt{2}, \sqrt{3})$. Then F is a finitely generated extension of **Q**. To find $[F: \mathbf{Q}]$ we introduce the intermediate field $E = \mathbf{Q}(\sqrt{2})$ of F over **Q**.

Since $\sqrt{2}$ is algebraic over **Q** with minimal polynomial $f = x^2 - 2$, $[E: \mathbf{Q}] = [\mathbf{Q}(\sqrt{2}): \mathbf{Q}] = 2$ and the set $B_1 = \{1, \sqrt{2}\}$ is a basis of E over **Q**.

Now $\sqrt{3}$ is algebraic <u>over E</u> with minimal polynomial $g = x^2 - 3$ (note that it must be verified that g is irreducible over E, or equivalently, that $\sqrt{3} \notin E$). Hence $[F: E] = [E(\sqrt{3}): E] = 2$ and the set $B_2 = \{1, \sqrt{3}\}$ is a basis of F over E.

It follows by (1.15) that

$$[F: \mathbf{Q}] = 4 \text{ and the set } B = \{1, \sqrt{2}, \sqrt{3}, \sqrt{6}\}$$

is a basis of F over **Q**. Thus every element w of F can be written uniquely as

$$w = a + b\sqrt{2} + c\sqrt{3} + d\sqrt{6} \quad (a,b,c,d \in \mathbf{Q})$$

(1.17.2) Let $F = \mathbf{Q}(w, i)$ where w is the positive real fourth root of 2. As above, to find the dimension of F over **Q** we introduce an intermediate field.

Let $E = \mathbf{Q}(w)$. Then by Example (1.9.2), $[E: \mathbf{Q}] = 4$ and the set $B_1 = \{1, w, w^2, w^3\}$ is a basis of E over **Q**.

Now i satisfies the monic irreducible polynomial $g = x^2 + 1$ in $E[x]$ (observe that the roots of g are complex and $E \subseteq \mathbf{R}$). Hence i is algebraic over E with minimal polynomial g. Since the degree of g is 2, $[F: E] = [E(i): E] = 2$ and the set $B_2 = \{1, i\}$ is a basis of F over E.

It follows that

$$[F: \mathbf{Q}] = 8 \text{ and } B = \{1, w, w^2, w^3, i, iw, iw^2, iw^3\}$$

is a basis of F over **Q**.

As an illustration of Corollary (1.16) we remark that $\mathbf{Q}(w, i) = \mathbf{Q}(w, wi)$ [we need only observe that $i = (1/2)w^3(wi)$]. The reader should also find $[F: \mathbf{Q}]$ as $[E(wi): E]\,[E: \mathbf{Q}]$. Observe that although wi has minimal polynomial f over $\mathbf{Q}$, f is not irreducible over E; in fact the minimal polynomial of wi over $\mathbf{Q}(w)$ must have degree 2 and divide f in $E[x]$. Using these facts we see that wi has minimal polynomial $h = x^2 + w^2$ over E.

(1.17.3) Let $F = \mathbf{Q}(w, \rho)$ where $w = \sqrt[3]{2}$ and ρ is a primitive cube root of unity. Since w has minimal polynomial $f = x^3 - 2$ over $\mathbf{Q}$, $[\mathbf{Q}(w): \mathbf{Q}] = 3$ and $3 \mid [F: \mathbf{Q}]$. Now

$$[F: \mathbf{Q}] = [F: \mathbf{Q}(w)]\,[\mathbf{Q}(w): \mathbf{Q}]$$

We observe that ρ has minimal polynomial $g = x^2 + x + 1$ over $\mathbf{Q}$ so that $[\mathbf{Q}(\rho): \mathbf{Q}] = 2$. Hence $2 \mid [F: \mathbf{Q}]$. To prove directly that g is irreducible over $\mathbf{Q}(w)$ would be difficult. However we do know that since ρ satisfies $g \in \mathbf{Q}(w)[x]$, $[F: \mathbf{Q}(w)] \le 2$. Thus $[F: \mathbf{Q}] \le (2)(3) = 6$. But since 2 and 3 both divide $[F: \mathbf{Q}]$ and $(2, 3) = 1$, $6 \mid [F: \mathbf{Q}]$; thus we conclude that $[F: \mathbf{Q}] = 6$. The set $B = \{1, w, w^2, \rho, \rho w, \rho w^2\}$ is a basis for F over $\mathbf{Q}$. This example is generalized in exercise (1.26).

EXERCISES

(1.1) Let F be an extension field of K and $u,v \in F$. Show that $(K(u))(v) = (K(v))(u)$.

(1.2) Let F be an extension field of K and $u,v \in F$ with u and v both transcendental over K. Show that there exists a unique isomorphism of rings from $K(u)$ to $K(v)$ which sends u to v and fixes K.

(1.3) Show that if F is an extension field of K and $u,v \in F$ are algebraic over K, then u and v have the same minimal polynomial over K <u>if and only if</u> there exists an isomorphism of rings from $K(u)$ to $K(v)$ which sends u to v and fixes K.

(1.4) Let E be an extension field of K and F an extension field of L. Show that if there are isomorphisms $\alpha: K \to L$ and $\tau: E \to F$ such that τ is an extension of α, then $[E: K] = [F: L]$ (Hint: show that if $B = \{u_i: 1 \le i \le n\}$ is a basis for E over K then $B_1 = \{T(u_i): 1 \le i \le n\}$ is a basis for F over L).

(1.5) (a) Is there an isomorphism of rings from $\mathbf{Q}(\sqrt{8})$ to $\mathbf{Q}(\sqrt{2})$ which fixes $\mathbf{Q}$ and sends $\sqrt{8}$ to $\sqrt{2}$?

(b) Is there a $\mathbf{Q}$-vector space isomorphism from $\mathbf{Q}(\sqrt{8})$ to $\mathbf{Q}(\sqrt{2})$ which fixes $\mathbf{Q}$ and sends $\sqrt{8}$ to $\sqrt{2}$?

(1.6) Let F be an extension field of K and $a, b \in K$ with $a \ne 0$. Show that if $u \in F$ then $K(au + b) = K(u)$.

(1.7) (a) Let $w = \sqrt[4]{2}$, $S = \{w, -w, iw, -iw\}$ and $T = \{i, -i\}$. Use the preceding exercise to show that if $u \in S$ and $v \in T$, then $\mathbf{Q}(u, v) = \mathbf{Q}(i, w)$.

(b) Show that the isomorphism $T: \mathbf{Q}(w) \to \mathbf{Q}(wi)$ defined by $w \mapsto wi$ can be extended uniquely to an isomorphism $T': \mathbf{Q}(w, i) \to \mathbf{Q}(w, i)$. What is the image of $w^4 - 5iw^2$ under this isomorphism?

(1.8) Let $u = \sqrt[3]{2}$ and $w = (-1 + \sqrt{3}i)/2$.

(a) Show that there is an isomorphism from $\mathbf{Q}(u)$ to $\mathbf{Q}(uw)$ which fixes $\mathbf{Q}$ and sends u to uw.

(b) Is $\mathbf{Q}(u) = \mathbf{Q}(uw)$?

(c) Is $\mathbf{Q}(u, w) = \mathbf{Q}(u, \sqrt{3}i)$?

(1.9) Suppose that F is an extension field of K and $u \in F$.

(a) Show that $K(u^2) \subseteq K(u)$.

(b) Show that if u is algebraic of odd degree over K, then $K(u^2) = K(u)$.

(1.10) In each case decide which of the given field extensions are equal.

(a) $\mathbf{Q}(\sqrt[3]{2})$, $\mathbf{Q}(\sqrt[3]{4})$, $\mathbf{Q}(\sqrt[3]{\sqrt{108}})$, $\mathbf{Q}(\sqrt{1 - \sqrt[3]{2}})$.

(b) $\mathbf{Q}(\sqrt{2})$, $\mathbf{Q}((1 + \sqrt{2})/3)$, $\mathbf{Q}(\sqrt{2}i)$, $\mathbf{Q}(\sqrt{2}, i)$.

(1.11) Let $u = \sqrt{1 - \sqrt[3]{2}}$.

(a) Show that $\mathbf{Q} \subseteq \mathbf{Q}(\sqrt[3]{2}) \subseteq \mathbf{Q}(u)$.

(b) Show that $[\mathbf{Q}(\sqrt[3]{2}): \mathbf{Q}] = 3$.

(c) Show that $[\mathbf{Q}(u): \mathbf{Q}(\sqrt[3]{2})] = 2$.

(d) Conclude that $[\mathbf{Q}(u): \mathbf{Q}] = 6$ and find the minimal polynomial of u over **Q**.

(1.12) Find the minimal polynomial of $\sqrt{2 + \sqrt{2}i}$ over each of the following fields.

(a) **R**.

(b) **Q**.

(c) $\mathbf{Q}(\sqrt{2})$.

(d) $\mathbf{Q}(i)$.

(1.13) Show that if F is algebraic over K and D is an integral domain such that $K \subseteq D \subseteq F$, then D is a field.

(1.14) Let $F = K(x)$ and $u = x^3/(x + 1)$. Show that x is algebraic over $K(u)$.

(1.15) Let $w = \sqrt[4]{2}$ and $F = \mathbf{Q}(w, i)$. Show that in fact F is the simple extension $F = \mathbf{Q}(i + w)$.

(1.16) Let F be an extension field of K, and $u,v \in F$ with v algebraic over $K(u)$ and v transcendental over K. Show that u is algebraic over $K(v)$.

(1.17) Let $a,b \in \mathbf{Q}$ such that $\sqrt{a}, \sqrt{b} \notin \mathbf{Q}$. Show that $\mathbf{Q}(\sqrt{a}) = \mathbf{Q}(\sqrt{b})$ if and only if $b = c^2a$ for some $c \in \mathbf{Q}$.

(1.18) Give examples of $a,b \in \mathbf{Q}$ such that $u = \sqrt{a + \sqrt{b}}$ has each of the following degrees over **Q**: 1, 2, 4.

(1.19) Give examples of $a,b \in \mathbf{Q}$ such that $u = \sqrt[3]{a + \sqrt{b}}$ has each of the following degrees over **Q**: 1, 3, 6.

(1.20) Find the minimal polynomial of each of the following values of u over **Q**.

(a) $\sqrt{1 + \sqrt{2 + \sqrt{3}}}$.

(b) $\sqrt[3]{5 + \sqrt[3]{4}}$.

(1.21) Let $u = \sqrt[3]{1 + \sqrt{3}}$.

(a) Show that $[\mathbf{Q}(u): \mathbf{Q}] = 6$.

(b) Show that $[\mathbf{Q}(\sqrt{3}): \mathbf{Q}] = 2$.

(c) Show that $\mathbf{Q}(u) = \mathbf{Q}(\sqrt{3}, u)$.

(d) Deduce that $[\mathbf{Q}(u): \mathbf{Q}(\sqrt{3})] = 3$ and find the minimal polynomial of u over $\mathbf{Q}(\sqrt{3})$.

(1.22) Suppose that $[F: K] = p$ where p is prime.

(a) Show that the only intermediate fields of F over K are F and K themselves.

(b) Show that F is a simple extension of K.

(1.23) Let F be a finite extension of K and $f \in K[x]$ irreducible of degree n. Show that if n and $[F: K]$ are relatively prime then f has no zeros in F.

(1.24) Suppose that $F = K(u, v)$ where u has minimal polynomial g over K of degree m, v has minimal polynomial h over K of degree n and $(n, m) = 1$. Show that $[F: K] = mn$ and g is also the minimal polynomial of u over $K(v)$.

(1.25) (a) Show that if ρ is a primitive 4-th root of unity, then $\mathbf{Q}(\rho) = \mathbf{Q}(i)$. What is the degree of ρ over $\mathbf{Q}$? What is the minimal polynomial of ρ over $\mathbf{Q}$?

(b) Show that if ρ is a primitive 8-th root of unity, then $\mathbf{Q}(\rho) = \mathbf{Q}(\sqrt{2}, i)$. What is the degree of ρ over $\mathbf{Q}$? Find the minimal polynomial of ρ over $\mathbf{Q}$.

(1.26) Let p be prime and ρ a primitive p-th root of unity [cf. Example (1.17.3)].

(a) Show that $[\mathbf{Q}(\rho): \mathbf{Q}] = p - 1$.

(b) Let $w = \sqrt[p]{2}$. Show that $[\mathbf{Q}(w): \mathbf{Q}] = p$.

(c) Show that $[\mathbf{Q}(w, \rho): \mathbf{Q}] = p(p - 1)$ and that w has minimal polynomial $x^p - 2$ over $\mathbf{Q}(\rho)$.

(1.27) (a) Let $n \in \mathbf{N}$ and ρ be a primitive n-th root of unity. Show that if $a \in \mathbf{Q}^+$ and $w = \sqrt[n]{a}$, then the roots of $x^n - a$ are $w, w\rho., \ldots, w\rho^{n-1}$.

(b) Find the roots of $x^6 - 2$.

(c) Let ρ be a primitive 6-th root of unity and $w = \sqrt[6]{2}$. Show that $[\mathbf{Q}(w, \rho): \mathbf{Q}] = 12$.

(1.28) Let $w = \sqrt[3]{3}$.

(a) Show that for each element u of $\mathbf{Q}(w)$, $u = a + bw + cw^2$ for unique $a,b,c \in \mathbf{Q}$.

(b) Find a,b,c for each of the following values of u: w^4, $1/w$, $(w - 2)/(w^2 + 1)$.

(1.29) Prove that $\mathbf{Q}(\sqrt{a}, \sqrt{b}) = \mathbf{Q}(\sqrt{a} + \sqrt{b})$ for all $a,b \in \mathbf{Q}$.

2. ALGEBRAIC EXTENSIONS

In this section we discuss the relation between finite dimensional and finitely generated extensions. We also show that if K is a field and $f \in K[x]$ then there is a simple extension field K(u) of K such that u is a root of f.

(2.1) THEOREM Let F be an extension field of K and $u \in F$. Then u is algebraic over K if and only if $[K(u): K]$ is finite.

PROOF By (1.8) if u is algebraic over K with minimal polynomial f of degree n, then $[K(u): K] = n$.

Now suppose that $[K(u): K]$ is finite; say $[K(u): K] = n$. Then since $\{1, u, u^2, \ldots, u^n\}$ is a subset of K(u) with n + 1 elements it must be linearly dependent over K. Hence there are elements $a_i \in K$ $(0 \le i \le n)$, not all zero, such that $a_0 + a_1u + \cdots + a_nu^n = 0$. Then u satisfies the nonzero polynomial $f = a_0 + a_1x + \cdots + a_nx^n \in K[x]$ and hence u is algebraic over K.

(2.2) THEOREM Let F be an extension field of K. Then the following statements are equivalent.

(i) F is finite dimensional over K.

(ii) F is finitely generated and algebraic over K.

(iii) There are elements $u_1,\ldots,u_n \in F$ with u_i algebraic over F $(i = 1, \ldots, n)$ and $F = K(u_1, \ldots, u_n)$.

PROOF (i) $\Rightarrow$ (ii) Suppose that $[F: K]$ is finite. We first show that F is algebraic over K. Let $u \in F$. Then, since $K \subseteq K(u) \subseteq F$, and $[F: K]$ is finite, $[K(u): K]$ is finite by (1.15). It follows by (2.1) that u is algebraic over K. Hence F is an algebraic extension of K.

Now let $\{v_1, \ldots, v_n\}$ be a basis of F over K. Then every element of F is a K-linear combination of the v_i $(i = 1, \ldots, n)$. Hence $F \subseteq K(v_1, \ldots, v_n) \subseteq F$ so that $F = K(v_1, \ldots, v_n)$. Thus F is finitely generated over K.

(ii)$\Longrightarrow$(iii) Follows by definition.

(iii)$\Longrightarrow$(i) Suppose that $F = K(u_1, \ldots, u_n)$ with each u_i algebraic over K. Let $F_0 = K$ and for each $i = 1, \ldots, n$, let

$$F_i = K(u_1, \ldots, u_i) = F_{i-1}(u_i)$$

Then for each i, u_i is algebraic over F_{i-1} (since it is algebraic over K and $K \subseteq F_{i-1}$) and hence by (2.1) $[F_i: F_{i-1}]$ is finite. Consider the following tower of extensions.

$$K = F_0 \subseteq F_1 \subseteq \cdots \subseteq F_{n-1} \subseteq F_n = F$$

Since each $[F_i: F_{i-1}]$ is finite, by (1.15), applied iteratively, $[F: K]$ is also finite. Hence F is a finite dimensional extension of K.

The next theorem shows that an algebraic extension of an algebraic extension is algebraic. The knowledge we now have of the relation between algebraic and finite dimensional extensions helps us to avoid the computations necessary in a proof using only the definition of algebraic.

(2.3) THEOREM Let F be an algebraic extension of E and E an algebraic extension of K. Then F is an algebraic extension of K.

PROOF Let $u \in F$. We wish to show that u is algebraic over K.

Since u is algebraic over E, there is a polynomial $f = a_0 + a_1x + \cdots + a_nx^n$ $(a_i \in E)$ such that $f(u) = 0$. Let $L = K(a_0, \ldots, a_n)$ and consider the following tower of extensions

$$K \subseteq L \subseteq L(u)$$

Since E is algebraic over K and $a_i \in E$ $(0 \le i \le n)$, by(2.2) [L: K] is finite.

Since u satisfies the nonzero polynomial $f \in L[x]$, u is algebraic over L. Hence by (2.1) [L(u): L] is finite.

It follows by (1.15) that [L(u): K] is finite. Hence by (2.2) u is algebraic over K as required.

(2.4) THEOREM Let F be an extension field of K and let

$$E = \{u \in F: u \text{ is algebraic over } K\}$$

Then E is an intermediate field of F over K which is algebraic over K.

PROOF If $k \in K$ then k satisfies the nonzero polynomial $x - k \in K[x]$. Hence $K \subseteq E$. To show that E is a subfield of F it suffices to prove that whenever $u, v \in E$, $u - v \in E$ and, provided $v \neq 0$, $u/v \in E$. Again, rather than working directly with the definition of algebraic we will appeal to (2.2).

Let $u, v \in E$. By (2.2), K(u, v) is an algebraic extension of K. Thus, since $u - v$ and u/v (provided $v \neq 0$) are elements of the field K(u, v), they are algebraic over K. Hence $u - v, u/v \in E$. The result now follows.

(2.5) EXAMPLES

(2.5.1) Since $\mathbf{C} = \mathbf{R}(i)$ and i is algebraic over **C** by (2.2), (ii), **C** is an algebraic extension of **R**. Thus every complex number satisfies a nonzero polynomial with coefficients in **R**. In particular the quadratic formula may be used to verify that if $a + bi \in \mathbf{C}$ $(a, b \in \mathbf{R})$ then $a + bi$ satisfies the nonzero polynomial $f = x^2 - 2ax + (a^2 + b^2)$ in $\mathbf{R}[x]$.

(2.5.2) Let $K = \{w \in \mathbf{C}: w \text{ is algebraic over } \mathbf{Q}\}$. Then by (2.4) K is an intermediate field of **C** over **Q** which is algebraic over **Q**. K is called the field of <u>algebraic numbers</u>. One may show that K is an infinite dimensional extension of **Q** by using Eisenstein's criterion to find irreducible polynomials in $\mathbf{Q}[x]$ of arbitrarily large degree [cf. exercise (2.12)].

Until now we have always assumed that we are working within a given field extension F of a field K - thus to adjoin an element u to K we assume that $K \subseteq F$ and $u \in F$. We then determine if u is algebraic over K and, if it is, we find the minimal polynomial $f \in K[x]$ of u; in this case, u is a root of the polynomial f.

In the next theorem we start with a field K and a polynomial $f \in K[x]$. We then find a simple extension field K(u) of K in which f has the root u without working inside a larger field F. In this case u is algebraic over K and the minimal polynomial of u over K divides f [since f(u) = 0; cf. (1.7)].

(2.6) THEOREM Let K be a field and $f \in K[x]$ a polynomial of degree greater than 0. Then there exists a simple extension field F = K(u) of K such that u is a root of the polynomial f.

PROOF Let g be a monic irreducible factor of f in K[x] and let

$$F = K[x]/I \quad \text{where} \quad I = (g)$$

Since g is irreducible, I is a maximal ideal by V, (3.8). Hence by IV, (3.3) F is a field.

Define a map

$$\alpha: K \to F \quad \text{by} \quad k \mapsto [k] = I + k \ (k \in K)$$

Then α is a ring homomorphism. If $h \in K[x]$, then $h \in I$ if and only if $g|h$. Hence, since deg $g \geq 1$, ker $\alpha = (0)$ so that α is an injective map. Thus F, together with the injective homomorphism of rings $\alpha: K \to F$, is an extension field of K.

Let $u = [x] = I + x \in F$. We claim that u is algebraic over K with minimal polynomial αg. [For purposes of clarity we temporarily continue to use the map α rather than identifying elements k of K with their image $\alpha(k) = [k]$ in F]. We wish to show that $\alpha g(u) = 0$. Suppose that $g = \sum_{i=0}^{n} a_i x^i$ with $a_i \in K$.

We calculate

$$\begin{aligned} \alpha g(u) &= \sum_{i=0}^{n} \alpha(a_i)u^i \\ &= \sum_{i=0}^{n} [a_i][x]^i \end{aligned}$$

$$= [\sum_{i=0}^{n} a_i x^i]$$
$$= [g]$$
$$= [0] \qquad \text{(since } g \in I)$$

It follows that the irreducible monic polynomial $g \in K[x]$ is the minimal polynomial of u [cf. (1.7)]. We now suppress the map α and identify $k \in K$ with its image $\alpha(k) = [k]$ in F. Under this identification $g(u) = 0$. Since g is a factor of the polynomial f, it is also true that $f(u) = 0$ so that u is a root of f.

Since F is an extension field of K and $u \in F$ is algebraic over K, the simple extension K(u) of K is now defined as a subfield of F; that is,

$$K(u) = \{\alpha h(u): h \in K[x]\}$$

We claim that in fact $K(u) = F$. Let $w \in F$. Then $w = [h]$ for some $h = \sum_{i=0}^{m} a_i x^i$ ($a_i \in K$, $m \in \mathbf{N} \cup \{0\}$) - hence

$$w = [\sum_{i=0}^{m} a_i x^i]$$
$$= \sum_{i=0}^{m} [a_i]\,[x]^i$$
$$= \sum_{i=0}^{m} \alpha(a_i) u^i$$
$$= \alpha h(u)$$

It follows that $w \in K(u)$ and $K(u) = F$ as desired.

(2.7) EXAMPLES [Illustrating (2.6)]

(2.7.1) We construct a simple extension $F = F_2(u)$ of F_2 (where F_2 is the field $\mathbf{Z}_2$) in which the irreducible monic polynomial $f = x^2 + x + 1 \in F_2[x]$ has the root u. (For ease of notation we write 0, 1 in place of [0], [1] for elements of F_2.)

Let $F = F_2[x]/I$ where $I = (f)$ and let $u = [x]$. Then by (2.6) $F = F_2(u)$ and u satisfies the polynomial $f \in F_2[x]$. Hence, since f is irreducible and monic, u is algebraic over F_2 with minimal polynomial f.

By (1.8), $[F: F_2] = \deg f = 2$ and the set $B = \{1, u\}$ is a basis for F over F_2. Then

$$F = \{a + bu: a,b \in F_2\} = \{0, 1, u, 1 + u\}$$

Note that F is a field with 4 elements. We use the fact that $f(u) = 0$ (so that $u^2 = -u - 1 = u + 1$) to construct Cayley tables for addition and multiplication in the field F.

+	0	1	u	1+u
0	0	1	u	1+u
1	1	0	1+u	u
u	u	1+u	0	1
1+u	1+u	u	1	0

•	0	1	u	1+u
0	0	0	0	0
1	0	1	u	1+u
u	0	u	1+u	1
1+u	0	1+u	1	u

For example,

$$\begin{aligned}(1+u)(1+u) &= 1 + u + u + u^2 \\ &= 1 + u^2 && [\text{since } u + u = (1 + 1)u = 0u = 0] \\ &= 1 + u + 1 && (\text{since } u^2 = u + 1) \\ &= u && (\text{since } 1 + 1 = 0)\end{aligned}$$

(2.7.2) Let $f = x^2 + x + 2 \in F_3[x]$. Then f is monic and irreducible in $F_3[x]$. Let

$$F = F_3[x]/I \text{ where } I = (f)$$

and let $u = [x]$. Then by (2.6) $F = F_3(u)$ and u satisfies the polynomial f. Hence u is algebraic over F_3 with minimal polynomial f, $[F: F_3] = 2$ and $\{1, u\}$ is a basis for F over F_3.

Then $F = \{a + bu: a,b \in F_3\}$ and, since F_3 contains 3 elements, the field F contains $3^2 = 9$ elements. The reader is urged to construct Cayley tables for addition and multiplication in the field F [see exercise (2.5)].

It will be shown later that in fact every finite field has order p^n for some prime p and $n \in \mathbf{N}$.

(2.8) DEFINITION A field K is said to be algebraically closed if it has no proper algebraic extensions.

By (2.6) K is algebraically closed if and only if every polynomial $f \in K[x]$ factors completely in $K[x]$ [cf. exercise (2.1)]. We will use the material developed in Chapter IX, Sections 1 and 2 to prove the following theorem.

THEOREM (The Fundamental Theorem of Algebra) The field **C** of complex numbers is algebraically closed.

We point out that although we may appeal to the preceding theorem in our examples, we do not use it to develop any material prior to its proof.

EXERCISES

(2.1) Let K be a field. Show that the following statements are equivalent.

(a) K is algebraically closed.

(b) Whenever $f \in K[x]$, f has a root in K.

(c) Whenever $f \in K[x]$, f factors into a product of linear factors in $K[x]$.

(2.2) Show that no finite field is algebraically closed. [Hint: Suppose that $F = \{u_1, \ldots, u_n\}$ with $u_1 \neq 0$ and consider $f(x) = u_1 - (x - u_1)(x - u_2) \cdots (x - u_n)$.]

(2.3) Let E be an intermediate field of F over K with E algebraic over K. Show that if $v \in F$ is transcendental over K then v is also transcendental over E.

(2.4) Let $F = F_2(u)$ where u is a root of the irreducible polynomial $f = x^2 + x + 1 \in F_2[x]$ [cf. Example (2.7.1)]. Find $v \in F$ such that $f = (x - u)(x - v)$.

(2.5) (a) Construct Cayley tables for addition and multiplication for the field $F = F_3(u)$ where u is a root of f =

$x^2 + x + 2$ [cf. Example (2.7.2)].

(b) Show that $F^\times$ is a cyclic group.

(c) Show that the polynomial f has two distinct roots in F.

(2.6) Let $E = F_3(v)$ where v is a root of $x^2 + 1$.

(a) Show that E is a field with 9 elements.

(b) Show that $E^\times$ is a cyclic group.

(2.7) Construct a field of order 27.

(2.8) (a) Show that $f = x^4 + x + 1 \in F_2[x]$ is irreducible. (Be careful! Besides showing that f has no roots in F_2, and hence has no linear factors in $F_2[x]$, it must also be shown that f does not factor as a product of two quadratics in $F_2[x]$.)

(b) Let u be a root of f and $F = F_2(u)$. Show that F is a field with 16 elements.

(c) Find a polynomial $g \in F[x]$ such that $f = (x - u)g$.

(2.9) Construct a field of order 81.

(2.10) Let $f = x^3 + x + 1 \in F_2[x]$, u a root of f and $F = F_2(u)$.

(a) Make Cayley tables for addition and multiplication for the field F.

(b) Find a polynomial $g \in F[x]$ such that $f = (x - u)g$.

(c) Show that in fact $f = (x - u)(x - v)(x - w)$ for distinct $u,v,w \in F$.

(2.11) (a) Show that the polynomial $f = x^4 - 4x^3 + 2$ is irreducible over **Q**.

(b) Let u be a root of f in **C** (recall that **C** is algebraically closed and hence f has a root in **C**). Find a basis for **Q**(u) over **Q** and express each of the following elements of **Q**(u) as a **Q**-linear combination of these basis elements: u^5, u^{-1}, $(u + 3)/(u - 2)$.

(2.12) Use Eisenstein's Criterion to find irreducible polynomials in **Q**[x] of arbitrarily large degree. Deduce that the field K of algebraic numbers is an infinite dimensional algebraic extension of **Q** [cf. (2.5.2)].

(2.13) An algebraic number u is said to be an algebraic integer if it satisfies a monic polynomial in $\mathbf{Z}[x]$. Let K be the field of algebraic numbers and let L be the set of algebraic integers.

(a) Show that if $u \in K$ then there is an $n \in \mathbf{N}$ such that $nu \in L$.

(b) Prove that $\mathbf{Q} \cap L = \mathbf{Z}$.

(c) Prove that if $u \in L$ and $m \in \mathbf{Z}$, then $u + m$ and $mu \in L$.

(d) Prove that L is a subring of K.

(2.14) (a) Prove that $\sin(1^\circ)$ is an algebraic number.

(b) Prove that $\sin(m^\circ)$ is an algebraic number for any $m \in \mathbf{Z}$.

(2.15) Let $L = \{u \in \mathbf{R}\colon u \text{ is an algebraic number}\}$. Prove that L is a subfield of $\mathbf{R}$.

(2.16) Let $u = a + bi$ with $a,b \in \mathbf{R}$.

(a) If u is an algebraic number are a and b necessarily algebraic numbers?

(b) If u is an algebraic integer are a and b necessarily algebraic integers?

CONSTRUCTIBILITY

The following exercises have to do with the classical problem of geometric constructions using a straightedge and compass alone. We will assume that we are given a unit length. The student should recall from high school geometry that we can construct with straightedge and compass a line perpendicular to and a line parallel to a given line through a given point. A real number c is said to be constructible if the point $(c, 0)$ can be constructed by using straightedge and compass alone.

(2.17) Prove each of the following facts.

(a) The set of constructible numbers is a subfield of $\mathbf{R}$ which contains $\mathbf{Q}$.

(b) A point (c, d) in the plane may be constructed using straightedge and compass alone if and only if both c and d are constructible numbers.

(c) If a nonnegative real number c is constructible, so is $\sqrt{c}$.

If F is a subfield of **R**, we define the <u>plane of F</u> to be the set

plane of F = $\{(c, d): c, d \in F\}$

If P and Q are in the plane of F, the unique line through P and Q is called a <u>line in F</u> and the circle with center P and radius the line segment PQ is called a <u>circle in F</u>.

(2.18) Let F be a subfield of **R**. Prove each of the following facts.

(a) If L is a line in F, then L has equation $ax + by + c = 0$ for some $a, b, c \in F$.

(b) If C is a circle in F, then C has equation $x^2 + y^2 + ax + by + c = 0$ for some $a, b, c \in F$.

(2.19) Let F be a subfield of **R**. Prove each of the following facts.

(a) If L_1 and L_2 are nonparallel lines in F, then $L_1 \cap L_2$ is a point in the plane of F.

(b) If L is a line in F and C is a circle in F and $L \cap C$ is not empty then $L \cap C$ consists of one or two points in the plane of $F(\sqrt{u})$ for some $u \in F$ with $u \geq 0$.

(c) If C_1 and C_2 are distinct circles in F and $C_1 \cap C_2$ is not empty, then $C_1 \cap C_2$ consists of one or two points in the plane of $F(\sqrt{u})$ for some $u \in F$ with $u \geq 0$.

(2.20) Let $c \in \mathbf{R}$. Prove that c is constructible if and only if there are $a_1, a_2, \ldots, a_n \in \mathbf{R}$ such that $c \in \mathbf{Q}(a_1, \ldots, a_n)$, $[\mathbf{Q}(a_1): \mathbf{Q}] \leq 2$ and $[\mathbf{Q}(a_1, \ldots, a_i): \mathbf{Q}(a_1, \ldots, a_{i-1})] \leq 2$ for $i = 2, \ldots, n$. Conclude that if c is constructible, then $[\mathbf{Q}(c): \mathbf{Q}] = 2^n$ for some nonnegative integer n.

(2.21) Show that an angle of 60° cannot be trisected by straightedge and compass constructions. [Hint: Show that since $\cos(3\alpha) = 4\cos^2(\alpha) - 3\cos(\alpha)$ for all α, $\alpha = 20^\circ$ satisfies the irreducible polynomial $f = 8x^3 - 6x - 1$ in $\mathbf{Q}[x]$.]

(2.22) Show that a cube of length 1 cannot be duplicated using straightedge and compass alone; that is, the side of a cube of volume 2 cannot be constructed.

FACTORIZATION IN COMPLEX QUADRATIC NUMBER FIELDS

The authors are indebted to Professor Hyman Bass of Columbia University for these exercises, which are adapted from his Number Theory course given during the 1972-73 academic year.

Throughout, d denotes a square-free integer; i.e., d is neither 0 nor 1 and d has no square factors. We write Δ for $\sqrt{d}$.

(2.23) Show that the extension $\mathbf{Q}(\Delta)$ of $\mathbf{Q}$ obtained by adjoining Δ to $\mathbf{Q}$ is given by

$$\mathbf{Q}(\Delta) = \{a + b\Delta\colon a,b \in \mathbf{Q}\}$$

and that $[\mathbf{Q}(\Delta)\colon \mathbf{Q}] = 2$.
An extension of $\mathbf{Q}$ of degree 2 is called a <u>quadratic number field</u>. (More generally, a finite extension of $\mathbf{Q}$ of arbitrary degree is called an <u>algebraic number field</u>.)

(2.24) Show that every quadratic number field takes the form $\mathbf{Q}(\Delta)$ for some Δ, the square root of a square-free integer, d. Show moreover that $\mathbf{Q}(\Delta) \subseteq \mathbf{R}$ if and only if $d > 0$. [In this case, $\mathbf{Q}(\Delta)$ is called a <u>real quadratic number field</u>; otherwise it is called a <u>complex quadratic number field</u>.]

(2.25) Define the map $\tau\colon \mathbf{Q}(\Delta) \to \mathbf{Q}(\Delta)$ by

$$\tau(a + b\Delta) = a - b\Delta$$

Note that in the complex quadratic case, τ is simply complex conjugation.

(a) Show that $\tau \circ \tau = 1_{\mathbf{Q}(\Delta)}$. Conclude that τ is bijective.

(b) Show that τ is in fact an automorphism of $\mathbf{Q}(\Delta)$ which restricts to the identity map on $\mathbf{Q}$.

(2.25) Let $u = a + b\Delta \in \mathbf{Q}(\Delta)$. Define the norm and trace of u by

$N(u) = u\tau(u)$

$tr(u) = u + \tau(u)$

Show that both the norm and the trace take values in $\mathbf{Q}$.

(2.26) Let $u, v \in \mathbf{Q}(\Delta)$. Show that the following assertions hold.

(a) $N(uv) = N(u)N(v)$.

(b) $N(u) = 0$ if and only if $u = 0$.

Conclude that the norm restricts to a homomorphism of groups $\mathbf{Q}(\Delta)^* \to \mathbf{Q}^*$.

(2.27) Let $u, v \in \mathbf{Q}(\Delta)$. Show that the following assertions hold.

(a) $tr(u + v) = tr(u) + tr(v)$.

(b) $tr(ku) = k\,tr(u)$ for all $k \in \mathbf{Q}$.

Conclude that the trace map is a homomorphism of $\mathbf{Q}$-vector spaces.

(2.28) Let $u \in \mathbf{Q}(\Delta)^*$. Show that $u^{-1} = \tau(u)/N(u)$.

An element u of $\mathbf{Q}(\Delta)$ is called an <u>algebraic integer</u> if both N(u) and tr(u) are in $\mathbf{Z}$. The set of all algebraic integers in $\mathbf{Q}(\Delta)$ is denoted A(d).

(2.29) Show that A(d) is a subring of $\mathbf{Q}(\Delta)$ containing $\mathbf{Z}[\Delta] = \mathbf{Z} \oplus \mathbf{Z}\Delta$. Show further that $\mathbf{Q} \cap A(d) = \mathbf{Z}$.

(2.30) Show that $u \in \mathbf{Q}(\Delta)$ is an algebraic integer as defined above if and only if u satisfies a monic polynomial in $\mathbf{Z}[x]$. Thus the notion of an algebraic integer given here is consistent with that given in exercise (2.13). [Hint: consider the product $(x - u)[x - \tau(u)]$.]

(2.31) Prove each of the following assertions.

(a) $A(d) = \mathbf{Z} \oplus \mathbf{Z}\Delta$ if $d \not\equiv 1 \pmod 4$.

(b) $A(d) = \mathbf{Z} \oplus \mathbf{Z}(1 + \Delta)/2$ if $d \equiv 1 \pmod 4$.

Illustrate both alternatives of the theorem in the complex case $(d < 0)$. [That is, show A(d) is a lattice of points in the complex plane.]

(2.32) Prove the following lemma. Suppose for every $u \in \mathbf{Q}(\Delta)$ there exists a $v \in A(d)$ such that $|N(u - v)| < 1$. Then A(d) is euclidean with respect to $|N|$.

Conclude that if $\mathbf{Q}(\Delta)$ is a complex quadratic number field, then $A(d)$ is euclidean with respect to $|N| = N$ if

(*) the open unit balls around the points of $A(d)$ cover $\mathbf{Q}(\Delta)$

Illustrate this as in exercise (2.31). (If $z \in \mathbf{C}$, the open unit ball around z is the set of all complex numbers whose distance from z is less than 1.)

(2.33) Suppose $d < 0$ and $d \not\equiv 1 \pmod 4$. Show that $A(d)$ satisfies (*) if and only if $d = -1,-2$. (Hint: use the pictures.)

(2.34) Suppose $d < 0$ and $d \equiv 1 \pmod 4$. Show that $A(d)$ satisfies (*) if and only if $d = -3,-7,-11$.

(2.35) Prove the following theorem. $A(d)$ is euclidean with respect to the norm map N for $d = -1,-2,-3,-7,-11$. Hence $A(d)$ is a principal ideal domain and a unique factorization domain for each of these values.

(2.36) Show that $A(-5)$ is not a unique factorization domain. [Hint: construct two essentially different prime factorizations in $A(-5)$ for the integer 6.]

3. SPLITTING FIELDS, NORMAL EXTENSIONS, AND FINITE FIELDS

In this section we show that if K is a field and $f \in K[x]$ then there is a minimal extension F of K which contains all the roots of f (such a field F will be called a splitting field over K of f). We then show that F is a finite field of characteristic p if and only if F is a splitting field over F_p of a polynomial $f \in F_p[x]$.

(3.1) DEFINITION Let F be an extension field over K and $f \in K[x]$. The polynomial f <u>splits</u> over F if f factors over F into a product of linear factors (not necessarily distinct). Thus f splits over F if and only if there are elements $c, u_1, u_2, \ldots, u_n \in F$ such that

$$f = c(x - u_1)(x - u_2) \cdots (x - u_n)$$

In this case the elements $u_1, u_2, \ldots, u_n \in F$ are roots of the polynomial f.

(3.2) DEFINITION An extension field F of K is a splitting field over K of the polynomial $f \in K[x]$ if F satisfies the following properties.

(i) f splits over F.

(ii) If $K \subseteq F' \subseteq F$ and f splits over F' then $F' = F$.

(3.3) PROPOSITION Let F be an extension field of K and $f \in K[x]$. Then F is a splitting field over K of f if and only if f splits over F and $F = K(u_1, \ldots, u_n)$ where $u_1, \ldots, u_n$ are the roots of f in F. In this case F is a finite dimensional algebraic extension of K.
PROOF The proof is routine and is left to the student [cf. exercise (3.1)].

Let f be a polynomial of degree n in K[x]. Then by (2.6) there exists a simple extension field K(u) of K (of degree at most n over K) in which f has a root u. The following theorem uses this fact to show the existence of a splitting field F over K of f such that $[F: K] \leq n!$

(3.4) THEOREM Let K be a field and $f \in K[x]$ a polynomial of degree $n \geq 1$. Then there exists a field F of dimension at most n! over K which is a splitting field of f over K.
PROOF We will induct on the degree n of f. If $n = 1$ then f splits over K and the result is trivial. Suppose that $n > 1$, f does not split over K, and the theorem is true for all polynomials of degree less than n.

Let g be a monic irreducible factor of f of degree greater than 1. By (2.6) there is a simple extension K(u) of K such that u is a root of g. Since g is then the minimal polynomial of u over K, $[K(u): K] = \deg g$ and hence $1 < [K(u): K] \leq n$.

Since u is a root of g and g is a factor of f, u is also a root of f. Hence by V, (1.10), $x - u$ divides f in the polynomial ring

$K(u)[x]$. Thus there is a polynomial $h \in K(u)[x]$ such that $f = (x - u)h$. Then $\deg h = n - 1$. Now by our induction hypothesis applied to the polynomial $h \in K(u)[x]$, there exists a splitting field F over $K(u)$ of h with $[F: K(u)] \le (n - 1)!$. Then by (1.15)

$$[F: K] = [F: K(u)]\,[K(u): K] \le (n - 1)!(n) = n!$$

Since F is a splitting field over $K(u)$ of h, and $f = (x - u)h$, F is a splitting field over K of f.

(3.5) PROPOSITION Let F be an extension field of K with F algebraically closed, and $f \in K[x]$. Then there exists a splitting field E of f over K such that $E \subseteq F$.

PROOF See exercise (3.2).

(3.6) EXAMPLES

(3.6.1) The polynomial $f = x^4 - 2 \in \mathbf{Q}[x]$ has roots $\pm w$, $\pm iw$ in $\mathbf{C}$ where $w = \sqrt[4]{2}$. Since $\mathbf{Q}(\pm w, \pm iw) = \mathbf{Q}(w, i)$, the field $F = \mathbf{Q}(w, i)$ is a splitting field over $\mathbf{Q}$ of f.

(3.6.2) Let $a \in \mathbf{Q}^+$, $m \in \mathbf{N}$ and $f = x^m - a$. Let $w = \sqrt[m]{a}$ and ρ be a primitive m-th root of unity. We recall that ρ is the complex number

$$\rho = e^{2\pi i/m} = \cos(2\pi/m) + i \sin(2\pi/m)$$

and $\rho^m = 1$ with $\rho^k \ne 1$ if $1 \le k < m$. Since $(w\rho^k)^m = w^m(\rho^m)^k = (a)(1) = a$, the m distinct elements $w\rho^k$ $(0 \le k < m)$ must comprise the roots of f. Hence the field $F = \mathbf{Q}(w, \rho)$ is a splitting field over $\mathbf{Q}$ of f. It is shown in exercise (1.26) that if $m = p$ is prime and $a = 2$, then $[F: \mathbf{Q}] = p(p - 1)$.

(3.6.3) Let $f = x^2 + x + 1 \in F_2[x]$. Then f is irreducible over F_2 and hence if u is a root of f and $F = F_2(u)$, F is a field with four elements; $F = \{0, 1, u, u + 1\}$ [cf. Example (2.7.1)]. The student may verify that $u + 1$ is also a root of f. Hence the field F is a splitting field over F_2 of the polynomial f.

(3.7) THEOREM (Extension Property for Splitting Fields) Let $\alpha: K \to L$ be an isomorphism of fields. If F is a splitting field over K of the polynomial $f \in K[x]$ and E is a splitting field over L of the corresponding polynomial $\alpha f \in L[x]$, then α can be extended to an isomorphism $\tau: F \to E$.

PROOF Since F is a splitting field over K of f, [F: K] is finite. We will induct on n = [F: K]. The astute reader may anticipate that we will use the extension property of simple extensions [cf. (1.12)] in our inductive argument.

If n = 1 then F = K so that f already splits over K. But then αf splits over L so that E = L and the result is trivial [observe that if u is a root of f then $\alpha(u)$ is a root of αf].

Suppose that n > 1 and the result is true for all extensions of dimension less than n. If f splits over K the result is again trivial. Otherwise, let $g \in K[x]$ be a monic irreducible factor of f of degree k > 1. Let u be a root of g in F (since f splits in F so does g) and let v be a root of αg in E. Then u has minimal polynomial g over K and v has minimal polynomial αg over L [cf. (1.10)]. Hence α extends to an isomorphism of fields $\alpha': K(u) \to L(v)$. The maps and relative dimensions are illustrated below.

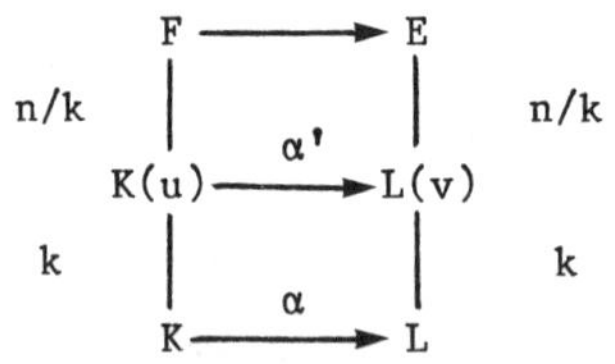

Since F is a splitting field over K of f, it is also a splitting field over K(u) of f [cf. Exercise (3.6)]. Then E is a splitting field over L(v) of $\alpha' f$ (note, that since $\alpha'|_K = \alpha$ and $f \in K[x]$, $\alpha' f = \alpha f$). Hence by our induction hypothesis (since [F: K(u)] < n), α' can be extended to an isomorphism of fields $\tau: F \to E$. Then τ also extends α and the proof is complete.

(3.8) COROLLARY (Uniqueness of Splitting Fields) Let F and E both be splitting fields over K of the polynomial $f \in K[x]$. Then there is an isomorphism of fields $\tau: F \to E$ such that $\tau|_K = 1_K$ and, in this case, $[F: K] = [E: K]$.

PROOF See (3.7) and exercise (1.4).

(3.9) COROLLARY Let F be a splitting field over K of $f \in K[x]$ and $u,v \in F[x]$ have the same minimal polynomial over K. Then there is an isomorphism of fields $\tau: F \to F$ such that $\tau(u) = v$ and $\tau|_K = 1_K$.

PROOF By (1.12) there is an isomorphism $\alpha: K(u) \to K(v)$ such that $\alpha(u) = v$ and $\alpha|_K = 1_K$. Then by the extension property for splitting fields, α extends to an isomorphism $\tau: F \to F$.

Let $f = a_0 + a_1x + a_2x^2 + \cdots + a_nx^n$ be an element of K[x]. The <u>formal derivative</u> Df of f is the polynomial

$$Df = a_1 + 2a_2x + \cdots + na_nx^{n-1} \in K[x]$$

Note that if $K = \mathbf{R}$, Df is the usual derivative of f. Let $f,g \in K[x]$, $k \in K$. Then the following facts may be routinely verified.

$$D(k) = 0$$
$$D(kf) = kDf$$
$$D(f + g) = Df + Dg$$
$$D(fg) = (Df)g + f(Dg)$$

REMARK If K is a field of characteristic 0 and $f \in K[x]$ has degree $n > 0$, then Df has degree $n - 1$.

(3.10) PROPOSITION Let K be a field, $f \in K[x]$ and $a \in K$. Then $(x - a)^2$ divides f in K[x] (that is, a is a <u>multiple root</u> of f) if and only if $(x - a)$ divides both f and Df in K[x].

PROOF ($\Longrightarrow$) Suppose that $f = (x - a)^2g$ with $g \in K[x]$. Then certainly $(x - a)$ divides f. But, since $Df = (x - a)^2Dg + 2(x - a)g$ and g and Dg are both elements of K[x], $(x - a)$ also divides Df.

($\Longleftarrow$) Suppose that $(x - a)$ divides both f and Df in K[x]. Since $(x - a)$ divides f, $f = (x - a)g$ for some $g \in K[x]$. Then Df =

$g + (x - a)Dg$. Hence, since $(x - a)$ divides both Df and $(x - a)Dg$, it also divides g. Thus $(x - a)^2$ divides f as required.

(3.11) DEFINITION Let K be a field and $f \in K[x]$ be irreducible. Then f is a separable polynomial if f has no multiple roots in any extension field of K.

(3.12) PROPOSITION Let K be a field of characteristic 0 and f an irreducible polynomial in $K[x]$. Then f is separable.

PROOF Suppose that a is a multiple root of f in an extension field F of K [so that $(x - a)^2$ divides f in $F[x]$].

We first observe that since K has characteristic 0 and $\deg f > 0$, Df is a nonzero polynomial in $K[x]$ of smaller degree than f. Without loss of generality we assume that f is monic; then f is the minimal polynomial over K of a. By (3.10) since $(x - a)^2$ divides f in $F[x]$, $(x - a)$ divides Df in $F[x]$. Hence $Df(a) = 0$. Thus, since f is the minimal polynomial over K of a, and a satisfies the nonzero polynomial $Df \in K[x]$, f divides Df [cf. (1.7)]. This implies that $\deg f \le \deg Df$ contradicting the fact that Df has smaller degree than f.

Exercise (3.13) shows that if K has characteristic p then an irreducible polynomial $f \in K[x]$ is separable if and only if f is <u>not</u> a polynomial in x^p.

In Section 1 we constructed some finite fields. In each case the order of the field was p^n for some prime p and $n \in \mathbf{N}$. We will now show that in fact if F is a finite field of characteristic p, then F has order p^n for some $n \in \mathbf{N}$ and, for each prime p and $n \in \mathbf{N}$ there is a unique field up to isomorphism having p^n elements.

Let F be a finite field. Then by IV, (4.8), F has characteristic p for some prime p (note that F cannot have characteristic 0 for then it would contain a copy of the infinite set $\mathbf{Z}$). Then $pu = 0$ for all $u \in F$ and F contains a subfield P which is isomorphic to the field F_p. The field P is generally called the <u>prime subfield</u> of F and F is then considered to be an extension of F_p. For future reference we also observe that any extension field of F has the same characteristic as F.

(3.13) PROPOSITION Let F be a field of nonzero characteristic p. Let $u, v \in F$, $n \in \mathbf{N}$ and $m = p^n$. Then

$$(u + v)^m = u^m + v^m \text{ and } (u - v)^m = u^m - v^m$$

PROOF We first show that $(u + v)^m = u^m + v^m$.

$$(u + v)^m = u^m + a_1u^{m-1}v + \cdots + a_iu^{m-i}v^i + \cdots + a_{m-1}uv^{m-1} + v^m$$

where a_i is the binomial coefficient $m!/i!(m-i)!$. If $1 \leq i \leq m-1$, then m divides a_i; hence (since $m = p^n$ and F has characteristic p) $a_iu^{m-i}v^i = 0$. Thus $(u + v)^m = u^m + v^m$.

Now, by the first paragraph, $(u - v)^m = u^m + (-v)^m = u^m + (-1)^mv^m$. If $p > 2$, m is odd so that $(-1)^m = -1$ and $(u - v)^m = u^m - v^m$. If $p = 2$, then $(-1)^m = 1 = -1$ and the result again follows.

(3.14) PROPOSITION Let F be a finite field of characteristic p. Then F contains p^n elements for some $n \in \mathbf{N}$.

PROOF Let P be the prime subfield of F. Then F is an extension field of P and, since F is finite, $[F: P] = n$ for some $n \in \mathbf{N}$. Let $\{u_1, \ldots, u_n\}$ be a basis of F over P. Then every element of F can be uniquely expressed in the form $a_1u_1 + \cdots + a_nu_n$ $(a_i \in P)$. Since $\text{card}(P) = \text{card}(F_p) = p$, there are p^n choices for such sums and it follows that $\text{card}(F) = p^n$.

(3.15) THEOREM Let p be a prime number and $n \in \mathbf{N}$. A field F has p^n elements if and only if it is a splitting field over some field K isomorphic to F_p of the polynomial

$$f = x^{p^n} - x$$

Hence for any prime p and $n \in \mathbf{N}$ there is a unique field (up to isomorphism) having p^n elements.

PROOF For ease of notation, let $m = p^n$.

($\Longrightarrow$) Suppose that the field F has m elements. Let K be the prime subfield of F. Then $K \cong F_p$. We will show that F is the splitting field over K of the polynomial $f = x^m - x$.

The group of units $F^\times$ of F has order m - 1; hence by II, (3.11), $u^{m-1} = 1$ for all $u \in F^\times$. It follows that, for all $u \neq 0$ in F, $u^m = u$, or in other words, u satisfies the polynomial $f = x^m - x$. Since 0 also satisfies f, f has m roots in F, namely every element of F. Thus, since f has degree m, f splits over F; hence F is a splitting field over K of f.

($\Longleftarrow$) Let K be a field isomorphic to F_p and F a splitting field over K of $f = x^m - x$. We wish to show that F has m elements.

Since F is an extension field of F_p, F has characteristic p. Hence $mu = p^n u = 0$ for all $u \in F$. Since $Df = mx^{m-1} - 1$, $Df(u) = -1$ for all $u \in F$. Thus Df and f have no common factors so that by (3.10) f has m distinct roots in F. Let

$$E = \{u \in F\colon u \text{ is a root of } f\} = \{u \in F\colon u^m = u\}$$

By the preceding paragraph, card (E) = m. By (3.13) if $u,v \in E$, $u - v \in E$. If $u,v \in E$ and $v \neq 0$, then $(uv^{-1})^m = u^m(v^m)^{-1} = uv^{-1}$; hence $uv^{-1} \in E$. Thus E is in fact a subfield of F. Since $u^m = u$ for all $u \in K$ (recall that $K \cong F_p$), $K \subseteq E$. Then since F is a splitting field over K of f, and E is a subfield of F containing K and the roots of f, F = E. The final statement of the theorem now follows from the uniqueness of splitting fields.

If $m = p^n$ (p prime, $n \in \mathbf{N}$) we will sometimes denote the unique field having m elements by F_m. The preceding proposition shows for example that the two fields of order 9 constructed in exercises (2.5) and (2.6) must be isomorphic.

(3.16) DEFINITION An extension field F of K is <u>normal</u> over K if whenever an irreducible polynomial $f \in K[x]$ has a root in F it splits in F.

(3.17) THEOREM Let F be an extension field of K. Then the following statements are equivalent.

(i) F is a finite normal extension of K.

(ii) F is a splitting field over K of some polynomial in K[x].

PROOF (i)$\Longrightarrow$(ii) Suppose that F is a finite normal extension of K. Then $F = K(u_1, \ldots, u_n)$ for some $u_i \in F$. By (2.2) each u_i is algebraic over K. Let f_i be the minimal polynomial of u_i over K. Then since f_i has a root in F (namely, u_i), f_i splits in F (since F is normal over K). Let $f = f_1 \cdots f_n$. Then, since F is generated by K and the zeros of f, F is a splitting field over K of f.

(ii)$\Longrightarrow$(i) Suppose that F is a splitting field over K of $f \in K[x]$. Then [F: K] is finite by (3.4). We must show that F is normal over K. Let g be an irreducible polynomial in K[x] with a root u in F. Let E be a splitting field over K of the polynomial fg (observe that since F is a splitting field of f over K we may assume that $F \subseteq E$).

We claim that if v_1, v_2 are roots of g in E then $[F(v_1): F] = [F(v_2): F]$ and hence for any root v of g in E, $[F(v): F] = [F(u): F] = 1$ so that $v \in F$. It then follows that g splits in F.

Consider the following Hasse diagram of field extensions.

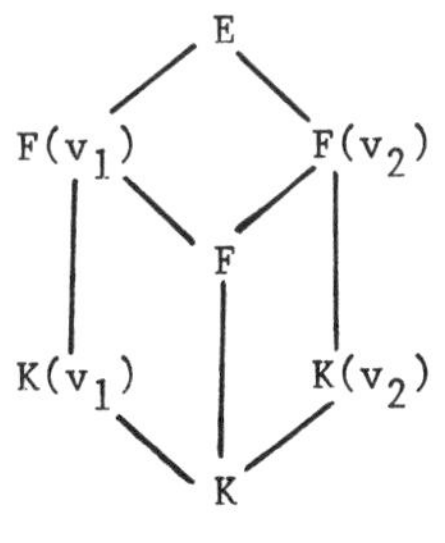

For i = 1 or 2 we have

$$[F(v_i): F]\ [F: K] = [F(v_i): K] = [F(v_i): K(v_i)]\ [K(v_i): K]$$

By (1.8) $[K(v_1): K] = \deg g = [K(v_2): K]$. By (1.13) there is an isomorphism $\alpha: K(v_1) \to K(v_2)$ such that $\alpha|_K = 1_K$. Then, by (3.7), since $F(v_1)$ is a splitting field for f over $K(v_1)$ and $F(v_2)$ is a splitting field for $\alpha f = f$ over $K(v_2)$, the map α extends to an isomorphism τ: $F(v_1) \to F(v_2)$. Thus $[F(v_1): K(v_1)] = [F(v_2): K(v_2)]$ [cf. exercise (1.4)], and it follows that $[F(v_1): F] = [F(v_2): F]$.

We remark that the preceding theorem shows that every finite field is a normal extension of its prime subfield.

(3.18) THEOREM Let K be a field and L a finite extension of K. Then there is a finite extension F of L satisfying the following properties.

(i) F is normal over K.

(ii) No proper subfield of F containing L is normal over K.

The field F is unique up to an isomorphism fixing K and is called the <u>normal closure</u> of L over K.

PROOF By (2.2) $L = K(u_1, \ldots, u_n)$ with u_i algebraic over K. Let $f = f_1 \cdots f_n$ where f_i is the minimal polynomial of u_i over K, and let F be a splitting field over K of f.

By (3.17) F is normal over K. Since each u_i is a root of f, $L = K(u_1, \ldots, u_n) \subseteq F$. We wish to show that no proper subfield of F containing L is normal over K. Suppose that $L \subseteq M \subseteq F$ and M is normal over K. Then M contains a root of $f_i \in K[x]$ (namely u_i); hence, since f_i is irreducible in K[x] and M is normal over K, f_i splits over M. Thus M contains the set S of roots of f, so that $F = K(S) \subseteq M$.

Suppose that E is also an extension of L satisfying (i) and (ii). We claim that E is also a splitting field over K of f so that by (3.8) there is an isomorphism from F to E which fixes K.

Since E is normal over K and E contains the root u_i of the irreducible polynomial $f_i \in K[x]$, f_i splits in E. Hence f splits in E so that E contains a splitting field E' of f over K. Since E' is then normal over K, by property (ii) L = E. The result now follows.

(3.19) EXAMPLES

(3.19.1) Let $w = \sqrt[4]{2}$ and $L = \mathbf{Q}(w)$. The minimal polynomial of w over $\mathbf{Q}$ is $f = x^4 - 2$ and the field $F = \mathbf{Q}(w, i)$ is a splitting field over **Q** of f. Hence F is the normal closure of **Q**(w) over **Q**.

(3.19.2) Let $m \in \mathbf{N}$ and $w = \sqrt[m]{2}$. Then the minimal polynomial of w over **Q** is $f = x^m - 2$ and the field $F = \mathbf{Q}(w, \rho)$ (where ρ is a primitive m-th root of unity) is a splitting field over **Q** of f. Hence F is the normal closure of **Q**(w) over **Q**.

EXERCISES

(3.1) Prove (3.3).

(3.2) Prove (3.5).

(3.3) Show that if $f \in K[x]$ is irreducible of degree n and F is a splitting field over K of f, then n divides [F: K]. Give an example to show that this is not necessarily true if f is reducible.

(3.4) Let F be an extension field of K, $g,h \in K[x]$ which split in F and f = gh. Show that if L is a splitting field for g over K and M is a splitting field for h over K, then $L \vee M$ is a splitting field for f over K (recall that $L \vee M$ is the smallest subfield of F containing $L \cup M$).

(3.5) Show that $(x^2 - 3)(x^3 + 1)$, $(x^2 - 2x - 2)(x^2 + 1)$ and $x^4 - 9$ all have the same splitting field F over **Q** and find [F: **Q**].

(3.6) Show that if F is a splitting field for f over K and $K \subseteq E \subseteq F$, then F is also a splitting field for f over E.

(3.7) Find a splitting field F over **Q** for each of the following polynomials and find [F: **Q**].

(a) $(x^2 - 2)(x^3 - 2)^2$.

(b) $x^4 + x^2 + 1$ [Hint: $x^4 + x^2 + 1 = (x^2 + 1)^2 - x^2$.]

(c) $x^4 + 4x^2 + 4$.

(d) $x^4 - 2x^2 - 2$ [Hint: $x^4 - 2x^2 - 2 = (x^2 - 1)^2 - 3$.]

(3.8) Give an example of a polynomial $f \in F_5[x]$ such that f has degree 8 and Df has degree 4.

(3.9) Give an example of a polynomial $f \in \mathbf{Q}[x]$ of degree 4 such that the splitting field F of f over $\mathbf{Q}$ has each of the following dimensions over $\mathbf{Q}$: 1, 2, 4, 8.

(3.10) Give an example of a polynomial $f \in \mathbf{Q}[x]$ of degree 3 such that the splitting field F of f over $\mathbf{Q}$ has each of the following dimensions over $\mathbf{Q}$: 1, 2, 6.

(3.11) In each case find the minimal polynomial f of the given element u over $\mathbf{Q}$, the splitting field F of f over $\mathbf{Q}$, $[\mathbf{Q}(u): \mathbf{Q}]$ and $[F: \mathbf{Q}]$.

(a) u is a primitive 6-th root of unity.

(b) u is a primitive 8-th root of unity.

(c) u is the positive real 6-th root of 2.

(d) u is the positive real 8-th root of 2.

(3.12) Let $I = ((x - 1)^2)$ be the principal ideal generated by $(x - 1)^2$ in $\mathbf{R}[x]$ and $J = (x^3 - 2x^2 + x)$ in $\mathbf{R}[x]$.

(a) Show that $\mathbf{R}[x]/I \cong \mathbf{R}^2$ as $\mathbf{R}$-vector spaces. [Hint: Consider the map $T: \mathbf{R}[x] \to \mathbf{R}^2$ by $f \mapsto (f(1), Df(1))$.]

(b) Show that $\mathbf{R}[x]/I$ and $\mathbf{R}^2$ are not isomorphic as rings. [Hint: Show that exactly one of the rings contains a nonzero nilpotent element - cf. IV, exercise (1.14).]

(c) Show that $\mathbf{R}[x]/J \cong \mathbf{R}^3$ as $\mathbf{R}$-vector spaces but not as rings.

(3.13) Let $f \in K[x]$ be irreducible.

(a) Show that f is separable if and only if $Df \neq 0$.

(b) Show that if K has characteristic $p > 0$, then f is <u>not</u> separable if and only if f is a polynomial in x^p.

(3.14) Let $f \in F_p[x]$ be irreducible of degree n, u a root of f and $F = F_p(u)$. Show that F is a field with p^n elements and is a finite normal extension of F_p. Conclude that F is a splitting field for f over F_p.

(3.15) Find a splitting field F for each of the following polynomials f and find the roots of f in F.

(a) $f = x^2 + x + 1 \in F_2[x]$ over F_2.

(b) $f = x^3 + x + 1 \in F_2[x]$ over F_2.

(3.16) (a) Construct a splitting field F for $f = x^3 + 2x + 1$ over F_3.

(b) Construct a splitting field E for $g = x^3 + x^2 + 2$ over F_3.

(c) Show that there is an isomorphism of fields from F to E which fixes F_3.

(3.17) Let p be a prime and $F = F_p(u)$ where u is transcendental over F_p. Let $f = x^p - u \in F[x]$.

(a) Show that f is not separable and that in fact if v is a zero of f, then $f = (x - v)^p$.

(b) Show that f is irreducible over F_p.

(3.18) Construct normal closures F for the following field extensions E of **Q**. Find [E: **Q**] and [F: **Q**].

(a) $E = \mathbf{Q}(w)$ where w is the real 5-th root of 7.

(b) $E = \mathbf{Q}(\sqrt{2}, \sqrt[3]{2})$.

(c) $E = \mathbf{Q}(w)$ where w is a root of $x^3 - 7$.

(3.19) Let F be a finite field. Show that the multiplicative group $F^\times$ of units of F is cyclic. [Hint: Use the fact that by Exercise VII, (2.24), $F^\times$ is isomorphic to a finite direct product of cyclic groups of order m_i $(1 \le i \le n, n \in \mathbf{N})$ with $m_1 | m_2 | \cdots | m_n$. Show that every element of $F^\times$ is a root of $x^k - 1$ where $k = m_n$ and hence a contradiction arises if $n > 1$.]

(3.20) Find generators for the group of units of the fields F_{25} and F_{27}. [Hint: Construct F_{25} as $F_5(u)$ where u is the root of an irreducible polynomial of degree 2 over F_5.]

(3.21) Show that if F is a finite field of characteristic p then every element of F has a unique p-th root.

(3.22) Show that every element in a finite field may be written as the sum of two squares.

9

Galois Theory

1. THE FUNDAMENTAL CORRESPONDENCE

Let F be an extension field of K. An automorphism τ of F is said to be a K-automorphism of F if $\tau|_K = 1_K$ (that is, $\tau(u) = u$ for all $u \in K$). In this case τ is also a K-vector space isomorphism of F. By exercise (1.1) the set of all K-automorphisms of F is a subgroup of the group Aut F of automorphisms of F.

(1.1) DEFINITION Let F be an extension field of K. The Galois group of F over K, denoted Gal(F/K), is the subgroup

$$\mathrm{Gal}(F/K) = \{\tau\colon \tau \text{ is a K-automorphism of } F\}$$

of Aut F.

Before presenting examples we make several observations.

(I) Suppose that $F = K(u_1, \ldots, u_n)$ and F is algebraic over K. Then by VIII, (1.16), F is generated as a vector space over K by products of the u_i. Hence if $\tau \in \mathrm{Gal}(F/K)$, since τ fixes K and is a homomorphism of rings, τ is uniquely determined by its action on the u_i $(1 \le i \le n)$.

(II) Let F be an extension field of K and $\tau \in \mathrm{Gal}(F/K)$. Let $f \in K[x]$ and $S = \{u \in F\colon u \text{ is a root of } f\}$. By VIII, (1.11), if $u \in S$, $\tau(u) \in S$. Then the map $\tau|_S\colon S \to S$ is an injective map. Since S is a finite set, $\tau|_S$ is also surjective. Thus τ merely permutes the roots of f. This fact will be used in Section 2 to show that the

Galois group of a finite extension is isomorphic to a subgroup of the group of permutations of the roots of a particular polynomial.

The reader is also advised to review the extension property for simple extensions [cf. VIII, (1.12)].

(1.1) EXAMPLES

(1.1.1) Let $G = \mathrm{Gal}(\mathbf{C}/\mathbf{R})$ be the Galois group of the field $\mathbf{C}$ of complex numbers over the field $\mathbf{R}$ of real numbers. We recall that $\mathbf{C} = \mathbf{R}(i)$ is a splitting field over $\mathbf{R}$ of $f = x^2 + 1$. Since f has roots i and -i in $\mathbf{C}$, by (I) and (II) any element τ of G is uniquely determined by its action on i and must take i to either itself or -i. Hence G contains at most two elements. By the extension property for simple extensions, if $u \in S = \{i, -i\}$, there is an isomorphism $\tau: \mathbf{C} \to \mathbf{R}(u)$ such that $\tau|_{\mathbf{R}} = 1_{\mathbf{R}}$ and $\tau(i) = u$. Since $\mathbf{C} = \mathbf{R}(u)$ for $u \in S$, $\tau \in G$. Hence G is a group of order 2 and the elements of G are the $\mathbf{R}$-automorphisms of $\mathbf{C}$ given below.

$i \mapsto i$ (the identity map on $\mathbf{C}$)
$i \mapsto -i$ $(a + bi \mapsto a - bi)$

(1.1.2) Let $G = \mathrm{Gal}(\mathbf{Q}(w)/\mathbf{Q})$ where w is the real cube root of 2. Since w is algebraic over $\mathbf{Q}$ with minimal polynomial $f = x^3 - 2$ of degree 3, $\{1, w, w^2\}$ is a basis for $\mathbf{Q}(w)$ over $\mathbf{Q}$. In particular, $\mathbf{Q}(w) \subseteq \mathbf{R}$. By (I) and (II) any element τ of G is determined by its action on w and must take w to a root of f in $\mathbf{Q}(w)$. But the only root of f in $\mathbf{Q}(w)$ is w itself (the other two roots of f are not real). Hence G is a group of order 1 whose only element is the identity map on $\mathbf{Q}(w)$.

(1.1.3) Let $G = \mathrm{Gal}(\mathbf{Q}(w)/\mathbf{Q})$ where w is the positive real fourth root of 2. Then as in Example VIII, (1.9.2), w is algebraic over $\mathbf{Q}$ with minimal polynomial $f = x^4 - 2$ of degree 4 so that $\{1, w, w^2, w^3\}$ is a basis of $\mathbf{Q}(w)$ over $\mathbf{Q}$. The roots of f are w, -w, iw, and -iw, and only w and -w are elements of $\mathbf{Q}(w)$. Hence, since any element τ of G is determined by its action on w and must take w to a root of f in $\mathbf{Q}(w)$, G has at most two elements.

By the extension property for simple extensions, if u is one of the two roots of f in $\mathbf{Q}(w)$, the identity map on $\mathbf{Q}$ extends to an isomorphism τ: $\mathbf{Q}(w) \to \mathbf{Q}(u)$ such that $\tau(w) = u$. Since $\mathbf{Q}(u) = \mathbf{Q}(w)$ for $u \in \{w, -w\}$, we may conclude that G is a group of order two whose elements are the $\mathbf{Q}$-automorphisms of $\mathbf{Q}(w)$ given below.

$w \mapsto w$ [the identity map on $\mathbf{Q}(w)$]
$w \mapsto -w$ $(a + bw + cw^2 + dw^3 \mapsto a - bw + cw^2 - dw^3)$

(1.1.4) Let $F = \mathbf{Q}(\sqrt{2}, \sqrt{3})$. Thus F is a splitting field over $\mathbf{Q}$ of $h = (x^2 - 2)(x^2 - 3)$. In Example VIII, (1.17.1) we established the following facts.

(i) The minimal polynomial of $\sqrt{2}$ over $\mathbf{Q}$ is $f = x^2 - 2$.

(ii) The minimal polynomial of $\sqrt{3}$ over both $\mathbf{Q}$ and $\mathbf{Q}(\sqrt{2})$ is $g = x^2 - 3$.

(iii) $\{1, \sqrt{2}, \sqrt{3}, \sqrt{6}\}$ is a basis for $\mathbf{Q}(\sqrt{2}, \sqrt{3})$ over $\mathbf{Q}$.

Since any element τ of G is determined by its action on $\sqrt{2}$ and $\sqrt{3}$ (by I) and must take $\sqrt{2}$ to $\pm\sqrt{2}$ and $\sqrt{3}$ to $\pm\sqrt{3}$ (by II), the order of G is at most 4.

To show that G has four elements we construct F via a tower of simple extensions. Let $S = \{\sqrt{2}, -\sqrt{2}\}$ and $T = \{\sqrt{3}, -\sqrt{3}\}$. By (i) and the extension property for simple extensions, if $u \in S$ there is an isomorphism α: $\mathbf{Q}(\sqrt{2}) \to \mathbf{Q}(u)$ with $\alpha(\sqrt{2}) = u$ and $\alpha|_{\mathbf{Q}} = 1_{\mathbf{Q}}$. Since α fixes $\mathbf{Q}$, $\alpha g = g$. If $u \in S$ and $w \in T$, w satisfies the monic polynomial $\alpha g = g$ in $\mathbf{Q}(u)[x]$. To use the extension property for simple extensions we must show that g is the minimal polynomial of w over $\mathbf{Q}(u)$. Rather than showing directly that g is irreducible over $\mathbf{Q}(u)$ (such a direct argument is possible in this case but in general is difficult), we present an indirect argument that will be more useful for future examples. The towers of field extensions are indicated in the illustration below.

$$\begin{array}{ccccc}
F = \mathbf{Q}(\sqrt{2}, \sqrt{3}) & \xrightarrow{\ \tau\ } & \mathbf{Q}(u, v) & & v \in T \\
| & & | & & \\
\mathbf{Q}(\sqrt{2}) & \xrightarrow{\ \alpha\ } & \mathbf{Q}(u) & & u \in S \\
| & & | & & \\
\mathbf{Q} & \xrightarrow{\ 1_{\mathbf{Q}}\ } & \mathbf{Q} & &
\end{array}$$

It is easily verified that if $u \in S$ and $v \in T$, then $\mathbf{Q}(u, v) = F$. Hence $[\mathbf{Q}(u, v): \mathbf{Q}(u)] = [F: \mathbf{Q}(u)] = 2$ so that since v satisfies the monic polynomial $g \in \mathbf{Q}(u)[x]$ and $\deg g = 2$, v has minimal polynomial g over $\mathbf{Q}(u)$. Hence, again by the extension property for simple extensions, α extends to an element $\tau \in G$ such that $\tau(\sqrt{3}) = v$.

Thus the order of G is 4 and the elements of G are the following four $\mathbf{Q}$-automorphisms of $\mathbf{Q}(\sqrt{2}, \sqrt{3})$.

1: $\sqrt{2} \mapsto \sqrt{2}$, $\sqrt{3} \mapsto \sqrt{3}$ (the identity map)

α: $\sqrt{2} \mapsto \sqrt{2}$, $\sqrt{3} \mapsto -\sqrt{3}$

τ: $\sqrt{2} \mapsto -\sqrt{2}$, $\sqrt{3} \mapsto \sqrt{3}$

$\alpha\tau$: $\sqrt{2} \mapsto -\sqrt{2}$, $\sqrt{3} \mapsto -\sqrt{3}$

The student should verify that $G \cong \mathbf{Z}_2 \times \mathbf{Z}_2$.

(1.1.5) Let $F = \mathbf{Q}(w, i)$ where w is the positive real fourth root of 2. Thus F is the splitting field over $\mathbf{Q}$ of the irreducible polynomial $f = x^4 - 2$. Let $G = \mathrm{Gal}(F/\mathbf{Q})$. The following facts were established in VIII, (1.17.2).

(i) The minimal polynomial of w over $\mathbf{Q}$ is f. The set of roots of f in F is $S = \{w, -w, iw, -iw\}$.

(ii) The minimal polynomial of i over both $\mathbf{Q}$ and $\mathbf{Q}(w)$ is $g = x^2 + 1$. The set of roots of g in F is $T = \{i, -i\}$.

Since F is a splitting field over $\mathbf{Q}$ which can be constructed as a tower of simple extensions, we will use a method similar to that of the preceding example.

Since w has minimal polynomial f over $\mathbf{Q}$ and i has minimal polynomial g over $\mathbf{Q}$, any element of G must take w to an element of S and i to an element of T. Hence the order of G is at most 8.

By (i) and the extension property for simple extensions, if $u \in S$, there is an isomorphism $\alpha: \mathbf{Q}(w) \to \mathbf{Q}(u)$ such that $\alpha(w) = u$. Verify that if $u \in S$ and $v \in T$ then $\mathbf{Q}(u, v) = F$. Hence, since v satisfies g and $[\mathbf{Q}(u, v): \mathbf{Q}(u)] = [F: \mathbf{Q}(u)] = \deg g$ [by (ii)], v has minimal polynomial $\alpha g = g$ over $\mathbf{Q}(u)$. Thus, again by the extension property for simple extensions, there is an isomorphism $\tau: F \to F$ such that τ is an extension of α and $\tau(i) = v$. Then $\tau \in G$ with $\tau(w) = u$ and $\tau(i) = v$. Hence G is a group of order 8.

We will see in (1.14.3) that G is isomorphic to the dihedral group D_4. Observe also that o(G) = [F: **Q**]. We will show in the next section that this follows from the fact that F is the splitting field over **Q** of a polynomial in **Q**[x] and hence a normal extension of **Q**.

(1.1.6) Let ρ be a primitive third root of unity and F = **Q**(ρ). We recall that F is a splitting field over **Q** of the polynomial $f = x^2 + x + 1$ and that the roots of f are ρ and ρ^2. If G = Gal(F/**Q**) then every element of G is determined by its action on ρ and must take ρ to itself or ρ^2; hence the order of G is at most 2. Since ρ^2 has the same minimal polynomial as ρ over **Q**, the identity map on **Q** extends to an isomorphism τ: F → F with $\tau(\rho) = \rho^2$. Thus G is a cyclic group of order 2 generated by the element τ.

(1.1.7) Let F be the splitting field of $f = x^3 - 2$ over **Q**. Thus F = **Q**(ρ, w) where ρ is a primitive third root of unity and $w = \sqrt[3]{2}$. We recall that ρ has minimal polynomial $g = x^2 + x + 1$ over **Q** and g has roots ρ and ρ^2, and that w has minimal polynomial f over **Q** and f has roots w, wρ, and $w\rho^2$. Then any element of G is determined by its action on w and ρ and must take w to a root of f and ρ to a root of g. Hence G has order at most 6. The student should use the method of the preceding examples to show that all six possibilities result in elements of G and hence G is a group of order 6. We again observe that o(G) = [F: **Q**].

We have now associated with each extension field F over K a related group, namely G = Gal(F/K). Our plan is to establish a connection between intermediate fields of the extension F over K and subgroups of the Galois group G.

(1.2) PROPOSITION Let F be an extension field of K and G = Gal(F/K) the associated Galois group. For each intermediate field L of F over K, let

L' = Gal(F/L) = {τ: τ is an L-automorphism of F}

Then the following assertions hold.

(i) $F' = (1)$ and $K' = G$.

(ii) If L and M are intermediate fields of F over K with $L \subseteq M$, then $M' \subseteq L'$. (Notice the reversal of inclusion.)

(iii) If L is an intermediate field of F over K, then L' is a subgroup of G.

PROOF (i) $F' = \mathrm{Gal}(F/F)$

$= \{\tau: \tau \text{ is an automorphism of F and } \tau|_F = 1_F\}$

$= (1)$

$K' = \mathrm{Gal}(F/K) = G$ by definition of G.

(ii) Let $\tau \in M' = \mathrm{Gal}(F/M)$. Then τ is an automorphism of F which fixes M. Since $L \subseteq M$, τ also fixes L. Thus $\tau \in \mathrm{Gal}(F/L) = L'$.

(iii) Since $K \subseteq L$, by (i) and (ii) $L' \subseteq K' \subseteq G$. Hence, since L' is a subgroup of Aut(F) which is contained in G, L' is a subgroup of G.

(1.3) PROPOSITION Let F be an extension field of K with associated Galois group $G = \mathrm{Gal}(F/K)$. For each subgroup H of G let

$$H' = \{u \in F: \tau(u) = u \text{ for all } \tau \in H\}$$

Then the following assertions hold.

(i) $(1)' = F$.

(ii) If H and J are subgroups of G with $H \subseteq J$, then $J' \subseteq H'$ (again notice the reversal of inclusion).

(iii) If H is a subgroup of G then H' is an intermediate field of F over K. H' is called the <u>fixed field of H</u> with respect to F over K.

PROOF (i) $(1)' = \{u \in F: 1_F(u) = u\} = F$.

(ii) Suppose that $H \subseteq J$. Let $u \in J'$. Then $\tau(u) = u$ for all $\tau \in J$. Since $H \subseteq J$, $\tau(u) = u$ for all $\tau \in H$. Hence $u \in H'$ and $J' \subseteq H'$.

(iii) Let H be a subgroup of G. Then, by definition, $H' \subseteq F$. Since $H \subseteq G = \mathrm{Gal}(F/K)$, every element of K is fixed by every element of H. Hence $K \subseteq H'$. It remains only to show that in fact H' is a subfield of F. Let $u,v \in H'$. Then $\tau(u) = u$ and $\tau(v) = v$. Since τ

is a homomorphism, $\tau(u - v) = \tau(u) - \tau(v) = u - v$ (so $u - v \in H'$) and, if $v \neq 0$, $\tau(uv^{-1}) = \tau(u)\tau(v)^{-1} = uv^{-1}$ (hence $uv^{-1} \in H'$). Thus H' is a subfield of F.

(1.4) PROPOSITION Let F be an extension field of K with associated Galois group $G = \text{Gal}(F/K)$. Let L be an intermediate field of F over K and H a subgroup of G. Then the following assertions hold.

(i) $L \subseteq L''$.

(ii) $H \subseteq H''$.

(iii) $L' = L'''$.

(iv) $H' = H'''$.

PROOF (i) Let $u \in L$. Then $\tau(u) = u$ for all $\tau \in L'$. Hence u is an element of the fixed field of L'; that is, $u \in L''$.

(ii) Let $\tau \in H$. Then τ is an automorphism of F. If $u \in H'$ (the fixed field of H) then, since $\tau \in H$, $\tau(u) = u$. Thus $\tau \in \text{Gal}(F/H') = H''$.

(iii) By (i), $L \subseteq L''$ and hence by (1.3), (ii), $L''' \subseteq L'$. On the other hand, L' is a subgroup of G and hence by (ii) $L' \subseteq L'''$.

(iv) The proof is similar to (iii) [cf. exercise (1.2)].

(1.5) DEFINITION Let F be an extension field of K with associated Galois group $G = \text{Gal}(F/K)$. An intermediate field L of F over K (or a subgroup H of G) is <u>closed</u> with respect to F over K if $L = L''$ (respectively $H = H''$).

Thus L is closed in F over K if and only if every element of F which is not in L is moved by some L-automorphism of F.

REMARK It suffices in the preceding definition to say that L is <u>closed in F</u> since L'' is the same in the extensions F over K and F over L [see exercise (1.3)].

(1.6) EXAMPLES

(1.6.1) Let $w = \sqrt[3]{2}$ and consider the extension $\mathbf{Q}(w)$ of $\mathbf{Q}$. By Example (1.1.2), $\text{Gal}(\mathbf{Q}(w)/\mathbf{Q}) = (1)$. Then $\mathbf{Q}'' = (1)' = \mathbf{Q}(w)$ so that $\mathbf{Q}$ is not closed in $\mathbf{Q}(w)$.

(1.6.2) Let ρ be a primitive third root of unity and consider the extension $\mathbf{Q}(\rho)$ over $\mathbf{Q}$. We have seen in (1.1.6) that the Galois group of $\mathbf{Q}(\rho)$ over $\mathbf{Q}$ is cyclic of order 2 with generator $\tau: \rho \mapsto \rho^2$ and that a basis for $\mathbf{Q}(\rho)$ over $\mathbf{Q}$ is $\{1, \rho\}$. If $u = a + b\rho \in \mathbf{Q}(\rho)$ and $u \notin \mathbf{Q}$, then $b \neq 0$ and $\tau(u) = a + b\rho^2 \neq u$. Hence $\mathbf{Q}$ is closed in $\mathbf{Q}(\rho)$.

(1.6.3) Let $w = \sqrt[4]{2}$ and $F = \mathbf{Q}(w)$. We consider the extension F over $\mathbf{Q}$. By Example (1.1.3), $\mathbf{Q}' = \mathrm{Gal}(F/\mathbf{Q}) = \{1, \tau)$ where τ is the $\mathbf{Q}$-automorphism of F which sends w to -w. Since the element w^2 of F is fixed by both 1 and τ, $w^2 \in \mathbf{Q}''$. Hence, since $w^2 \notin \mathbf{Q}$, $\mathbf{Q}'' \neq \mathbf{Q}$. Thus $\mathbf{Q}$ is not closed in $\mathbf{Q}(w)$.

We will show in Corollary (1.13) that in fact if F is a finite dimensional extension of K then K is closed in F if and only if $[F: K] = o(\mathrm{Gal}(F/K))$. In the next section we will also show that if K is a field of characteristic 0 then K is closed in F if and only if F is a splitting field over K of a polynomial $f \in K[x]$. For example, in part (ii) of Example (1.6.1), the field $F = \mathbf{Q}(\rho)$ is a splitting field over $\mathbf{Q}$ of the polynomial $f = x^2 + x + 1$.

(1.7) THEOREM Let F be an extension field of K with associated Galois group $G = \mathrm{Gal}(F/K)$. Then the mapping

$$\theta: L \mapsto L' \text{ (L a closed intermediate field of F over K)}$$

is an order reversing bijection from the set of closed intermediate fields of F over K to the set of closed subgroups of G. The mapping

$$T: H \mapsto H' \text{ (H a closed subgroup of G)}$$

is an inverse for θ.

PROOF By (1.4) L' and H' are closed for all intermediate fields L of F over K and all subgroups H of G. We now show that θ and T are inverse maps.

$$T\theta(L) = T(L') = L'' = L \text{ (since L is closed in F)}$$

$\theta T(H) = \theta(H') = H'' = H$ (since H is closed in G)

We will show in (1.12) that if F is a finite dimensional extension of K with associated Galois group G = Gal(F/K) and K is closed in F then all intermediate fields of F over K and all subgroups of G are closed.

Let F be an extension field of K with associated Galois group G = Gal(F/K) and L and M intermediate fields of F over K with $L \subseteq M$. Then the inclusion of fields $K \subseteq L \subseteq M \subseteq F$ produces a reverse inclusion of groups $G = K' \supseteq L' \supseteq M' \supseteq F' = (1)$. We wish to establish conditions under which [M: L], the dimension of the extension M over L, equals (L': M'), the index of M' in L' [cf. II, (3.6)]. (Observe the difference between the field notation and the group notation.)

(1.8) PROPOSITION Let F be an extension field of K with associated Galois group G = Gal(F/K). Then the following assertions hold.

(i) Let M be an intermediate field of F over K and $\alpha,\tau \in G$. Then $\alpha M' = \tau M'$ if and only if $\alpha|_M = \tau|_M$.

(ii) Let L be an intermediate field of F over K, $u \in F$ algebraic over L, M = L(u) and $\alpha,\tau \in L'$. Then $\alpha M' = \tau M'$ if and only if $\alpha(u) = \tau(u)$.

(iii) Let H be a subgroup of G and $\alpha,\tau \in G$. If $\alpha H = \tau H$ then $\alpha|_{H'} = \tau|_{H'}$.

PROOF (i) The two left cosets $\alpha M'$ and $\tau M'$ of M' in G are equal if and only if $\tau^{-1}\alpha \in M'$. Since M' = Gal(F/M), $\tau^{-1}\alpha \in M'$ if and only if $\tau^{-1}\alpha(u) = u$ for all $u \in M$; that is, if and only if $\tau(u) = \alpha(u)$ for all $u \in M$.

(ii) Let $\alpha,\tau \in L'$. Since u is algebraic over L, L(u) = L[u]. Hence if $v \in M = L(u)$, v is an L-linear combination of powers of u. Then, since α and τ are both homomorphisms of rings which fix L, α and τ agree if and only if $\alpha(u) = \tau(u)$.

(iii) Suppose that $\alpha,\tau \in G$ are such that $\alpha H = \tau H$. Then $\tau^{-1}\alpha \in H$. Since $H \subseteq H''$ by (1.4), $\tau^{-1}\alpha \in H''$; that is, $\alpha H'' = \tau H''$. It now follows by (i) that $\alpha|_{H'} = \tau|_{H'}$.

(1.9) THEOREM Let F be an extension field of K and L and M intermediate fields with $L \subseteq M$. If [M: L] is finite then

$$(L': M') \leq [M: L]$$

In particular, if [F: K] is finite and G = Gal(F/K), then G is a finite group and $o(G) \leq [F: K]$.

PROOF We shall prove the result by induction on the dimension of M over L. Let n = [M: L].

If n = 1, then M = L; hence M' = L' and (L': M') = 1 = [M: L].

Suppose that n > 1 and the result is true for all extensions of dimension less than n. Our plan is to use the induction hypothesis to reduce to the case M = L(u) and to then make use of (1.8), (ii). Since [M: L] = n > 1, there is an element $u \in M$ such that $u \notin L$. Then $L \subseteq L(u) \subseteq M$. Since [M: L] is finite, by VIII, (1.15), [M: L] = [M: L(u)] [L(u): L].

Let m = [L(u): L]. Then m > 1 and [M: L(u)] = n/m < n. If m < n, then by our induction hypothesis applied to the extensions L(u) over L and M over L(u),

$$(L(u)': M') \leq [M: L(u)] = n/m \text{ and } (L': L(u)') \leq [L(u): L] = m$$

and hence $(L': M') = (L': L(u)')\,(L(u)': M') \leq m(n/m) = n = [M: L]$.

Now suppose that m = n. Then M = L(u) and u is algebraic over L with minimal polynomial $f \in L[x]$ of degree n [cf. VIII, (1.8)]. In order to show that $(L': M') \leq [M: L] = \deg f$, it then suffices to construct an injective map from the set S of left cosets of M' in L' to the set T of roots of f in F.

If $\tau \in L' = \mathrm{Gal}(F/L)$, then since u has minimal polynomial $f \in L[x]$, $\tau(u)$ is also a root of f. Hence $\tau(u) \in T$. Let $\theta: S \to T$ be defined by $\tau M' \mapsto \tau(u)$ $(\tau \in L')$. By (1.8), (ii), if $\tau, \alpha \in L'$ then $\tau M' = \alpha M'$ if and only if $\tau(u) = \alpha(u)$. Hence the map θ is both well-defined and injective.

Finally, if [F: K] is finite and G = Gal(F/K), then, since F' = (1) and K' = G, $o(G) = (G: (1)) \leq [F: K]$.

(1.10) THEOREM Let F be an extension field of K and let H and J be subgroups of the Galois group $G = Gal(F/K)$ with $H \subseteq J$. If $(J: H)$ is finite, then

$$[H': J'] \leq (J: H)$$

PROOF Let $(J: H) = n$ and $\tau_1, \tau_2, \ldots, \tau_n$ be a complete system of (left) coset representatives of H in J [cf. II, (3.2)].

Suppose that $[H': J'] > n$. Then there are $n + 1$ elements $u_1, u_2, \ldots, u_{n+1} \in H'$ that are linearly independent over J'. Let $a_{ij} = \tau_i(u_j) \in F$ $(1 \leq i \leq n;\ 1 \leq j \leq n+1)$ and consider the following system of n homogeneous linear equations in $n + 1$ unknowns with coefficients in F.

$$\text{(I)} \qquad \begin{aligned} a_{11}x_1 + a_{12}x_2 + \cdots + a_{1,n+1}x_{n+1} &= 0 \\ a_{21}x_1 + a_{22}x_2 + \cdots + a_{2,n+1}x_{n+1} &= 0 \\ &\vdots \\ a_{n1}x_1 + a_{n2}x_2 + \cdots + a_{n,n+1}x_{n+1} &= 0 \end{aligned}$$

Since the homogeneous system in (I) has more unknowns than equations it must have a nontrivial solution in F^{n+1} [cf. VII, exercise (3.15)]. Among all nontrivial solutions to (I) we may choose a nontrivial solution $a = (a_1, a_2, \ldots, a_{n+1}) \in F^{n+1}$ with a <u>minimal</u> number r of nonzero entries. We may assume, by reindexing if necessary, that $a_i \neq 0$ for $1 \leq i \leq r$ and $a_i = 0$ for $r \leq i \leq n + 1$. Since any constant multiple of a in F is again a solution to (I) and $a_1 \neq 0$ we may assume that $a_1 = 1$.

Since a is a solution to (I), for each $i = 1, \ldots, n$.

$$(*) \qquad \sum_{j=1}^{n+1} a_{ij}a_j = 0$$

Our plan is to contradict the minimality of r as follows. We will use the independence of the u_j $(1 \leq j \leq n + 1)$ over J' to find a $\tau \in J$ satisfying the following conditions.

(i) $\tau(a_k) \neq a_k$ for some k with $1 \leq k \leq n + 1$.

(ii) The element $b = (b_1, \ldots, b_{n+1}) \in F^{n+1}$ with $b_j = \tau(a_j)$ $(1 \leq j \leq n + 1)$ is also a solution to (I).

Then if $c = b - a$, c is also a solution to (I). By (i), c is nonzero. Since $a_1 = 1$, $b_1 - a_1 = \tau(a_1) - a_1 = \tau(1) - 1 = 0$. If $j > r$, $a_j = 0$; hence $b_j - a_j = \tau(a_j) - a_j = \tau(0) - 0 = 0$. Thus c is a solution to (I) with at most r - 1 nonzero entries. Since this contradicts the minimality of r the result then follows.

We proceed to find such a $\tau \in J$. Since $\tau_1, \ldots, \tau_n$ is a complete system of coset representatives of H in J, there is an i with $1 \leq i \leq n$ such that $\tau_i \in H$. Since $u_j \in H'$ $(1 \leq j \leq n + 1)$, the fixed field of H, and $\tau_i \in H$, $a_{ij} = \tau_i(u_j) = u_j$. Thus, for this fixed i,

$$0 = \sum_{j=1}^{n+1} a_{ij}a_j \quad [\text{by } (*)]$$

$$= \sum_{j=1}^{n+1} u_j a_j$$

Since the u_j are independent <u>over J'</u>, there must be some index k with $1 \leq k \leq n + 1$ such that $a_k \notin J'$. Now, since J' is the fixed field of J, there is a $\tau \in J$ such that $\tau(a_k) \neq a_k$.

Let $b_i = \tau(a_i)$ $(1 \leq i \leq n + 1)$ and $b = (b_1, b_2, \ldots, b_{n+1})$. It remains only to show that b is also a solution to (I). We do this by introducing a second system (II) of equations. We then show that b is a solution to (II) and that the equations in (II) are merely a reordering of those in (I). Thus b is also a solution to (I).

Let $b_{ij} = \tau(a_{ij}) \in F$ $(1 \leq i \leq n;\ 1 \leq j \leq n + 1)$ and consider the following system of equations.

$$\text{(II)} \quad \begin{aligned} b_{11}x_1 + b_{12}x_2 + \cdots + b_{1,n+1}x_{n+1} &= 0 \\ b_{21}x_1 + b_{22}x_2 + \cdots + b_{2,n+1}x_{n+1} &= 0 \\ &\vdots \\ b_{n1}x_1 + b_{n2}x_2 + \cdots + b_{n,n+1}x_{n+1} &= 0 \end{aligned}$$

To show that the element $b \in F^{n+1}$ defined above is a solution to (II) we substitute b into an arbitrary i-th row of (II).

$$\sum_{j=1}^{n+1} b_{ij}b_j = \sum_{j=1}^{n+1} \tau(a_{ij})\tau(a_j)$$

$$= \tau\left(\sum_{j=1}^{n+1} a_{ij}a_j\right)$$

$$= \tau(0) \qquad \text{[by (*)]}$$

$$= 0$$

Claim: The equations in (II) are just a reordering of those in (I). Hence the element $b \in F^{n+1}$ is also a solution to (I) and the proof is complete.

Proof of Claim: If $i \neq j$, then $\tau_iH \neq \tau_jH$ and hence $\tau\tau_iH \neq \tau\tau_jH$. It follows that the n cosets $\tau\tau_iH$ are merely a reordering of the n cosets τ_iH ($1 \leq i \leq n$). Suppose that $\tau\tau_kH = \tau_mH$ for some k and m between 1 and n. Then by (1.8), for each j with $1 \leq j \leq n + 1$, since $u_j \in H'$ and $\tau\tau_kH = \tau_mH$, $\tau\tau_k(u_j) = \tau_m(u_j)$. Hence, for $1 \leq j \leq n + 1$,

$$b_{kj} = \tau(a_{kj}) = \tau\tau_k(u_j) = \tau_m(u_j) = a_{mj}$$

Thus the k-th equation in (II) is identical to the m-th equation in (I).

(1.11) PROPOSITION Let F be an extension field of K with Galois group $G = \mathrm{Gal}(F/K)$. Then the following assertions hold.

(i) Let L and M be intermediate fields of F over K with $L \subseteq M$. If L is closed and $[M: L]$ is finite, then M is closed and $(L': M') = [M: L]$.

(ii) Let H and J be subgroups of G with $H \subseteq J$. If H is closed and $(J: H)$ is finite, then J is closed and $[H': J'] = (J: H)$.

In particular, if G is finite, then all subgroups of G are closed.

PROOF (i) Since $[M: L]$ is finite,

(*) $(L': M') \leq [M: L]$ [by (1.9)]

Thus $(L': M')$ is finite and it follows by (1.10) that

(**) $[M'': L''] \leq (L': M')$

Hence

$$\begin{aligned}
[M: L] &\le [M'': L] && \text{(since } L \subseteq M \subseteq M''\text{)}\\
&= [M'': L''] && \text{(since L is closed)}\\
&\le (L': M') && \text{[by (**)]}\\
&\le [M: L] && \text{[by (*)]}
\end{aligned}$$

It therefore follows that $[M: L] = (L': M') = [M'': L]$. Since M is a subspace of the vector space M'' over L with the same dimension, $M = M''$ and hence M is closed in L.

(ii) Using a proof similar to that of (i) we may show that

$$(J: H) = [H': J'] = (J'': H)$$

[cf. exercise (1.5)]. It then follows that $J = J''$ and J is closed.

(1.12) THEOREM (The Fundamental Correspondence) Let F be a finite extension of K with K closed in F and let $G = \mathrm{Gal}(F/K)$. Then the following assertions hold.

(i) All intermediate fields of F over K and all subgroups of G are closed and the mapping

$$L \mapsto L' \text{ (L an intermediate field of F over K)}$$

is an order reversing bijection from the set of all intermediate fields of F over K to the set of all subgroups of G.

(ii) If L and M are intermediate fields of F over K with $L \subseteq M$, then $[M: L] = (L': M')$. In particular, $o(G) = [F: K]$.

PROOF We first show that all intermediate fields of F over K are closed. Let L be an intermediate field of F over K. Since $[F: K]$ is finite and $K \subseteq L \subseteq F$, $[L: K]$ is also finite. Hence, by (1.11), since K is closed, L is also closed.

Now let L and M be intermediate fields of F over K with $L \subseteq M$. Since $K \subseteq L \subseteq M \subseteq F$ and $[F: K]$ is finite, $[M: L]$ is also finite. Hence, since L is closed, it follows by (1.11) that $(L': M') = [M: L]$.

In particular, $o(G) = (G: (1)) = (K': F') = [F: K]$.

Now, since G is a finite group, all subgroups of G are closed. The result therefore follows by (1.7).

The following corollary indicates one method of determining if K is closed in F (where F is a finite extension of K).

(1.13) COROLLARY Let F be a finite extension of K with Galois group $G = \mathrm{Gal}(F/K)$. Then K is closed in F if and only if $o(G) = [F: K]$.

PROOF By the preceding theorem if F is a finite extension of K and K is closed in F then $o(G) = [F: K]$.

Now suppose that $o(G) = [F: K]$. Then $K \subseteq K''$, K'' is closed in F [by (1.4)] and $\mathrm{Gal}(F/K'') = \mathrm{Gal}(F/K) = G$ [again by (1.4)]. Thus

$$\begin{aligned} [F: K''] &= o(\mathrm{Gal}(F/K'')) && \text{[by (1.12) since } K'' \text{ is closed in F]} \\ &= o(G) && \text{[since } G = \mathrm{Gal}(F/K'')\text{]} \\ &= [F: K] && \text{(by hypothesis)} \end{aligned}$$

It follows that $[F: K''] = [F: K]$ and hence, since $K \subseteq K'' \subseteq F$, $K = K''$. Thus K is closed in F as required.

(1.14) EXAMPLES

(1.14.1) We first consider the extension $F = \mathbf{Q}(\sqrt{2}, \sqrt{3})$ of **Q**. By Example (1.1.4) the Galois group $G = \mathrm{Gal}(F/\mathbf{Q})$ contains the four **Q**-automorphisms of F listed below (we observe that an element of G is determined by its action on $\sqrt{2}$ and $\sqrt{3}$ but for reasons which will soon be apparent our table will include the images of all four of the basis elements of F over **Q**).

	1	$\sqrt{2}$	$\sqrt{3}$	$\sqrt{6}$
1	1	$\sqrt{2}$	$\sqrt{3}$	$\sqrt{6}$
α	1	$\sqrt{2}$	$-\sqrt{3}$	$-\sqrt{6}$
τ	1	$-\sqrt{2}$	$\sqrt{3}$	$-\sqrt{6}$
$\alpha\tau$	1	$-\sqrt{2}$	$-\sqrt{3}$	$\sqrt{6}$

Since $o(G) = 4 = [F: \mathbf{Q}]$, by Corollary (1.13) **Q** is closed in F. Thus, since F is also finite over **Q**, by the Fundamental Correspondence Theorem (1.12), the mapping θ: $L \mapsto L'$ (L an intermediate field of F over $\mathbf{Q}$) is an order reversing bijection from the set of intermediate fields of F over $\mathbf{Q}$ to the set of subgroups of G.

Since the set of subgroups of G is perhaps more familar to us than the set of intermediate fields of F over $\mathbf{Q}$, instead of θ we consider the inverse mapping T: $H \mapsto H'$ (H a subgroup of G). Thus given a subgroup H of G, we wish to determine the fixed field H' of H. By part (i) of the Fundamental Correspondence Theorem, $H = H''$ and therefore, by part (ii), $[H': \mathbf{Q}] = (G: H)$. Thus, if $m = (G: H)$, to find H' it suffices to find an intermediate field L of F over $\mathbf{Q}$ of dimension at least m over $\mathbf{Q}$ which is fixed by H; for in this case, $L \subseteq H'$, $m \leq [L: \mathbf{Q}] \leq [H':\mathbf{Q}] = m$, and hence $L = H'$.

The group G has five subgroups: (1), $H_1 = \langle\alpha\rangle$, $H_2 = \langle\tau\rangle$, $H_3 = \langle\alpha\tau\rangle$ and G itself.

By (1.3), $(1)' = F$. Since $\mathbf{Q}$ is closed in F, $G' = \mathbf{Q}$. Each of the remaining three subgroups of G has index two in G. Since the element $\sqrt{2}$ of F is fixed by H_1 and $[\mathbf{Q}(\sqrt{2}): \mathbf{Q}] = 2$, $H_1' = \mathbf{Q}(\sqrt{2})$. Similary, $H_2' = \mathbf{Q}(\sqrt{3})$. Finally, since $\alpha\tau(\sqrt{6}) = \sqrt{6}$, and $[\mathbf{Q}(\sqrt{6}): \mathbf{Q}] = 2$, $H_3' = \mathbf{Q}(\sqrt{6})$.

The correspondence is indicated in the Hasse diagrams below. Note that the Galois correspondence is order reversing so that the fields near the top correspond to the groups near the bottom.

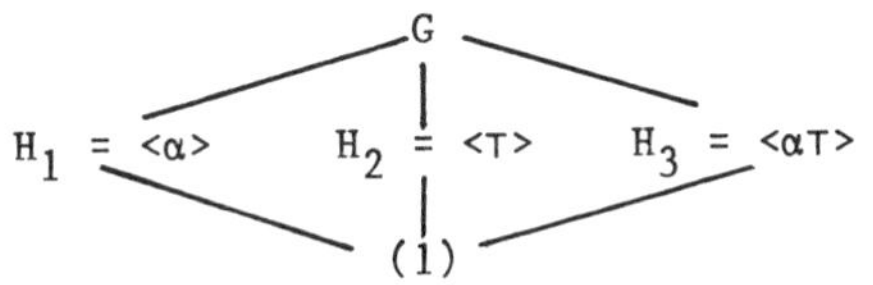

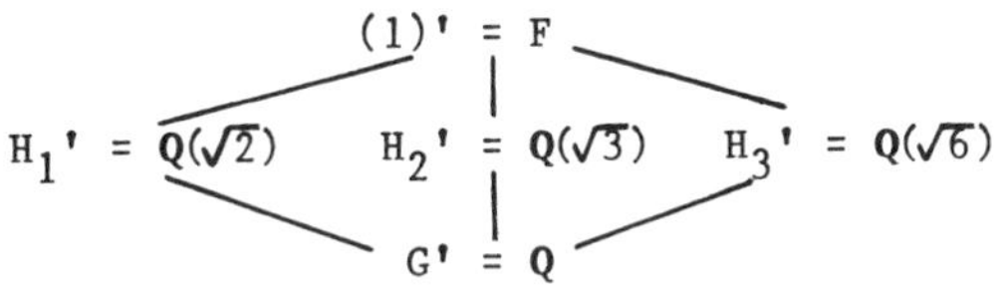

(1.14.2) Let F be a splitting field for $f = x^3 - 2$ over $\mathbf{Q}$. Thus $F = \mathbf{Q}(\rho, w)$ where $w = \sqrt[3]{2}$ and ρ is a primitive cube root of unity. We have seen in Example (1.1.7) that if $\tau \in G = Gal(F/\mathbf{Q})$ then τ is one of the six $\mathbf{Q}$-automorphisms of F determined by $\rho \mapsto \rho, \rho^2$ and $w \mapsto w, w\rho, w\rho^2$. We first show that G is isomorphic to the dihedral group D_3.

Let α: $w \mapsto w\rho$, $\rho \mapsto \rho$, and τ: $w \mapsto w$, $\rho \mapsto \rho^2$. Then $\alpha^2(w) = \alpha(\alpha(w)) = \alpha(w\rho) = \alpha(w)\alpha(\rho) = w\rho^2$ and $\alpha^3(w) = \alpha(w\rho^2) = \alpha(w)[\alpha(\rho)]^2 = w\rho^3 = w$. Since we also have $\alpha^3(\rho) = \rho$, it follows that $o(\alpha) = 3$. In a similar manner we find that τ has order 2. We now show that $\tau\alpha = \alpha^2\tau$. It will then follow that $G \cong D_3$. It suffices to show that $\tau\alpha$ and $\alpha^2\tau$ agree on ρ and w.

$$\tau\alpha(w) = \tau(w\rho) = \tau(w)\tau(\rho) = w\rho^2$$
$$\alpha^2\tau(w) = \alpha^2(w) = w\rho^2$$
$$\tau\alpha(\rho) = \tau(\rho) = \rho^2$$
$$\alpha^2\tau(\rho) = \alpha^2(\rho^2) = \alpha([\alpha(\rho)]^2) = \alpha(\rho^2) = \rho^2$$

The following table lists the images under elements of G of all six basis elements of F over $\mathbf{Q}$.

	1	w	w^2	ρ	$w\rho$	$w^2\rho$
1	1	w	w^2	ρ	$w\rho$	$w^2\rho$
α	1	$w\rho$	$w^2\rho$	ρ	$w\rho^2$	w^2
α^2	1	$w\rho^2$	$w^2\rho$	ρ	w	$w^2\rho^2$
τ	1	w	w^2	ρ^2	$w\rho^2$	$w^2\rho^2$
$\alpha\tau$	1	$w\rho$	$w^2\rho$	ρ^2	w	$w^2\rho$
$\alpha^2\tau$	1	$w\rho^2$	$w^2\rho^2$	ρ^2	$w\rho$	w^2

We observe for example that the images of w under elements of G must also be roots of the minimal polynomial f of w. Similarly the images of $w^2\rho$ must be roots of the minimal polynomial of $w^2\rho$; hence the polynomial $h = (x - w^2\rho)(x - w^2)(x - w^2\rho^2) = x^3 - 4$ must divide the minimal polynomial of $w^2\rho$. But, since $h \in \mathbf{Q}[x]$ and $h(w^2\rho) = 0$, we see in fact that h is the minimal polynomial of $w^2\rho$ over $\mathbf{Q}$ and it follows that $[\mathbf{Q}(w^2\rho): \mathbf{Q}] = 3$.

The Hasse diagrams below indicate the correspondence between the subgroups of G and the intermediate fields of F over **Q**.

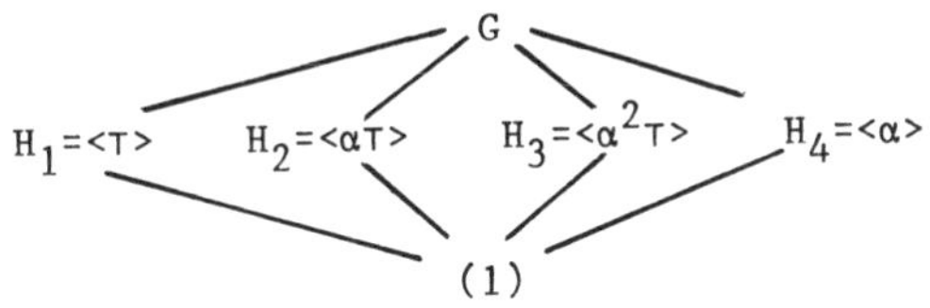

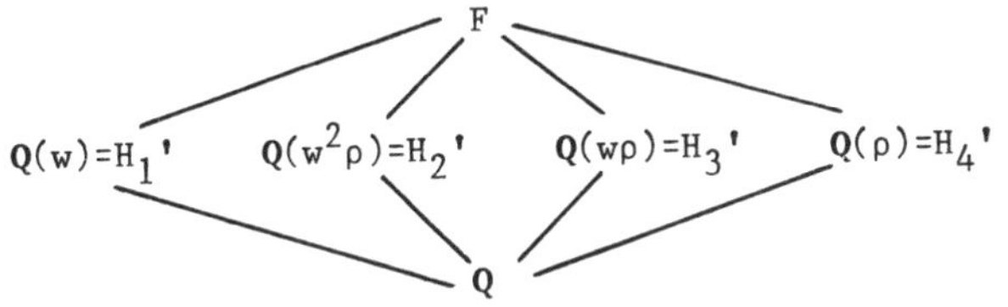

We point out that since $\rho = \cos(2\Pi/3) + i \sin(2\Pi/3) = -(1/2) + i(\sqrt{3}/2)$, F may also be realized as $\mathbf{Q}(w, \sqrt{3}i)$. We urge the student to construct G by considering the minimal polynomials of w over **Q** and $\sqrt{3}i$ over $\mathbf{Q}(w)$. The group G and corresponding subgroups and intermediate fields should be the same as the ones that we determined above [cf. exercise (1.16)].

(1.14.3) Let $F = \mathbf{Q}(w, i)$ where $w = \sqrt[4]{2}$, and $G = \mathrm{Gal}(F/\mathbf{Q})$. By Example (1.1.5), $o(G) = 8 = [F: \mathbf{Q}]$ and hence, by (1.13), **Q** is closed in F. Since F is finite over **Q** and **Q** is closed in F, the extension F over **Q** satisfies the hypotheses of the Fundamental Correspondence Theorem. The student should verify that if α: $w \mapsto iw$, $i \mapsto i$, and τ: $w \mapsto w$, $i \mapsto -i$, then α has order 4, τ has order 2 and $\tau\alpha = \alpha^3\tau$; hence G is isomorphic to D_4.

To illustrate the Fundamental Correspondence Theorem, we first list the images under elements of G of all eight basis elements of F over **Q**.

	1	w	w^2	w^3	i	iw	iw^2	iw^3
1	1	w	w^2	w^3	i	iw	iw^2	iw^3
α	1	iw	$-w^2$	$-iw^3$	i	$-w$	$-iw^2$	w^3
α^2	1	$-w$	w^2	$-w^3$	i	$-iw$	iw^2	$-iw^3$
α^3	1	$-iw$	$-w^2$	iw^3	i	$-w$	$-iw^2$	$-w^3$
τ	1	w	w^2	w^3	$-i$	$-iw$	$-iw^2$	$-iw^3$
$\alpha\tau$	1	iw	$-w^2$	$-iw^3$	$-i$	$-w$	iw^2	$-w^3$
$\alpha^2\tau$	1	$-w$	$-w^2$	$-w^3$	$-i$	iw	$-iw^2$	iw^3
$\alpha^3\tau$	1	$-iw$	$-w^2$	iw^3	$-i$	$-w$	iw^2	w^3

The Hasse diagrams below indicate the correspondence between the subgroups of G and the intermediate fields of F over **Q**.

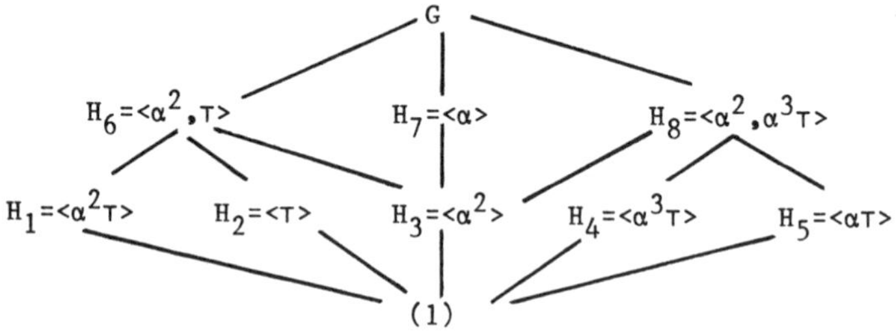

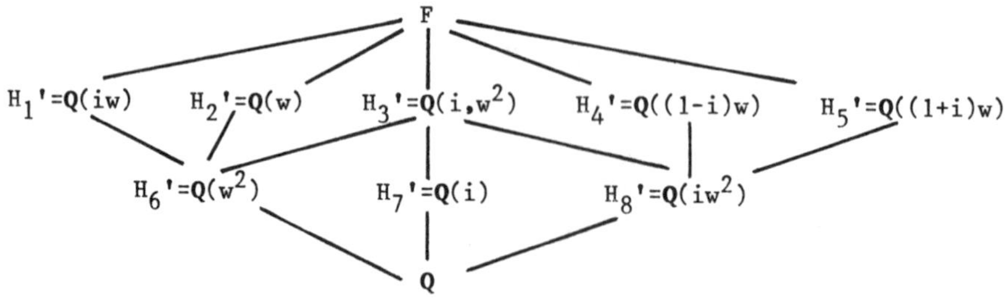

REMARK If H is a subgroup of G an element u of F is in the fixed field H' of H if and only if it is fixed by each of the generators of H [cf. exercise (1.7)].

(i) We first find H_8'. Since $H_8 = \langle\alpha^2, \alpha^3\tau\rangle$, by the above remark if $u \in F$ then $u \in H_8'$ if and only if it is fixed by both α^2 and $\alpha^3\tau$. From the table we see that iw^2 is fixed by both α^2 and $\alpha^3\tau$; thus $\mathbf{Q}(iw^2) \subseteq H_8'$. By the Fundamental Correspondence Theorem, $[H_8': \mathbf{Q}] = (G: H_8) = 2$. Hence, since $\mathbf{Q}(iw^2) \subseteq H_8'$ and $[\mathbf{Q}(iw^2): \mathbf{Q}] = 2$, $H_8' = \mathbf{Q}(iw^2)$.

(ii) We next find H_4'. By (1.12), since $(G: H_4) = 4$, $[H_4': \mathbf{Q}] = 4$. Now since iw^2 is the only basis element of F over $\mathbf{Q}$ fixed by H_4 and $[\mathbf{Q}(iw^2): \mathbf{Q}] = 2$, we must look further for other elements of H_4'.

Let $u \in H_4'$. Then, since $u \in F$, u can be written uniquely as a $\mathbf{Q}$-linear combination of the eight basis elements of F over $\mathbf{Q}$. Hence

$$u = a_1 + a_2w + a_3w^2 + a_4w^3 + a_5i + a_6iw + a_7iw^2 + a_8iw^3$$

Since $u \in H_4'$,

$$\begin{aligned} u &= \alpha^3\tau(u) \\ &= a_1 - a_2iw - a_3w^2 + a_4iw^3 - a_5i - a_6w + a_7iw^2 + a_8w^3 \end{aligned}$$

Comparing coefficients of like terms (we are using here the uniqueness of representation of u as a $\mathbf{Q}$-linear combination of basis elements) we obtain the following results.

$$a_2 = -a_6;\ a_4 = a_8;\ a_3 = a_5 = 0;\ a_1,\ a_7 \text{ arbitrary}$$

An arbitrary element u of H_4' can then be written as

$$\begin{aligned} u &= a + bw + cw^3 - biw + diw^2 + ciw^3 \\ &= a + b(1 - i)w + c(1 + i)w^3 + diw^2 \quad (a,b,c,d \in \mathbf{Q}) \end{aligned}$$

Thus $H_4' = \mathbf{Q}((1 - i)w, (1 + i)w^3, iw^2)$. However, since $iw^2 = -(1/2)[(1 - i)w]^2$ and $(1 + i)w^3 = -(1/2)[(1 - i)w]^3$, we see that in fact H_4' can be written more simply as $\mathbf{Q}((1 - i)w)$.

EXERCISES

(1.1) Let F be an extension field of K. Prove that Gal(F/K) is a subgroup of Aut F.

(1.2) Prove (1.4), (iv).

(1.3) Let L be an intermediate field of F over K. Show that L'' is the same in the extensions F over K and F over L [cf. Remark (1.5)].

(1.4) Let $w = \sqrt[4]{2}$. Show that **Q** is not closed in **Q**(wi).

(1.5) Let G = Gal(F/K) and $H \subseteq J \subseteq G$ be subgroups of G such that (J: H) is finite. Show that (J: H) = [H': J'] = (J'': H) [see (1.11), (ii)].

(1.6) Find the remaining fixed fields in the extension **Q**(w, i) over **Q** where $w = \sqrt[4]{2}$ as in Example (1.14.2).

(1.7) Let G = Gal(F/K) and H be a finitely generated subgroup of G. Prove that if $u \in F$, then $u \in H'$ if and only if $\tau(u) = u$ for each of the generators τ of H [cf. Example (1.14.2)].

(1.8) Let ρ be a primitive third root of unity and $w = \sqrt[3]{2}$.

(a) Find the minimal polynomial f of w over **Q**.

(b) Show that the roots of f are w, $w\rho$ and $w\rho^2$.

(c) Find [**Q**(w, ρ): **Q**] and list the basis elements.

(d) Let G = Gal(**Q**(w, ρ)/**Q**). Show that o(G) = 6 and hence **Q** is closed in **Q**(w, ρ).

(e) Show that $G \cong D_3$.

(f) Use the Galois correspondence to find all subgroups of G and corresponding fixed fields. Which fixed fields correspond to normal subgroups of G?

(1.9) Show that Gal(**Q**($\sqrt{2}$, $\sqrt{3}$, $\sqrt{5}$)/**Q**) $\cong \mathbf{Z}_2 \times \mathbf{Z}_2 \times \mathbf{Z}_2$. Determine all subgroups and corresponding fixed fields.

(1.10) Prove that Gal(**R**/**Q**) = (1). [Hint: Show that if $\tau \in$ Gal(**R**/**Q**) then τ sends positives to positives and hence preserves the order in **R**.]

(1.11) Prove that if [F: K] = 2 then K is closed in F.

(1.12) Let E be an intermediate field of F over K such that K is closed in E, E is closed in F and every $\tau \in$ Gal(E/K) extends to an isomorphism of F. Show that K is closed in F.

(1.13) Let L and M be intermediate fields of F over K. Recall that $L \vee M$ is the smallest subfield of F containing $L \cup M$.

(a) Show that $L \vee M = L\,M$ where

$$L\,M = \{v \in F\colon v = \sum_{i=1}^{n} u_i v_i,\ u_i \in L,\ v_i \in M,\ n \in \mathbf{N}\}$$

(b) Show that $(L \vee M)' = L' \cap M'$.

(1.14) Suppose that $[F\colon K]$ is finite. Show that if $\tau\colon F \to F$ is a nonzero homomorphism such that $\tau(u) = u$ for all $u \in K$, then $\tau \in \mathrm{Gal}(F/K)$.

(1.15) If $f \in K[x]$, the Galois group of f over K is the group Gal(F/K) where F is a splitting field of f over K. Describe the Galois groups of each of the polynomials f over the field K.

(a) $f = x^2 - 5$, $K = \mathbf{Q}$.

(b) $f = x^3 - 1$, $K = \mathbf{Q}$.

(c) $f = (x^2 - 5)(x^3 - 1)$, $K = \mathbf{Q}$.

(d) $f = (x^5 - 2)$, $K = \mathbf{Q}(\rho)$ where ρ is a primitive 5-th root of unity.

(1.16) Let $F = \mathbf{Q}(w, \sqrt{3}i)$ where $w = \sqrt[3]{2}$. Show that $\mathrm{Gal}(F/\mathbf{Q})$ is the same group as that determined in (1.14.2).

(1.17) A <u>Galois connection</u> between two partially ordered sets X and Y is a pair of order reversing maps $f\colon X \to Y$ and $g\colon Y \to X$ satisfying $x \le (g \circ f)(x)$ for all $x \in X$ and $y \le (f \circ g)(y)$ for all $y \in Y$. For $x \in X$ and $y \in Y$, let $x' = f(x)$ and $y' = g(y)$. Then x is said to be a closed (or Galois) element of X if $x'' = x$. Similarly y is called a closed or Galois element of Y if $y'' = y$. Suppose that f and g define a Galois connection between X and Y. Let $X' = \mathrm{im}\, f$ and $Y' = \mathrm{im}\, g$.

(a) Show that $Z' = Z'''$ for all Z in X or Y.

(b) Show that X' and Y' consist precisely of all closed elements of X and Y respectively.

(c) Let $f'\colon X' \to Y'$ and $g'\colon Y' \to X'$ be the restriction of f and g to the set of Galois objects. Show that f' and g' are inverse bijections to one another.

2. THE FUNDAMENTAL THEOREM OF GALOIS THEORY

We continue now with our study of the interrelationship between intermediate fields of an extension F over K and subgroups of the associated Galois group G = Gal(F/K). We recall from Chapter VIII that an extension L over K is said to be normal over K if every irreducible polynomial over K which has a root in L splits in L. In VIII, (3.17) it was shown that if L is a finite extension of K then L is normal over K if and only if L is a splitting field over K of a polynomial $f \in K[x]$.

Let F be a finite extension of K such that K is closed in F. Our goal is to prove the following third part of the Fundamental Correspondence Theorem for fields of characteristic 0.

(III) If L is an intermediate field of F over K then L is normal over K if and only if L' is a normal subgroup of G. In this case the mapping

$$\theta: G \to \mathrm{Gal}(L/K) \text{ by } \tau \mapsto \tau|_L$$

is a surjective homomorphism of groups with $\ker \theta = L'$ and hence θ induces an isomorphism of groups $\theta': G/L' \to \mathrm{Gal}(L/K)$.

In order that θ be a map into Gal(L/K) it is necessary that, for each $\tau \in G$, $\tau|_L$ be a K-automorphism of L. In particular, we require that for each $\tau \in G$, $\tau(L) \subseteq L$. We are thus motivated to make the following definition.

(2.1) DEFINITION Let F be an extension field of K with Galois group G = Gal(F/K) and L an intermediate field of F over K. Then L is <u>stable</u> (relative to F over K) if $\tau(L) \subseteq L$ for every $\tau \in G$.

(2.2) PROPOSITION Let F be an extension field of K with Galois group

$G = \mathrm{Gal}(F/K)$ and L a stable intermediate field of F over K. Then the mapping

$$\theta: G \to \mathrm{Gal}(L/K) \text{ by } \tau \mapsto \tau|_L$$

is a homomorphism of groups with $\ker \theta = L'$. Hence L' is a normal subgroup of G and θ induces an injective homomorphism of groups

$$\theta': G/L' \to \mathrm{Gal}(L/K) \text{ by } [\tau] \mapsto \tau|_L$$

PROOF We must first show that if $\tau \in G$, $\tau|_L \in \mathrm{Gal}(L/K)$. Let $\tau \in G$. Since τ and τ^{-1} are both elements of G and L is stable in F over K, τ and τ^{-1} both map L into itself. Hence (since τ and τ^{-1} also fix K), $\tau|_L$ is a K-automorphism of L with inverse $\tau^{-1}|_L$.

If $\alpha, \tau \in G$ then

$$\theta(\alpha\tau) = \alpha\tau|_L = \alpha|_L\tau|_L = \theta(\alpha)\theta(\tau)$$

and hence θ is a homomorphism of groups. Finally, if $\tau \in G$, then $\tau \in \ker \theta$ if and only if $\tau|_L = 1_L$; that is, if and only if $\tau \in L'$. Thus $\ker \theta = L'$ and the result now follows from group theory.

(2.3) PROPOSITION Let F be an extension field of K with Galois group $G = \mathrm{Gal}(F/K)$. If H is a normal subgroup of G then H' is a stable intermediate field of F over K.

PROOF Since H is a normal subgroup of G, $\alpha^{-1}\tau\alpha \in H$ for all $\alpha \in G$ and $\tau \in H$. Let $\alpha \in G$ and $u \in H'$. We must show that $\alpha(u) \in H'$; that is, $\tau\alpha(u) = \alpha(u)$ for all $\tau \in H$. But since $u \in H'$ and $\alpha^{-1}\tau\alpha \in H$, $\alpha^{-1}\tau\alpha(u) =$ u or, equivalently, $\tau\alpha(u) = \alpha(u)$ as required.

The next two propositions show that if F is a finite extension of K and K has characteristic 0 then F is normal over K if and only if K is closed in F.

(2.4) PROPOSITION Let K be closed in F and f be an irreducible polynomial in K[x] which has a root u in F. Then

$$f = a(x - u_1)(x - u_2) \cdots (x - u_m) \quad (a \in K)$$

where $u_1 = u, u_2, \ldots, u_m$ are the distinct images of u under elements of Gal(F/K). Hence, in particular, F is normal over K.

PROOF Without loss of generality we assume that f is monic. Then u is algebraic over K with minimal polynomial f.

Let G = Gal(F/K). If $\tau \in G$ then $\tau(u) \in F$ is a root of f by VIII, (1.11). Hence the set of images of u under elements of G is a subset of the set of roots of f. Let $u_1 = u, u_2, \ldots, u_m$ be the distinct images of u under elements of G (in which case $m \leq \deg f$) and define

$$g = (x - u_1)(x - u_2) \cdots (x - u_m)$$

Then $g \in F[x]$. We will show that in fact $g \in K[x]$. It will then follow by VIII, (1.7) that, since g(u) = 0 and f is the minimal polynomial of u over K, $f|g$ and hence $\deg f \leq \deg g$. Thus deg f = deg g and, since f and g are monic of the same degree and $f|g$, f = g.

We proceed to show that $g \in K[x]$. Suppose that g = $a_0 + a_1x + \cdots + a_mx^m$ with $a_i \in F$. Since K is closed in F, to show that $a_i \in K$, it suffices to show that a_i is fixed by every element of G. Let $\tau \in G$. Then, since τ merely permutes the u_i, $\tau g = g$. Hence coefficients of like terms in g and τg are equal; that is, $a_i = \tau(a_i)$ for each i. The result therefore follows.

Since the polynomial f had distinct roots in the preceding proposition, it is not surprising that some conditions on K are required in order that the converse hold. In the next proposition we require that K be a field of characteristic 0. In fact, it may be observed from the proof that a weaker condition suffices - that is, that F be a splitting field over K of a separable polynomial [cf. VIII, (3.11)].

(2.5) PROPOSITION Let K be a field of characteristic 0 and F a finite extension of K. Then if F is normal over K, K is closed in F.

PROOF By VIII, (3.17), F is a splitting field over K of a polynomial $f \in K[x]$. Let G = Gal(F/K). We shall prove that K is closed in F by showing that o(G) = [F: K] [cf. (1.13)]. The counting argument will use the fact that, since K has characteristic 0, any irreducible polynomial $g \in K[x]$ splits over F into a product of distinct linear factors. We will induct on $n = [F: K]$.

If $[F: K] = 1$ (or equivalently, if f splits over K), then $F = K$ and the result is trivial. Otherwise let g be a monic irreducible factor of f having degree m greater than 1. Let u be a root of g (so that g is the minimal polynomial of u over K) and $L = K(u)$. Then by VIII, (1.8), $[L: K] = m > 1$. Thus by our induction hypothesis applied to the extension F over L, L is closed in F (recall that F is normal over L since F is also a splitting field over L of f). It follows by (1.12) that $o(L') = o(\mathrm{Gal}(F/L)) = [F: L]$.

Claim: $(G: L') = [L: K]$. It then follows that

$$o(G) = o(L')\,(G: L') = [F: L]\,[L: K] = [F: K]$$

and the proof is complete.

Proof of Claim: Let S be the set of distinct cosets of L' in G and T the set of distinct roots of g in F. Since f splits in F so does g. Thus, since g is irreducible in K[x] and K has characteristic 0, card $(T) = \deg g = [L: K]$. It therefore suffices to construct a bijective map from the set S to the set T.

Let $\theta: S \to T$ be defined by $\tau L' \mapsto \tau(u)$ $(\tau \in G)$. By (1.8) if $\alpha, \tau \in G$, then $\alpha L' = \tau L'$ if and only if $\alpha(u) = \tau(u)$. Hence the map θ is both well-defined and injective. By VIII, (3.9) if $v \in T$, then since u and v both have minimal polynomial g over K and F is a splitting field over K there is an element $\tau \in G$ such that $\tau(u) = v$. Thus θ is also surjective.

(2.6) COROLLARY Let K be a field of characteristic 0 and F an finite extension of K. Then the following statements are equivalent.

(i) F is normal over K.

(ii) K is closed in F.

(iii) F is a splitting field over K of some $f \in K[x]$.

(2.7) THEOREM Let F be a finite normal extension of K with Galois group $G = Gal(F/K)$ and L an intermediate field of F over K. Suppose that K has characteristic 0. Then the following statements are equivalent.

(i) K is closed in L.

(ii) L is a splitting field over K of some $f \in K[x]$.

(iii) L is normal over K.

(iv) L is stable (relative to F over K).

(v) L' is a normal subgroup of G.

PROOF The equivalence of (i), (ii) and (iii) follows by the preceding corollary.

(iii)$\Longrightarrow$(iv) Suppose that L is normal over K. Let $\tau \in G$ and $u \in L$. We must show that $\tau(u) \in L$. Since F is a finite extension of K, F is algebraic over K by VIII, (2.2). By VIII, (1.11), if f is the minimal polynomial of u over K, then $\tau(u)$ is also a root of f. Then, since L is normal over K, $\tau(u) \in L$ as required.

(iv)$\Longrightarrow$(v) If L is stable then L' is normal in G by (2.2).

(v)$\Longrightarrow$(i) Suppose that L' is a normal subgroup of G. To show that K is closed in L we must show that for each $u \in L - K$ there is a $\tau \in Gal(L/K)$ such that $\tau(u) \neq u$.

Let $u \in L - K$. Since K is closed in F there is an $\alpha \in Gal(F/K)$ such that $\alpha(u) \neq u$. By (2.3), since L' is normal in G, L'' is stable in F over K. By the Fundamental Correspondence Theorem (1.12), L is closed in F; hence $L'' = L$. Thus L is stable in F over K and the map $\tau = \alpha|_L$ is an element of $Gal(L/K)$ by (2.2). Since $\tau(u) = \alpha(u) \neq u$, we have produced the required element τ of $Gal(L/K)$.

We are now ready to prove the Fundamental Theorem of Galois Theory. Although parts (i) and (ii) have been proved in (1.12) we state them again for the sake of completeness.

(2.8) THEOREM (The Fundamental Theorem of Galois Theory) Let F be a finite extension of K with K closed in F and $G = Gal(F/K)$. Then the following assertions hold.

(i) All intermediate fields of F over K and all subgroups of G are closed and the mapping

$$L \mapsto L' = \mathrm{Gal}(F/L)$$

is an order reversing bijection from the set of all intermediate fields L of F over K to the set of all subgroups of G.

(ii) If M and L are intermediate fields of F over K with $L \subseteq M$, then $[M: L] = (L': M')$. In particular, $o(G) = [F: K]$.

(iii) Suppose that K has characteristic 0. If L is an intermediate field of F over K then L is normal over K if and only if L' is a normal subgroup of G.

$$\begin{array}{ccl} F & \longleftrightarrow & (1) \\ \cup & & \cap \\ L & \longleftrightarrow & L' = \mathrm{Gal}(F/L) \\ \cup & & \cap \\ K & \longleftrightarrow & G = \mathrm{Gal}(F/K) \end{array}$$

In this case the mapping

$$\theta: G \to \mathrm{Gal}(L/K) \text{ by } \tau \mapsto \tau|_L \ (\tau \in G)$$

is a surjective homomorphism of groups with $\ker \theta = L'$. Hence θ induces an isomorphism of groups

$$G/L' \cong \mathrm{Gal}(L/K)$$

and there is an exact sequence of groups [cf. II, (4.12)]

$$1 \to \mathrm{Gal}(F/L) \to \mathrm{Gal}(F/K) \to \mathrm{Gal}(L/K) \to 1$$

We say that the exact sequence given above is induced by the Galois correspondence.

PROOF (iii) By (2.7) L is normal over K if and only if L' is a normal subgroup of G.

Suppose now that L is normal over K. Then L is stable in F over K by (2.3). Hence, by (2.2) the mapping θ defined in (iii) is a homomorphism of groups with $\ker \theta = L'$ and θ induces an injective homomorphism of groups

$$\theta': G/L' \to \mathrm{Gal}(L/K) \text{ by } [\tau] \mapsto \tau|_L \ (\tau \in G)$$

It then suffices to show that θ' is a surjective map.

Since L is a finite normal extension of K, by (ii) Gal(L/K) is a finite group of order [L: K]. Then card (Im θ') = o(G/L') = (G: L') = (K': L') = [L: K] = o(Gal(L/K)) and hence Im θ = Gal(L/K).

(2.9) EXAMPLE We illustrate the preceding theorems by considering the finite extension F = **Q**(w, i) over **Q** where $w = \sqrt[4]{2}$. We recall that F is a splitting field over **Q** of the irreducible polynomial f = $x^4 - 2$. Thus F is a finite normal extension of **Q** and we may conclude by the Fundamental Theorem of Galois Theory, without explicitly calculating the Galois group G = Gal(F/K), that o(G) = [F: K] = 8.

As in Example (1.1.5), if $\tau \in G$, then τ must take w to one of w, -w, iw or -iw and i to i or -i. This results in 8 possibilities, and since o(G) = 8, all 8 of these possibilities must indeed define elements of G. The reader is referred to Example (1.14.3) for the correspondence between the subgroups of the Galois group G = Gal(F/K) and the intermediate fields of F over K.

ILLUSTRATION OF (2.4) Theorem (2.4) gives us a method for determining the minimal polynomial over **Q** of any element $u \in F$. We illustrate by considering the element $u = (1 - i)w \in F$. Lef f be the minimal polynomial of u over **Q** (note that since F is a finite extension of **Q** it is also algebraic over **Q**). Then f has a root in F, namely u. By (2.4), since F is normal over **Q**, f = $(x - u_1)(x - u_2) \cdots (x - u_m)$ where $u_1, u_2, \ldots, u_m$ are the <u>distinct</u> images of u under elements of G. Referring to the table in (1.14.3) we find that the distinct images of u under elements of G are the four elements

$$u_1 = u,\ u_2 = \bar{u},\ u_3 = (-1 + i)w,\ u_4 = \bar{u}_3$$

Hence

$$\begin{aligned} f &= (x - u)(x - \bar{u})(x - u_3)(x - \bar{u}_3) \\ &= (x^2 - 2wx + 2w^2)(x^2 + 2wx + 2w^2) \\ &= x^4 + 8 \end{aligned}$$

ILLUSTRATION OF (2.2) We saw in Example (1.6.3) that the intermediate field $\mathbf{Q}(w)$ of F over $\mathbf{Q}$ is not normal over $\mathbf{Q}$. Recall that $[\mathbf{Q}(w): \mathbf{Q}] = 4$ and $\{1, w, w^2, w^3\}$ is a basis for $\mathbf{Q}(w)$ over $\mathbf{Q}$. Let α be the element of G defined by $w \mapsto iw$, $i \mapsto i$. Then $\alpha(\mathbf{Q}(w))$ is not contained in $\mathbf{Q}(w)$ since $\alpha(w) \notin \mathbf{Q}(w)$; we have therefore verified that $\mathbf{Q}(w)$ is not a stable intermediate field of F over $\mathbf{Q}$. We also observe that $\mathbf{Q}(w)'$ is not a normal subgroup of G since $\mathbf{Q}(w)' = H_2 = \langle\tau\rangle$ and $\alpha^{-1}\tau\alpha = \alpha^3\tau\alpha = \alpha^2\tau \notin \langle\tau\rangle$.

ILLUSTRATION OF (2.6) Since $H_3 = \langle\alpha^2\rangle$ is a normal subgroup of G, the corresponding intermediate field $H_3' = \mathbf{Q}(i, w^2)$ must be normal over $\mathbf{Q}$. By (2.6) $\mathbf{Q}(i, w^2)$ must be a splitting field over $\mathbf{Q}$ of some polynomial $g \in \mathbf{Q}[x]$. The student may verify that $\mathbf{Q}(i, w^2)$ is a splitting field over $\mathbf{Q}$ of $g = (x^2 + 1)(x^2 - 2) \in \mathbf{Q}[x]$.

ILLUSTRATION OF (2.8) Consider the intermediate field $L = \mathbf{Q}(i)$ of F over $\mathbf{Q}$. Since $L' = \langle\alpha\rangle$ has index 2 in G, L' is a normal subgroup of G by II, Exercise (4.2). Hence by the Galois correspondence L must be normal over $\mathbf{Q}$. We observe that in fact L is the splitting field over $\mathbf{Q}$ of $g = x^2 + 1$ and $\mathrm{Gal}(L/\mathbf{Q}) = \{1, \rho\}$ where $\rho: i \mapsto -i$.

As in (2.8) we define a map

$$\theta: G \to \mathrm{Gal}(L/\mathbf{Q}) \text{ by } \tau \mapsto \tau|_L \quad (\tau \in G)$$

Referring to the table of images of the elements of G given in Example (1.14.3) we see that the elements $1, \alpha, \alpha^2, \alpha^3$ all have images 1 under θ [that is, they all restrict to the identity on $L = \mathbf{Q}(i)$] and $\tau, \alpha\tau, \alpha^2\tau, \alpha^3\tau$ all have images ρ under θ [that is, they all restrict to complex conjugation on $L = \mathbf{Q}(i)$]. Hence θ is a surjective homomorphism of groups with $\ker\theta = \langle\alpha\rangle = L'$ and θ induces an isomorphism $G/L' \cong \mathrm{Gal}(L/\mathbf{Q})$.

(2.10) DEFINITION Let K be a field and $f \in K[x]$. The Galois group of the polynomial f is the group $\mathrm{Gal}(F/K)$ where F is a splitting field over K of f.

Since by VIII, (3.8), any two splitting fields over K of f are isomorphic by an isomorphism fixing K, the Galois group of f over K is independent of the choice of splitting field F [cf. exercise (2.2)].

(2.11) THEOREM Let K be a field of characteristic 0, $f \in K[x]$ and $G = Gal(F/K)$ the Galois group of f (where F is a splitting field over K of f). Then the following assertions hold.

(i) If f has m distinct roots in F then G is isomorphic to a subgroup of the symmetric group S_m.

(ii) If f is an irreducible polynomial of degree n then $n|o(G)$ and G is a subgroup of S_n.

PROOF (i) Let $u_1, \ldots, u_m$ be the distinct roots of f in F. Consider S_m as the group of permutations of $\{u_1, \ldots, u_m\}$. If $\tau \in G$ then τ merely permutes the roots of f and hence the map τ' defined by $u_i \mapsto \tau(u_i)$ is an element of S_m. Let $\theta: G \to S_m$ be defined by $\theta(\tau) = \tau'$. Then θ is a homomorphism of groups. Since F is a splitting field over K of f, $F = K(u_1, \ldots, u_m)$ and hence an element of G is uniquely determined by its action on the u_i. It follows that θ is an injective map.

(ii) Suppose that f is irreducible of degree n. Then by VIII, (3.12), f has n distinct roots in F. Hence by (i) G is isomorphic to a subgroup of S_n. Let u be a root of f in F. Then $[K(u): K] = n$. Since F is a finite normal extension of K, the subgroup $K(u)'$ of G must have index n in G. Hence $n|o(G)$.

The preceding theorems provide valuable insight into the structure of finite normal extensions. If K has characteristic 0 then a finite extension F of K is normal over K, or equivalently, K is closed in F, if and only if F is a splitting field over K of some polynomial $f \in K[x]$. In this case the Galois group $G = Gal(F/K)$ is the Galois group of the polynomial f and can be thought of as a subgroup of the group of permutations of the roots of f.

In the next section we will apply this theory to show that if a polynomial equation $f(x) = 0$ is 'solvable by radicals' (a term we will define precisely in Section 3) then the Galois group of f is a solvable group. We will then use this fact to produce a fifth degree equation which is not solvable by radicals over **Q**.

(2.12) EXAMPLES

(2.12.1) Let $F = \mathbf{Q}(\sqrt[4]{2}, i)$ and $G = \mathrm{Gal}(F/\mathbf{Q})$. Since F is a splitting field over **Q** of the irreducible polynomial $f = x^4 - 2$, G is isomorphic to a subgroup of the group of permutations of the roots of f; that is, the group of permutations of the set $\{w, -w, iw, -iw\}$. For example, the element τ of G determined by $w \mapsto iw$, $i \mapsto i$ results in the permutation

$$w \mapsto iw,\ -w \mapsto -iw,\ iw \mapsto -w,\ -iw \mapsto w$$

(2.12.2) Let G be the Galois group of the polynomial $f = x^6 - 2$ over **Q**. To find a splitting field F over **Q** of f, let $w = \sqrt[6]{2}$ and ρ be a primitive 6-th root of unity. Then $\rho = \cos(2\Pi/6) + i\sin(2\Pi/6) = (1 + \sqrt{3}i)/2$ and $F = \mathbf{Q}(w, \rho)$. Since w has minimal polynomial f over **Q**, $[\mathbf{Q}(w): \mathbf{Q}] = 6$. Then, since $\mathbf{Q}(w) \subseteq \mathbf{R}$, $\rho \notin \mathbf{Q}(w)$; then ρ has minimal polynomial $g = x^2 - x + 1$ over $\mathbf{Q}(w)$ and $[F: \mathbf{Q}(w)] = 2$. Hence $[F: \mathbf{Q}] = 12$ and it follows by the Fundamental Theorem of Galois Theory that $o(G) = 12$. The reader should verify that if $\alpha: w \mapsto w\rho$, $\rho \mapsto \rho$ and $\tau: w \mapsto w$, $\rho \mapsto \rho^2$ then α has order 6, τ has order 2 and $\tau\alpha = \alpha^5\tau$ and hence $G \cong D_6$.

Let H be the subgroup of G generated by α^3. Then H is a normal subgroup of G of index 6 and it follows by the Galois correspondence that the fixed field H' of H is a normal extension of **Q** of dimension 6. Since w^2 and ρ^2 are fixed by H and $[\mathbf{Q}(w^2, \rho^2): \mathbf{Q}] = 6$, it follows that $H' = \mathbf{Q}(w^2, \rho^2)$. By (2.6) H' must be a splitting field over **Q** of some polynomial $h \in \mathbf{Q}[x]$. In fact it may be verified that H' is a splitting field over **Q** of the polynomial $h = x^3 - 2$.

We conclude this section with a proof of the Fundamental Theorem of Algebra which states that the field **C** of complex numbers is algebraically closed. Our proof is based on the Galois theory we have developed. We assume that every positive element of **R** has a square root in **R** and that every polynomial in **R**[x] of odd degree has a root in **R**. The first fact follows from the construction of the real numbers from the rationals and the second fact is a consequence of the Intermediate Value Theorem of elementary calculus.

(2.13) PROPOSITION Let p be a prime number and K a field of characteristic 0 satisfying the property that every finite extension F of K has dimension [F: K] divisible by p. If F is a finite extension of K then $[F: K] = p^n$ for some $n \in \mathbf{N}$.

PROOF Let $F \neq K$ be a finite extension of K. If E is the normal closure of F over K [cf. VIII, (3.18)], then $K \subseteq F \subseteq E$, E is normal over K and [E: K] = [E: F] [F: K]. Hence if the dimension of E over K is a power of p so is the dimension of F over K. We thus assume without loss of generality that F is normal over K.

Let G = Gal(F/K). Since F is a finite normal extension of K, by the Fundamental Theorem of Galois Theory o(G) = [F: K]; hence p divides o(G). By the first Sylow theorem [III, (1.17)], G contains a p-Sylow subgroup P; that is, a subgroup P of G satisfying $o(P) = p^n$ (some $n \in \mathbf{N}$) and p does not divide (G: P). Then since F is normal over K, if P' is the fixed field of P (relative to F over K), [P': K] = (G: P). But then p does not divide [P': K]. Therefore, by hypothesis, P' = K and consequently P = G. Thus $[F: K] = o(G) = o(P) = p^n$.

(2.14) PROPOSITION Every finite extension of **R** has dimension 2^n over **R** for some $n \in \mathbf{N} \cup \{0\}$.

PROOF Suppose that **R** has a finite extension F with $[F: \mathbf{R}] = r > 1$ with r odd. Then there is an element $u \in F$ which is not in **R**. Let f be the minimal polynomial of u over **R**. Since deg f = [**R**(u): **R**] which divides [F: **R**], f has odd degree. But then f has a root in **R** which contradicts the fact that f is irreducible over **R**. We conclude therefore that every finite extension of **R** has even dimension over **R** and hence, by the preceding proposition every finite extension of **R** has dimension 2^n over **R** for some $n \in \mathbf{N}$.

(2.15) PROPOSITION There are no extension fields of dimension 2 over **C**.

PROOF We first show that every element of **C** has a square root in **C**. It will then follow, using the quadratic formula, that every second degree polynomial in **C**[x] splits over **C**.

Let a + bi be a nonzero element of **C** (a,b $\in$ **R**). Then the positive real numbers $|(a + \sqrt{a^2 + b^2})/2|$ and $|(-a + \sqrt{a^2 + b^2})/2|$ have real positive square roots c and d respectively. The student should verify that with a proper choice of signs $(\pm c \pm di)^2 = a + bi$. Hence every element of **C** has a square root in **C**.

Now, if $f = ax^2 + bx + c \in$ **C**[x], the element $d = b^2 - 4ac \in$ **C** has a square root $\sqrt{d}$ in **C** as shown above. Then (by the quadratic formula) f has roots $(-b \pm \sqrt{d})/2$ in **C**. Thus any second degree polynomial f $\in$ **C**[x] splits over **C**.

Suppose that F is an extension field of dimension 2 over **C**. Since F $\neq$ **C**, there is an element u $\in$ F which is not in **C**. Then [**C**(u): **C**] = 2 and hence u is a root of some irreducible monic polynomial f $\in$ **C**[x]. But by the previous paragraph all such polynomials split over **C** contradicting the fact that u $\notin$ **C**.

(2.16) THEOREM (The Fundamental Theorem of Algebra) The field **C** of complex numbers is algebraically closed.

PROOF Let f $\in$ **C**[x] and let F be a splitting field over **C** of f. We wish to show that F = **C** and hence f splits over **C**.

Since F is a splitting field over **C**, F is a finite normal extension of **C** by (2.7). Since [F: **R**] = [F: **C**][**C**: **R**] = [F: **C**] (2) and by (2.14) [F: **R**] = 2^n for some n $\in$ **N**, [F: **C**] = 2^{n-1}.

Suppose that F $\neq$ **C**. Then n > 1. Let G = Gal(F/**C**). By the Fundamental Theorem of Galois Theory, o(G) = [F: **C**] = 2^{n-1}. Since n > 1, G is a 2-group; hence by III, (1.6) G has a subgroup H of index 2. But then, under the Galois correspondence, the fixed field H' of H (relative to F over **C**) has dimension 2 over **C** contradicting (2.15). We conclude therefore that F = **C** and **C** is algebraically closed as claimed.

EXERCISES

(2.1) Let $w = \sqrt[4]{2}$ and $G = \mathrm{Gal}(\mathbf{Q}(w, i)/\mathbf{Q})$.

(a) Determine the minimal polynomial of $(1 + i)w$ over $\mathbf{Q}$.

(b) Verify that the fixed fields corresponding to the normal subgroups of G are precisely the intermediate fields of $\mathbf{Q}(w, i)$ which are normal over $\mathbf{Q}$.

(2.2) Let $f \in K[x]$ and E and F both be splitting fields of f over K. Let $G = \mathrm{Gal}(F/K)$ and $H = \mathrm{Gal}(E/K)$. Show that $G \cong H$.

(2.3) Let F be a finite normal extension of K and L and M intermediate fields which are normal over K.

(a) Show that $(L \vee M)' = (L\, M)' = L' \cap M'$ [cf. exercise (1.13)].

(b) Show that $(L \cap M)' = L' \vee M' = L'M'$.

(c) Show that the map

$$\theta\colon \mathrm{Gal}(F/K) \to \mathrm{Gal}(L/K) \times \mathrm{Gal}(M/K) \quad \text{by} \quad f \mapsto (f|_L, f|_M)$$

is a homomorphism of groups with $\ker \theta = L' \cap M'$.

(d) Show that θ is injective if and only if $L \vee M = F$.

(e) Show that θ is surjective if and only if $L \cap M = K$.

(2.4) Let K be a field of characteristic 0, $g, h \in K[x]$ and $f = gh$. Let G_1 be the Galois group of g over K, G_2 the Galois group of h over K and G the Galois group of f over K. Show that G is isomorphic to a subgroup of $G_1 \times G_2$ and, if the splitting fields for g and h over K intersect in K, then $G \cong G_1 \times G_2$.

(2.5) Show that $\mathrm{Gal}(\mathbf{Q}(\sqrt{3}, \sqrt{5})/\mathbf{Q}) \cong \mathrm{Gal}(\mathbf{Q}(\sqrt{3})/\mathbf{Q}) \times \mathrm{Gal}(\mathbf{Q}(\sqrt{5})/\mathbf{Q}) \cong \mathbf{Z}_2 \times \mathbf{Z}_2$.

(2.6) Let ρ be a primitive cube root of unity and $w = \sqrt[3]{2}$. Recall that $F = \mathbf{Q}(w, \rho)$ is a normal extension of $\mathbf{Q}$ and that if $G = \mathrm{Gal}(F/\mathbf{Q})$ then $G \cong D_3$ [cf. exercise (1.8)].

(a) Verify that H is a normal subgroup of G if and only if H' is normal over K.

(b) Verify that $\mathbf{Q}(\rho)$ is a normal extension of $\mathbf{Q}$.

(c) Find $(\mathbf{Q}(\rho))'$ and verify that $\mathrm{Gal}(\mathbf{Q}(\rho)/\mathbf{Q}) \cong G/\mathbf{Q}(\rho)'$.

(2.7) Show that if G is the Galois group of the polynomial $(x^4 - 2)(x^3 - 2)$ over $\mathbf{Q}$, then $G \cong D_4 \times D_3$.

(2.8) Let $f = x^3 + x + 1 \in F_2[x]$, u be a root of f and $F = F_2(u)$.

(a) Show that F is a normal extension of F_2 and $\mathrm{Gal}(F/F_2) \cong \mathbf{Z}_3$.

(b) Find the minimal polynomial of $u + 1$ over F_2.

(2.9) Let G be the Galois group of $f = x^6 - 2$ over $\mathbf{Q}$. Find all subgroups of G and corresponding intermediate fields of F over $\mathbf{Q}$ where F is the splitting field of f over $\mathbf{Q}$. Determine which are normal. For the normal intermediate fields E of F over $\mathbf{Q}$ find a polynomial g such that E is the splitting field of g over $\mathbf{Q}$.

(2.10) Let p be a prime and ρ a primitive p-th root of unity. Let $G = \mathrm{Gal}(\mathbf{Q}(\rho)/\mathbf{Q})$. Show that $G \cong \mathbf{Z}_p^\times$. [Hint: Show that if $\tau \in G$, then $\tau(\rho) = \rho^k$ for some unique k with $1 \le k \le p - 1$.]

(2.11) Let p be a prime, $a \in \mathbf{Q}^+$, $w = \sqrt[p]{a}$ and suppose that $w \notin \mathbf{Q}(\rho)$ where ρ is a primitive p-th root of unity. Let $F = \mathbf{Q}(w, \rho)$ and $G = \mathrm{Gal}(F/\mathbf{Q}(\rho))$. Show that $G \cong \mathbf{Z}_p$. [Hint: show that if $\tau \in G$, then $\tau(w) = w\rho^k$ for some unique k with $0 \le k \le p - 1$.]

(2.12) Let p be a prime, $a \in \mathbf{Q}^+$, and G the Galois group of $f = x^p - a$ over $\mathbf{Q}$. Show that G is a solvable group.

(2.13) Let G be the Galois group of $f = (x^3 - 5)(x^7 - 1)^2$ over $\mathbf{Q}$. Show that G is a solvable group.

(2.14) For each of the following polynomials f find the splitting field F of f over $\mathbf{Q}$, $[F: \mathbf{Q}]$, and describe $G = \mathrm{Gal}(F/\mathbf{Q})$.

(a) $f = (x^2 - 2)(x^3 - 2)^2$.

(b) $f = x^4 + x^2 + 1$.

(c) $f = x^4 + 4x^2 + 4$.

(d) $f = x^4 - 6x^2 - 3$.

(2.15) Let ρ be a primitive 5-th root of unity, $F = \mathbf{Q}(\rho)$ and $G = \mathrm{Gal}(F/\mathbf{Q})$.

(a) Show that $G \cong \mathbf{Z}_5^\times \cong \mathbf{Z}_4$ has a unique proper nontrivial subgroup H.

(b) Show that $H' = \mathbf{Q}(\rho^2 + \rho^3) = \mathbf{Q}(\sqrt{5})$ is the unique proper intermediate field of F over **Q**.

(c) Show that the splitting field of $g = x^2 - 5$ is contained in F.

(d) Describe the Galois group of the polynomial $(x^5 - 1)(x^2 - 5)$ over **Q**.

(e) Describe the Galois group of the polynomial $(x^5 - 1)(x^2 - 3)$ over **Q**.

(2.16) (a) Show that if L is a splitting field of $x^4 - 2$ over **Q** and M is a splitting field of $x^5 - 1$ over **Q**, then $L \cap M = \mathbf{Q}$.

(b) Describe the Galois group of the polynomial $f = (x^4 - 2)(x^5 - 1)$ over **Q**.

(2.17) (a) Use the fact that the Galois group G of $f = x^7 - 1$ over **Q** is isomorphic to $\mathbf{Z}_7^\times \cong \mathbf{Z}_6$ to find all subgroups of G and corresponding intermediate fields of of the splitting field F of f over **Q**.

(b) For each of the intermediate fields E of F over **Q** find a polynomial $h \in \mathbf{Q}[x]$ such that E is the splitting field for h over **Q**.

(c) Show that the unique intermediate field of F over **Q** of dimension 2 over **Q** is $\mathbf{Q}(\sqrt{7}i)$.

(d) Describe the Galois group of the polynomial $(x^7 - 1)(x^2 + 7)$.

(e) Describe the Galois group of the polynomial $(x^7 - 1)(x^2 - 7)$.

(2.18) (a) Prove that if F is the splitting field of $f = x^{11} - 1$ over **Q**, then there is an intermediate field L of F over **Q** such that L is normal over **Q** and $[L\colon \mathbf{Q}] = 5$.

(b) Use (a) to find a polynomial $g \in \mathbf{Q}[x]$ such that the Galois group of g over **Q** is isomorphic to $\mathbf{Z}_5$.

(2.19) In each case find a polynomial f such that the Galois group of f over **Q** is isomorphic to the given group.

(a) D_3.

(b) $\mathbf{Z}_2 \times D_4$.

(c) $\mathbf{Z}_2 \times \mathbf{Z}_4$.

(d) $\mathbf{Z}_2 \times \mathbf{Z}_7^{\times}$.

(e) $\mathbf{Z}_7$ [Hint: Use the methods of the preceding exercise.]

(2.20) Describe the Galois group of the polynomial $f = x^3 + x + 1 \in F_2[x]$ over F_2. Find all intermediate fields of the splitting field of f over F_2.

(2.21) Construct normal closures F for the following field extensions E of $\mathbf{Q}$. Describe Gal(E/$\mathbf{Q}$) and Gal(F/$\mathbf{Q}$).

(a) $E = \mathbf{Q}(w)$ where w is the real 5-th root of 7.

(b) $E = \mathbf{Q}(\sqrt{2}, \sqrt[3]{2})$.

(c) $E = \mathbf{Q}(w)$ where w is a root of $x^3 - 7$.

(2.22) If X is a set, a subgroup G of the group of bijections B(X) is said to be <u>transitive</u> if for any $x, y \in X$ there is a $g \in G$ such that $g(x) = y$.

(a) Show that if $f \in K[x]$ is irreducible then the Galois group of f over K can be identified with a transitive subgroup of B(X) where X is the set of roots of f.

(b) Is the hypothesis that f is irreducible necessary in (a)?

(c) Find the transitive subgroups of S_3 and S_4.

(2.23) Let $m \in \mathbf{N}$, ρ a primitive m-th root of unity, $K = \mathbf{Q}(\rho)$, $a \in K$, and G be the Galois group of $f = x^m - a$ over K.

(a) Show that G is isomorphic to a subgroup of $\mathbf{Z}_m$.

(b) Give an example of a polynomial $f = x^m - a \in K[x]$ such that G is isomorphic to a proper subgroup of $\mathbf{Z}_m$.

(2.24) Suppose that K is an infinite field and [F: K] is finite (we do not assume that K has characteristic 0). Show that F is a simple extension of K if and only if there are only finitely many intermediate fields of F over K. [Hint: Choose $u \in F$ such that [K(u): K] is maximal; if $v \in F - K(u)$ consider $\{K(u + av): a \in K\}$.]

(2.25) Use the preceding exercise to show that if K has characteristic 0 and F is a finite normal extension of K, then F is a simple extension of K.

(2.26) Show that if F is a field with p^n elements and G = $Gal(F/F_p)$ then o(G) = n and G is cyclic with generator the Froebenius automorphism $\tau: F \to F$ by $u \mapsto u^p$.

(2.27) Show that if F is any finite dimensional extension of a finite field K, then F is a finite field, F is normal over K and G = Gal(F/K) is cyclic. (Hint: If char K = p, then $F_p \subseteq K \subseteq F$.)

(2.28) Let $f = x^4 + x + 1 \in F_2[x]$, u be a root of f and F = $F_2(u)$.

(a) Show that F is a field with 16 elements.

(b) Find a generator for $F^\times$.

(c) Find a generator for $Gal(F/F_2)$.

(2.29) Let F be a field of order p^n.

(a) Show that for each r with $1 \le r \le n$ there is a unique subfield E of F containing F_p such that $E \cong F_m$ where $m = p^r$.

(b) Show that every intermediate field of F over F_p is of this form.

(c) Show that if $m = p^r$, then $Gal(F/F_m)$ is cyclic with generator $\tau: F \to F$ by $u \mapsto u^m$.

(2.30) Show that if $m = p^n$, then the additive group of F_m is a direct product of n cyclic groups of order p.

(2.31) Show that if F is a finite field of characteristic p, then F is a simple extension of F_p.

3. SOLVABILITY BY RADICALS

In this section all fields have characteristic 0.

The solutions to the general quadratic equation f(x) = 0 where $f = ax^2 + bx + c \in \mathbf{Q}[x]$ are given by the quadratic formula as $r = (-b \pm \sqrt{d})/2a$ where $d = b^2 - 4ac \in \mathbf{Q}$. Hence the field $F = \mathbf{Q}(\sqrt{d})$ is a splitting field of f over $\mathbf{Q}$. By letting $u = \sqrt{d} \in \mathbf{C}$, we see that we may write F as $F = \mathbf{Q}(u)$ where $u^2 \in \mathbf{Q}$.

The solutions to the general cubic equation $x^3 + ax^2 + bx + c = 0$ $(a,b,c \in \mathbf{Q})$ are given explicitly by Cardan's formulas [see exercise (3.10)]. The solutions involve rational operations (that is, addition, subtraction, multiplication and division) and the taking of certain m-th roots $(m \in \mathbf{N})$.

More generally, one might ask whether, given an $n \in \mathbf{N}$, there exists a formula for the solutions of an n-th degree equation which involves only rational operations and the taking of roots. In fact such a solution does exist for the general fourth degree equation. We will show in this section that the existence of such a 'radical solution' to the equation $f(x) = 0$ $(f \in K[x])$ implies that the Galois group of f over K is a solvable group. Using this fact we then produce a fifth degree equation which does <u>not</u> have a radical solution over $\mathbf{Q}$, thereby showing that no general radical solution exists for fifth degree equations over $\mathbf{Q}$.

(3.1) DEFINITION An extension field F of K is a <u>radical extension</u> of K if there exist elements $u_1,\ldots,u_n \in F$ and $m(1),m(2),\ldots,m(n) \in \mathbf{N}$ satisfying the following properties.

(i) $F = K(u_1, \ldots, u_n)$.

(ii) $u_1^{m(1)} \in K$ and $u_i^{m(i)} \in K(u_1, \ldots, u_{i-1})$ for $2 \leq i \leq n$.

In this case we say that $(u_1, \ldots, u_n; m(1), \ldots, m(n))$ forms a <u>radical sequence</u> for the extension F over K.

We observe that if F is a radical extension of K the associated radical sequence is by no means unique. For example, if $F = K(u)$ with $u^{12} \in F$ [and hence radical sequence $(u; 12)$], we may also write F as a radical extension of K with associated radical sequence $(u^6, u^3, u; 2, 2, 3)$. In fact, by inserting extra powers of u_i (if necessary), we may always assume that the exponents $m(i)$ are prime [cf. exercise (3.2)].

(3.2) PROPOSITION If F is a radical extension of K then $[F: K]$ is finite.

PROOF Let F be a radical extension of K with associated radical sequence $(u_1, \ldots, u_n; m(1), \ldots, m(n))$. Let $F_0 = K$ and $F_i = F_{i-1}(u_i) = K(u_1, \ldots, u_i)$ for $1 \leq i \leq n$. Then we have the following chain of field extensions.

$$K = F_0 \subseteq F_1 \subseteq \cdots \subseteq F_{i-1} \subseteq F_i \subseteq \cdots \subseteq F_n = F$$

For each i, $u_i^{m(i)} \in F_{i-1}$ and hence u_i is a root of the polynomial $x_i^{m(i)} - u_i^{m(i)} \in F[x]$. Therefore u_i is algebraic over F_{i-1} and, since $F_i = F_{i-1}(u_i)$, $[F_i: F_{i-1}]$ is finite [cf. VIII, (2.1)]. It follows by VIII, (1.15), that [F: K] is finite.

(3.3) DEFINITION Let K be a field, $f \in K[x]$ and F a splitting field over K of f. The equation $f(x) = 0$ is solvable by radicals over K if F is contained in a radical extension E of K.

We again observe that since all splitting fields over K of f are isomorphic by an isomorphism fixing K, Definition (3.3) is independent of the choice of splitting field F.

EXAMPLE Let us consider the fourth degree equation $f(x) = 0$ where $f = x^4 - 4x^2 + 2 \in \mathbf{Q}[x]$. Using the quadratic formula we see that f has roots $\pm\sqrt{2 \pm \sqrt{2}}$. Since the solutions to the equation $f(x) = 0$ involve only rational operations and the taking of square roots, it seems reasonable to expect that the equation $f(x) = 0$ be solvable by radicals over **Q**. Verifying this fact, we see that $F = \mathbf{Q}(\sqrt{2 + \sqrt{2}}, \sqrt{2 - \sqrt{2}})$ is a splitting field for f over **Q** and F is contained in the radical extension $E = \mathbf{Q}(\sqrt{2}, \sqrt{2 + \sqrt{2}}, \sqrt{2 - \sqrt{2}})$ of **Q** [which has associated radical sequence $(\sqrt{2}, \sqrt{2 + \sqrt{2}}, \sqrt{2 - \sqrt{2}}; 2, 2, 2)$].

We recall that if m is a positive integer the complex number $\rho = e^{(2\pi i/m)} = \cos(2\pi/m) + i \sin(2\pi/m)$ is a primitive m-th root of unity; that is $\rho^m = 1$ and $\rho^i \neq 1$ for $1 \leq i \leq m - 1$. We observe that ρ is a generator for the multiplicative group of roots of the polynomial $x^m - 1$ in **C**. Now let K be any field of characteristic 0 and p a prime. Since the polynomial $f = x^p - 1$ and its derivative $Df = px^{p-1}$

have no roots in common, the roots of f (in any splitting field) are distinct. Since the set of roots of f forms a multiplicative group of prime order, it must be cyclic. Thus there is an element ρ of the splitting field of f such that $\rho^p = 1$ and $\rho^m \neq 1$ for any $1 \leq m \leq p - 1$. Such a ρ is a primitive p-th root of unity.

(3.4) PROPOSITION Let K be a field, p a prime, ρ a primitive p-th root of unity, and $F = K(\rho)$. Then the following assertions hold.

(i) F is a splitting field over K of the polynomial $f = x^p - 1 \in K[x]$ and hence F is a finite normal extension of K.

(ii) If $G = \mathrm{Gal}(F/K)$ then G is a finite abelian group.

PROOF Since ρ is a primitive p-th root of unity the p distinct elements $1, \rho, \rho^2, \ldots, \rho^{p-1}$ of F are precisely the roots of f and hence $F = K(\rho)$ is a splitting field of f over K. Hence by VIII, (3.17) F is a finite normal extension of K and by (2.8) G is a finite group.

We wish to show that G is abelian. Any element of G is determined by its action on ρ and must take ρ to ρ^i for some i with $0 \leq i \leq p - 1$. Hence if α and τ are elements of G, $\alpha(\rho) = \rho^k$ and $\tau(\rho) = \rho^i$ for some integers k and i between 0 and p - 1. But then $\alpha\tau$ and $\tau\alpha$ both take ρ to ρ^{ki} so that $\alpha\tau = \tau\alpha$ as required and the proof is complete.

(3.5) PROPOSITION Let $m \in \mathbb{N}$ and K a field over which $x^m - 1$ splits (that is, K contains all the m-th roots of unity). Let $F = K(u)$. If $u^m \in K$ then the following assertions hold.

(i) F is a splitting field over K of the polynomial $f = x^m - u^m \in K[x]$ and hence F is a finite normal extension of K.

(ii) If $G = \mathrm{Gal}(F/K)$ then G is a finite abelian group.

PROOF Since K has characteristic 0 the polynomial $x^m - 1$ has m distinct roots. If ρ is a root of $x^m - 1$, then $(u\rho)^m = u^m\rho^m = u^m$. Hence the set of roots of $f = x^m - u^m$ is precisely the set of $u\rho$ such that ρ is an m-th root of unity. Thus, since K contains the m-th roots of unity, the field $F = K(u)$ is a splitting field for f over K.

Now, since $F = K(u)$, if α and $\tau \in G$ they are determined by their action on u and must take u to a root of f. Suppose that $\alpha(u) = u\rho$ and $\tau(u) = uw$ where ρ and w are both m-th roots of unity. Then, since α and τ are both homomorphisms which fix K (and hence ρ and w), $\alpha\tau$ and $\tau\alpha$ both take u to $u\rho w$. Hence $\alpha\tau = \tau\alpha$ and G is abelian as claimed.

REMARKS Before proving the main result of this section we recall two important facts.

(1) Let $1 \to K \to G \to H \to 1$ be an exact sequence of groups. Then G is solvable if and only if both K and H are solvable [cf. III, (2.10)].

(2) If F is a finite normal extension of K and E is an intermediate extension which is normal over K, then by (2.7) the Galois correspondence induces an exact sequence of groups

$$1 \to \mathrm{Gal}(F/E) \to \mathrm{Gal}(F/K) \to \mathrm{Gal}(E/K) \to 1$$

(3.6) THEOREM Let F be a radical extension of K and $G = \mathrm{Gal}(F/K)$. Then G is a solvable group.

PROOF We first show that we may assume, without loss of generality, that F is normal over K.

Let K'' be the fixed field of G (relative to F over K). Then $K \subseteq K'' \subseteq F$, F is normal over K'', and $\mathrm{Gal}(F/K'') = K''' = K' = \mathrm{Gal}(F/K)$. Since $K \subseteq K'' \subseteq F$ and F is a radical extension of K, F is also a radical extension of K'' [cf. exercise (3.1)]. Hence the hypotheses of the theorem are unchanged by replacing K by K''.

We therefore assume that F is a normal radical extension of K with Galois group $G = \mathrm{Gal}(F/K)$. We wish to show that G is a solvable group.

Since F is a radical extension of K, there is an associated radical sequence $(u_1, \ldots, u_n; p(1), \ldots, p(n))$ of F over K with p(i) prime [cf. Exercise (3.2)]. We will induct on n. Let $p = p(1)$ and let ρ be a primitive p-th root of unity. We will consider the following three towers of field extensions.

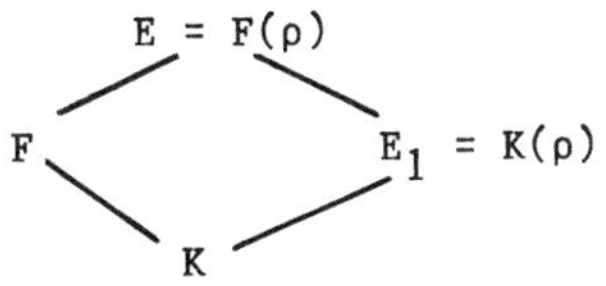

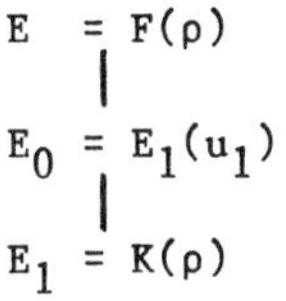

(I) (II) (III)

(I) Since F is normal over K, F is the splitting field of some polynomial $f \in K[x]$. Then E is the splitting field over K of the polynomial $(x^p - 1)f$ and hence E is a finite normal extension of K by (3.17). The Galois correspondence therefore induces the following exact sequence of groups.

$$1 \to \mathrm{Gal}(E/F) \to \mathrm{Gal}(E/K) \to \mathrm{Gal}(F/K) \to 1$$

By Remark (1) to prove that Gal(F/K) is solvable it suffices to prove that Gal(E/K) is solvable.

(II) By (3.4) E_1 is normal over K and $\mathrm{Gal}(E_1/K)$ is abelian and hence solvable. Since E is a finite normal extension of K we have a second exact sequence of groups.

$$1 \to \mathrm{Gal}(E/E_1) \to \mathrm{Gal}(E/K) \to \mathrm{Gal}(E_1/K) \to 1$$

Now, since $\mathrm{Gal}(E_1/K)$ is solvable, Gal(E/K) will be solvable provided $\mathrm{Gal}(E/E_1)$ is solvable [again by Remark (1)].

(III) Since E is finite and normal over K it is also finite and normal over E_1. By (3.5) E_0 is normal over E_1 and $\mathrm{Gal}(E_0/E_1)$ is abelian (and hence solvable). We now have a third exact sequence of groups.

$$1 \to \mathrm{Gal}(E/E_0) \to \mathrm{Gal}(E/E_1) \to \mathrm{Gal}(E_0/E_1) \to 1$$

Using Remark (1) again the proof is reduced to showing that $\mathrm{Gal}(E/E_0)$ is a solvable group. If $n = 1$ then $E = F(\rho) = K(u_1, \rho) = E_0$ and the result is trivial. Suppose that $n > 1$ and the result is true for all radical extensions which can be written with prime exponents and fewer than n terms. Then $E = F(\rho) = K(u_1, \ldots, u_n, \rho) =$

$K(\rho)(u_1, \ldots, u_n) = E_0(u_2, \ldots, u_n)$ and hence E is radical over E_0 with associated radical sequence $(u_2, \ldots, u_n; p(2), \ldots, p(n))$. It therefore follows by our induction hypothesis that $Gal(E/E_0)$ is a solvable group and the proof is now complete.

(3.7) COROLLARY Let f be an element of K[x]. If the equation $f(x) = 0$ is solvable by radicals then the Galois group of f over K is a solvable group.
PROOF Since the equation $f(x) = 0$ is solvable by radicals, there is a splitting field F of f over K and an extension field E of F such that E is a radical extension of K. Then by the previous theorem Gal(E/K) is a solvable group. Now since F is a splitting field over K, F is a finite normal extension of K and hence the Galois correspondence induces the following exact sequence of groups.

$$1 \to Gal(E/F) \to Gal(E/K) \to Gal(F/K) \to 1$$

Finally, since Gal(E/K) is a solvable group, it follows that Gal(F/K) is also a solvable group.

We recall from Chapter II that if $n \in \mathbf{N}$ and a subgroup G of the symmetric group S_n contains a transposition and an n-cycle, then in fact $G = S_n$ [cf. II, exercise (1.45)]. This fact is used in the following theorem to show that under certain circumstances a polynomial f of degree n has Galois group over **Q** isomorphic to S_n. Since S_n is <u>not</u> a solvable group for $n > 4$ we will then be able to use the preceding Corollary to produce a fifth degree equation which is not solvable by radicals over **Q**.

(3.8) THEOREM If p is a prime and f is an irreducible polynomial of degree p over **Q** which has exactly two nonreal roots in **C** then the Galois group of f over **Q** is isomorphic to S_p.
PROOF Let G be the Galois group of f over **Q**. Thus $G = Gal(F/\mathbf{Q})$ where F is a splitting field of f over **Q**. Since f splits in **C** [**C** is algebraically closed - cf. (2.15)] we may assume that $F \subseteq \mathbf{C}$ [see Remark (3.3)]. By (2.11), G can be considered to be a subgroup of S_p.

Since $p | o(G)$ by (2.11), G contains an element τ of order p by Cauchy's Theorem, III, (1.6). Hence G contains the p-cycle τ.

Let α: $\mathbf{C} \to \mathbf{C}$ be defined by $\alpha(a + bi) = a - bi$. Then $\alpha \in \text{Gal}(\mathbf{C}/\mathbf{Q})$. Since F is a splitting field over $\mathbf{Q}$ and $\mathbf{Q}$ has characteristic 0, F is normal over $\mathbf{Q}$. Hence if $\alpha_1 = \alpha|_F$, $\alpha_1 \in \text{Gal}(F/\mathbf{Q}) = G$. But α_1 is an element of G which interchanges the two nonreal roots of f and fixes the other roots; that is, α_1 is a transposition. Hence G contains both a transposition and a p-cycle and it follows that $G = S_p$.

(3.9) EXAMPLE Let $f = x^5 + 5x^4 - 5 \in \mathbf{Q}[x]$. By Eisenstein's criterion [cf. V, (3.18)], f is irreducible over $\mathbf{Q}$. Using elementary calculus we see that $Df = 5x^4 + 20x^3 = 5x^3(x + 4)$ and f has a local maximum when $x = -4$ and $f(x) = 251$ and a local minimum when $x = 0$ and $f(x) = -5$. A rough sketch of the graph of f then indicates that f has exactly three real roots and hence exactly two nonreal roots. By the previous theorem the Galois group of f over $\mathbf{Q}$ is isomorphic to S_5. Since S_5 is not a solvable group by III, (2.11), the fifth degree equation $x^5 + 5x^4 - 5 = 0$ is not solvable by radicals [cf. Corollary (3.6)].

Our goal now is to show that the converse to Corollary (3.7) also holds; that is, a polynomial f in K[x] is solvable by radicals <u>if and only if</u> the Galois group of f over K is a solvable group. We first require some definitions and technical propositions.

(3.10) DEFINITION Let F be a field and S a nonempty set of automorphisms of F. S is said to be linearly independent provided whenever $a_1, \ldots, a_n \in F$ and $\tau_1, \ldots, \tau_n \in S$ are such that

$$a_1\tau_1(u) + \cdots + a_n\tau_n(u) = 0 \text{ for } \underline{\text{all}}\ u \in F$$

then $a_k = 0$ for all k.

(3.11) PROPOSITION (Dedekind) If F is a field and S is any set of distinct automorphisms of F, then S is linearly independent.

PROOF Suppose the statement is false. Then there exist $a_k \in F$ (not all zero) and distinct $\tau_k \in S$ such that

(i) $\sum_{k=1}^{n} a_k \tau_k(u) = 0$ for all $u \in F$

Among all such dependence relations we may choose one with the number n of terms minimal. Clearly $n > 1$. Since $\tau_1 \neq \tau_2$ there is a $v \in F$ such that $\tau_1(v) \neq \tau_2(v)$. Then applying (i) to the element uv of F we get the following equation (observe that we need the fact that the functions τ_k in question are automorphisms).

(ii) $\sum_{k=1}^{n} a_k \tau_k(u)\tau_k(v) = 0$

Multiplying (i) by $\tau_1(v)$ produces the following equation.

(iii) $\sum_{k=1}^{n} a_k \tau_k(u)\tau_1(v) = 0$

Let $b_k = a_k[\tau_k(v) - \tau_1(v)]$ for $2 \leq k \leq n$. Then $b_2 \neq 0$ and subtracting (iii) from (ii) produces a dependence relation

$$\sum_{k=2}^{n} b_k \tau_k(u) = 0 \text{ for all } u \in F$$

with fewer than n terms, contradicting the minimality of n.

(3.12) DEFINITION Let F be a finite normal extension of K and G = Gal(F/K). If $u \in F$, the norm of u is

$$N(u) = \tau_1(u) \cdots \tau_n(u) \quad \text{where } G = \{\tau_1, \ldots, \tau_n\}$$

REMARKS (i) If $\tau \in G$ then $\{\tau\tau_k : 1 \leq k \leq n\}$ is merely a rearrangement of the elements of G and hence N(u) is fixed by every element of G. This implies, since F is normal over K, that $N(u) \in K$ for every $u \in F$.

(ii) If $u \in K$, then every element of G fixes u and hence $N(u) = u^n$

(3.13) THEOREM (Hilbert's Theorem 90) Suppose that F is a finite normal extension of K and G = Gal(F/K) is cyclic with generator τ. Then if $v \in F$, $N(v) = 1$ if and only if $v = u/\tau(u)$ for some nonzero u in F.

PROOF Suppose that $o(G) = n$. Then $\tau^n = 1$ [cf. II, (3.11)]. If $v = u/\tau(u)$ for some $u \in F$ then the proof that $N(v) = 1$ is relatively straightforward and is left to the reader [cf. exercise (3.13)].

Suppose now that $v \in F$ is such that $N(v) = 1$. We wish to produce a nonzero $u \in F$ such that $v = u/\tau(u)$. We will apply Dedekind's Lemma to the collection $\tau, \tau^2, \ldots, \tau^{n-1}, \tau^n = 1$, of distinct elements of G. Let

$$\begin{aligned} a_1 &= v\tau(v) \\ a_{k+1} &= a_k\tau^{k+1}(v) \\ &= v\tau(v) \cdots \tau^{k+1}(v) \quad \text{for } 1 \le k \le n-1 \end{aligned}$$

Then

$$\begin{aligned} a_{n-1} &= v\tau(v) \cdots \tau^{n-1}(v) = N(v) = 1 \text{ and hence} \\ a_n &= a_{n-1}\tau^n(v) = v \text{ (recall that } \tau^n = 1) \end{aligned}$$

We observe the following facts.

(i) $\tau(a_k) = \tau(v)\tau^2(v) \cdots \tau^{k+1}(v) = a_{k+1}/v$ for $1 \le k \le n-1$

(ii) $\tau(a_n) = \tau(v) = a_1/v$

Since not all $a_k = 0$ (in particular $a_n = v \neq 0$), by Dedekind's Lemma there is an element $w \in F$ such that

$$u = \sum_{k=1}^{n} a_k\tau^k(w) \neq 0.$$

We then calculate

$$\begin{aligned} \tau(u) &= \sum_{k=1}^{n} \tau(a_k)\tau^{k+1}(w) \\ &= \sum_{k=1}^{n-1} (a_{k+1}/v)\tau^{k+1}(w) + \tau(a_n)\tau(w) \quad \text{[by (i)]} \\ &= \sum_{k=2}^{n} (a_k/v)\tau^k(w) + (a_1/v)\tau(w) \quad \text{[by (ii)]} \\ &= u/v \end{aligned}$$

Hence $v = \tau(u)/u$ and the proof is complete.

(3.14) THEOREM Suppose that F is a finite normal extension of K, $G = \mathrm{Gal}(F/K)$ has prime order p and K contains the p-th roots of unity. Then the following assertions hold.

(i) There is an element $u \in F$ such that $F = K(u)$ and $u^p \in K$. Hence F is a radical extension of K.

(ii) F is the splitting field over K of the polynomial $f = x^p - u^p$.

PROOF Let ρ be a primitive p-th root of unity. Then since $o(G) = p$ and $\rho \in K$ (so that every element of G fixes ρ), $N(\rho) = \rho^p = 1$. Therefore, by Hilbert's Theorem 90, $\rho = u/\tau(u)$ for some nonzero $u \in F$ where $G = \langle\tau\rangle = \{\tau, \ldots, \tau^{p-1}, \tau^p = 1\}$. Then $\tau(u^p) = [\tau(u)]^p = (u/\rho)^p = u^p/\rho^p = u^p$ and hence u^p is fixed by every element of G. Since F is normal over K, this implies that $u^p \in K$.

Since $\tau(u) = u/\rho \neq u$, $u \notin K$ and hence $K(u) \neq K$. Then, since $[K(u): K]$ divides $[F: P] = p$ and does not equal 1, $[K(u): K] = p$ so that $F = F(u)$ as claimed. Property (ii) now follows from Lemma (3.5).

Before proving the next theorem we make the following observation.

If F is an extension of K and E is an intermediate field with E a radical extension of K and F a radical extension of E, then F is also a radical extension of K [cf. exercise (3.3)].

(3.15) THEOREM Let F be a finite normal extension of K with solvable Galois group $G = \text{Gal}(F/K)$. Then F is contained in a radical extension E of K.

PROOF Our goal is to use the properties of solvable groups and induction to reduce consideration to a field extension satisfying the hypotheses of the previous theorem.

Since G is a finite group it has a maximal normal subgroup H (that is, a normal subgroup H which is not properly contained in any other normal subgroup of G). Then G/H is a simple, solvable group [cf. III, (2.9)] and therefore is cyclic of prime order for some prime p.

Let ρ be a primitive p-th root of unity. We shall consider the following field extensions.

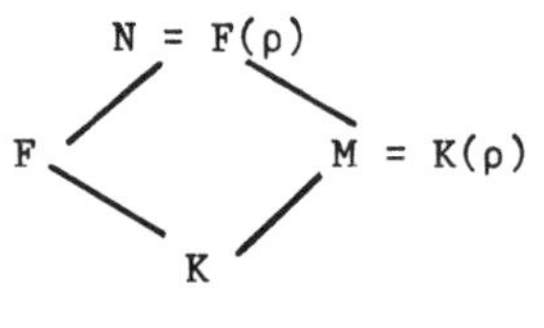

As in the proof of (3.6) N is finite dimensional and normal over K (and therefore over M as well). Since M is a radical extension of K, it suffices to show that there exists a radical extension of M that contains N (since this extension will also be radical over K by the observation preceding this theorem).

Since F is normal over K and therefore stable and $K \subseteq M$, if $\tau \in \mathrm{Gal}(N/M)$, then $\tau|_F \in \mathrm{Gal}(F/K)$. Define

$$\alpha\colon \mathrm{Gal}(N/M) \to \mathrm{Gal}(F/K) \text{ by } \tau \mapsto \tau|_F \quad [\tau \in \mathrm{Gal}(N/M)]$$

Then α is a homomorphism of groups. If $\tau \in \ker(\alpha)$, then $\tau|_F = 1$ and τ fixes ρ (since $\rho \in M$) and hence τ fixes $N = F(\rho)$. Thus $\ker(\alpha) = 1$ and α is an injective map.

We will now induct on $n = [F\colon K]$. We consider two cases.

Case (i) If α is not a surjective map, then N is a finite normal extension of M [cf. (3.5)], Gal(N/M) is isomorphic to a subgroup of Gal(F/K) and is therefore solvable, and $[N\colon M] = o(\mathrm{Gal}(N/M)) < o(\mathrm{Gal}(F/K)) = [F\colon K]$. Hence by the induction hypothesis there is a radical extension E of M which contains N and the result follows.

Case (ii) Suppose now that α is surjective (and therefore an isomorphism). Let $J = \alpha^{-1}(H)$. Then J is a normal subgroup of Gal(N/M) of index p. Hence the fixed field J' of J is a normal extension of M of dimension p as illustrated below.

$$\begin{array}{ccc} N & \to & (1) \\ \cup & & \cap \\ J' & \to & J = \mathrm{Gal}(N/J') \\ \cup & & \cap \\ M & \to & \mathrm{Gal}(N/M) \end{array}$$

By the Fundamental Theorem of Galois Theory, $o(\mathrm{Gal}(J'/M)) = [J'\colon M] = p$. Since M contains the p-th roots of unity, (3.14) implies that J' is a radical extension of M. It therefore suffices to show that there is a radical extension E of J' containing N.

Now $\mathrm{Gal}(N/J') = J$ is a solvable group (since it is a subgroup of a solvable group) and $[N: J'] = o(\mathrm{Gal}(N/J')) < [N: M] = [F: K] = n$. Therefore, by the induction hypothesis, there is a radical extension E of J' containing N. The proof is now complete.

(3.16) COROLLARY Let $f \in K[x]$. Then f is solvable by radicals over K if and only if the Galois group of f over K is a solvable group.

EXERCISES

(3.1) Prove that if $K \subseteq E \subseteq F$ and F is a radical extension of K, then F is also a radical extension of E.

(3.2) (a) Prove that if F is a radical extension of K there is an associated radical sequence $(u_1, \ldots, u_n; p(1), \ldots, p(n))$ of F over K with p(i) prime.

(b) Find such a radical sequence for the extension $\mathbf{Q}(\sqrt[4]{2}, \sqrt[6]{2 + \sqrt{3}})$ over **Q**.

(3.3) Suppose that E is an intermediate field of F over K such that E is a radical extension of K and F is a radical extension of E. Prove that F is a radical extension of K.

(3.4) Suppose that F is a radical extension of K. Show that if E is an intermediate field then E need not be a radical extension of K.

(3.5) Find a radical extension of **Q** containing the splitting field of each of the following polynomials and give the corresponding radical sequence.

(a) $x^4 + x^2 + 1$.

(b) $x^4 + 3x^2 + 1$.

(c) $x^3 - 2x + 1$.

(d) $x^5 + 4x^3 + x$.

(e) $x^7 - 2$.

(3.6) Show that if $f, g \in K[x]$, then fg is solvable by radicals over K if and only if both f and g are solvable by radicals over K.

(3.7) Show that if $f \in K[x]$ is reducible of degree 5 then f is solvable by radicals over K.

(3.8) (a) Show that if $a \in \mathbf{Q}^+$ and $m \in \mathbf{N}$, then $f = x^m - a$ is solvable by radicals over $\mathbf{Q}$.

(b) Find a solvable series for the Galois group of $f = x^p - 2$ over $\mathbf{Q}$ where p is a prime. [Hint: use exercises (2.10) and (2.11).]

(3.9) Verify the following formulas (Cardan's formulas) for the solution of a cubic in $K[x]$.

Let $a,b,c \in K$ and consider the equation $x^3 + ax^2 + bx + c = 0$. Let $p = b - a^2/3$, $q = 2a^3/27 - ab/3 + c$, $P = \sqrt[3]{-q/2 + \sqrt{p^3/27 + q^2/4}}$, and $Q = \sqrt[3]{-q/2 - \sqrt{p^3/27 + q^2/4}}$. Then the solutions of the given equation are $P + Q -a/3$, $wP + w^2Q - a/3$ and $w^2P + wQ - a/3$ where w is a primitive cube root of unity.

(3.10) Prove that if F is a finite normal extension of K and $G = \mathrm{Gal}(F/K)$ is cyclic with generator τ and $v = u/\tau(u)$, $u \in F$, then $N(v) = 1$ [cf. (3.13)].

(3.11) Prove that every polynomial equation over $\mathbf{R}$ is solvable by radicals over $\mathbf{R}$. (Hint: $\mathbf{C}$ is algebraically closed.)

(3.12) Prove that if K is a field (of characteristic 0), p is prime and ρ is a primitive p-th root of unity then the Galois group of $K(\rho)$ over K is a cyclic group [Hint: cf. exercise VIII, (3.19)].

Appendix A
Zorn's Lemma

We state Zorn's Lemma and give an example of its application. First recall the following elementary notions.

Let S be a set. We say S is <u>partially ordered</u> with respect to the relation $\leqslant$ if for all $s,t,u \in S$, the following assertions hold:

(i) $s \leqslant s$ (Reflexivity)

(ii) If $s \leqslant t$ and $t \leqslant s$, then $s = t$ (Antisymmetry)

(iii) If $s \leqslant t$ and $t \leqslant u$, then $s \leqslant u$ (Transitivity)

A subset T of S is called a <u>chain</u> (or a <u>totally ordered subset</u>) if for all $s,t \in T$, either $s \leqslant t$ or $t \leqslant s$.

For any subset T of S, an element $b \in S$ is called an <u>upper bound</u> for T if $t \leqslant b$ for all $t \in T$.

An element $x \in S$ is called a <u>maximal element</u> if $x \leqslant s$ implies $x = s$ for all $s \in S$.

THEOREM (Zorn's Lemma) Let S be a nonempty, partially ordered set and suppose that every chain in S has an upper bound in S. Then S has a maximal element.

The proof of the theorem is conveniently beyond the scope of this book. (See, for instance, Stoll's <u>Set Theory and Logic</u> for a good treatment.) Mathematicians, however, are quite unable to mention Zorn's Lemma without noting its equivalence to the well-grilled Axiom of Choice. (And usually in the next breath.)

Here is a modest application.

EXAMPLE Let A be a commutative ring and let X be a subset of A which is closed under multiplication and does not contain 0. Then there is an ideal J of A maximal with respect to the set of all ideals of A not intersecting X. Moreover, J is prime.

PROOF Consider the set S of all ideals in A which have empty intersection with X. S is nonempty (why?) and partially ordered by inclusion. Moreover, any chain of ideals in S has an upper bound in S given by the union of all ideals in the chain. (Check this!) Thus, by Zorn's Lemma, X has a maximal element J as required.

To show that J is indeed prime, let $a,b \in A$ and assume that neither a nor b lies in J, but the product ab does. Then the ideals $(a) + J$ and $(b) + J$ both properly contain J and so, by the maximality of J, must intersect X. Therefore there exist $x,y \in A$ and $p,q \in J$ such that the sums $ax + p$ and $by + q$ lie in X. Since X is closed under multiplication, the product

$$(ax + p)(by + q) = abxy + axq + pby + pq$$

likewise lies in X. But consider the four terms on the right: the second, third, and fourth lie in J since p and q do. The first lies in J by the assumption that the product ab does. Thus the sum lies both in J and in X, contradicting the disjointness of these two sets. Consequently $a,b \notin J$ implies $ab \notin J$, whence by definition J is prime.

Appendix B
Categories and Functors

In this appendix we will introduce the basic terminology of category theory. The intent is not to prove, or even state, theorems about categories. Instead we include only basic definitions and examples. We urge the student to work through all the examples and fill in all missing details. Much of the richness of mathematics is enhanced in this very general setting. We begin with the definition of a category.

(1.1) DEFINITION A category **K** is a pair (**O**, **M**) of classes. **O** is called the class of objects of the category and **M** is called the class of morphisms. **K** satisfies the following axioms.

(a) For any $A, B \in \mathbf{O}$ there is a subset $\mathbf{K}(A, B)$ of **M** called the set of morphisms from A to B. If $f \in \mathbf{K}(A, B)$ we write $f: A \to B$.

(b) If $A, B, C, D \in \mathbf{O}$ are such that $\mathbf{K}(A, B) \cap \mathbf{K}(C, D) \neq \emptyset$ then $A = C$ and $B = D$.

(c) For any $A, B, C \in \mathbf{O}$ there is a mapping

$$\mathbf{K}(B, C) \times \mathbf{K}(A, B) \to \mathbf{K}(A, C)$$
$$(g, f) \mapsto gf$$

called the rule of composition for **K**.

(d) Associativity of composition. For any $A, B, C, D \in \mathbf{O}$, $f \in \mathbf{K}(A, B)$, $g \in \mathbf{K}(B, C)$, $h \in \mathbf{K}(C, D)$ we have $(hg)f = h(gf)$.

(e) Existence of identities. For any $A \in \mathbf{O}$ there is a $1_A \in \mathbf{K}(A, A)$ such that for any $B, C \in \mathbf{O}$, $f \in \mathbf{K}(B, A)$, $g \in \mathbf{K}(A, C)$ we have $1_A f = f$ and $g 1_A = g$.

(1.2) EXAMPLES We will now consider some examples of categories. The reader is urged to construct many others.

(1.2.1) The Category of Sets. Let $\mathbf{S} = (\mathbf{O}, \mathbf{M})$ where $\mathbf{O}$ is the class of all sets and $\mathbf{M}$ is the class of all functions, and the rule of composition is composition of functions. The resulting structure is clearly a category which we will call the category of sets.

(1.2.2) Categories contained in the category of sets. Many categories are in some sense contained in the category of sets, more specifically categories in which the objects are sets with some kind of structure, e.g. groups, and the morphisms are functions which preserve this structure, e.g. group homomorphisms. In these cases we will always take function composition to be the rule of composition. The following table lists some examples which we have studied in this book.

Objects	Morphisms
1. Sets	1. Functions
2. Groups	2. Group homomorphisms
3. Additive groups	3. Group homomorphisms
4. Rings	4. Ring homomorphisms
5. R-modules	5. R-homomorphisms
6. Commutative rings	6. Ring homomorphisms

To show that each of these examples is a category we need only observe that the composition of morphisms is a morphism and that the identity mapping is a morphism.

(1.2.3) The category of complexes of additive groups. In II, Exercise (4.34) we defined a complex of additive groups to be a sequence A of additive groups and group homomorphisms

$$\cdots \longrightarrow A_{i-1} \xrightarrow{d_{i-1}} A_i \xrightarrow{d_i} A_{i+1} \longrightarrow \cdots$$

such that $d_i d_{i-1} = 0$ for each $i \in \mathbf{Z}$.

If A,B are complexes of additive groups then a morphism $f: A \to B$ is a sequence $\{f_i: i \in \mathbf{Z}$ of homomorphisms such that $f_i: A_i \to B_i$ and

$$\begin{array}{ccc} A_i & \xrightarrow{d_i} & A_{i+1} \\ {\scriptstyle f_i}\downarrow & & \downarrow{\scriptstyle f_{i+1}} \\ B_i & \xrightarrow{d_i} & B_{i+1} \end{array}$$

is commutative for each $i \in \mathbf{Z}$.

A category may be formed by taking objects to be the class of complexes of additive groups, and morphisms to be the class of morphisms of complexes. The rule of composition will be defined as componentwise composition of morphisms. This category is called the category of complexes of additive groups. Note that the objects in this category are not sets and the morphisms are not functions.

(1.2.4) The category $\mathbf{K}^I$. Let $\mathbf{K}$ be any category and let I be a set; then a category may be formed by taking families $\{A_i: i \in I\}$ of objects of $\mathbf{K}$ as objects in $\mathbf{K}^I$. A morphism from $\{A_i: i \in I\}$ to $\{B_i: i \in I\}$ is a family $\{f_i: i \in I\}$ where $f_i \in \mathbf{K}(A_i, B_i)$ for each $i \in I$. The rule of composition in $\mathbf{K}^I$ is 'componentwise composition.'

(1.3) DEFINITION Let $f: A \to B$ be a morphism in a category $\mathbf{K}$. Then

(i) f is said to be a monomorphism, or monic, if for any pair of morphisms $g,h \in \mathbf{K}(C,A)$ such that $fg = fh$, we have $g = h$.

(ii) f is said to be an epimorphism, or epic, if for any pair of morphisms $g,h \in \mathbf{K}(B,C)$ such that $gf = hf$, we have $g = h$.

(1.4) REMARK In all the categories in example (1.2) in which objects are sets and morphisms are functions it is easy to show that injective mappings are monic and surjective mappings are epic. The converse to this statement is not true in general.

(1.5) DEFINITION Sums and Products. Let $\{A_i: i \in I\}$ be an indexed family of objects in a category **K**. Then $\{P; p_i; i \in I\}$ is called a product of $\{A_i: i \in I\}$ in **K** if

(a) P is an object in **K**.

(b) $p_i: P \to A_i$ is a morphism in **K** which is called the i-th projection.

(c) If B is an object in **O** and $\{f_i: i \in I\}$ is a family of morphisms in **K** such that $f_i: B \to A_i$ for each $i \in I$, then there is a unique $f: B \to P$ rendering commutative the following family of diagrams.

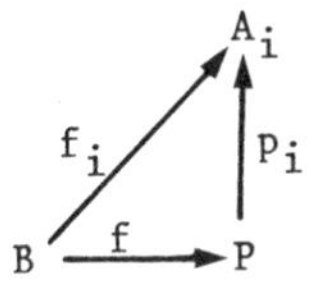

We will call $\{S; u_i; i \in I\}$ a sum of $\{A_i: i \in I\}$ in **K** if

(a) S is an object in **K**.

(b) $u_i: A_i \to S$ is a morphism in **K** which is called the i-th injection.

(c) If B is an object in **K** and $\{f_i: i \in I\}$ is a family of morphisms in **K** such that $f_i: A_i \to B$ for each $i \in I$, then there is a unique $f: S \to B$ rendering commutative the following family of diagrams.

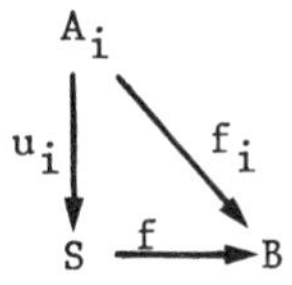

$(i \in I)$

(1.6) EXAMPLES We will now give some examples of products and sums in some categories.

(1.6.1) Sets and functions. Let **S** be the category of sets and $\{A_i: i \in I\}$ be a family of sets. The product set P is defined in a similar way to the product of a family of modules, see VII section 2.

The i-th projection mappings are also defined in a similar way. With this definition, $\{P;\ p_i;\ i \in I\}$ is a product of the family $\{A_i:\ i \in I\}$ in the category of sets.

Define $S = \{(a_i, i):\ a_i \in A_i,\ i \in I\}$ and define $u_i: A_i \to S$ by $u_i(a_i) = (a_i, i)$. Then $\{S;\ u_i;\ i \in I\}$ is a sum of $\{A_i:\ i \in I\}$ in the category **S**. The set S is called the disjoint union of the family $\{A_i:\ i \in I\}$.

(1.6.2) The category of R-modules. We have seen in VII Proposition (2.2) that $\{\prod_{i\in I} M_i;\ p_i;\ i \in I\}$ is a product of the family $\{M_i:\ i \in I\}$ of R-modules in the category of R-modules. Likewise, from VII Proposition (2.5) $\{\bigoplus_{i\in I} M_i;\ u_i;\ i \in I\}$ is a sum in the category of R-modules.

(1.6.3) The category of groups. Let **G** be the category of groups and let $\{G_i:\ i \in I\}$ be a family of groups. Define $\prod_{i\in I} G_i$ in a similar manner as defined in the category of R-modules. Let $p_i: \prod_{i\in I} G_i \to G_i$ be the i-th projection mapping for each $i \in I$. Then $\{\prod_{i\in I} G_i;\ p_i;\ i \in I\}$ is a product of the family $\{G_i:\ i \in I\}$ in the category **G**.

The description of the sum of a family of groups in **G** is beyond the scope of this work.

(1.7) DEFINITION Let **K** and **K'** be categories. Then a functor F from **K** to **K'** is a function from **O** ∪ **M** to **O'** ∪ **M'** such that

(a) For any $A \in$ **O** we have $F(A) \in$ **O'**.

(b) For any $f \in$ **K**(A, B), we have $F(f) \in$ **K'**$(F(A), F(B))$.

(c) For any $A, B, C \in$ **O** and for any $f \in$ **K**(A, B), $g \in$ **K**(B, C), we have $F(gf) = F(g)F(f)$.

(d) For any $A \in$ **O**, we have $F(1_A) = 1_{F(A)}$.

(1.8) EXAMPLES We will now give some examples of functors.

(1.8.1) The forgetful functor from the category of groups to the category of sets. Define F from the category of groups to the category of sets as follows: $F(G, *) = G$ and $F(f) = f$. Then F is a functor. F is sometimes called the forgetful functor from the category of groups to the category of sets. The name is very suggestive since F simply forgets the group structure.

(1.8.2) The functor $\mathbf{K}(A, \ )$. Let $\mathbf{K}$ be any category and let A be an object in $\mathbf{K}$. Define a functor $\mathbf{K}(A, \ )$ from $\mathbf{K}$ to $\mathbf{S}$ as follows. For each object B in $\mathbf{K}$,

$\mathbf{K}(A, B)$ is the set of all morphisms from A to B in $\mathbf{K}$ as usual

If $f: B \to C$ is a morphism in $\mathbf{K}$, then $\mathbf{K}(A, f): \mathbf{K}(A, B) \to \mathbf{K}(A, C)$ is defined by

$$\mathbf{K}(A, f)(g) = f \circ g$$

It is easy to verify that $\mathbf{K}(A, \ )$ is a functor from $\mathbf{K}$ to $\mathbf{S}$.

In the case that $\mathbf{K}$ is the category of R-modules we have seen previously that $\mathrm{Hom}_R(M, N)$ is an additive group. The student can also verify that if $f: N \to K$ is an R-homomorphism, [cf. VII(1.8.10)] then $\mathrm{Hom}_R(M, f)$ is a homomorphism of additive groups. Hence, $\mathrm{Hom}_R(M, \ )$ can be regarded as a functor from the category of R-modules to the category of additive groups.

Bibliography

Artin, E., Galois Theory, Notre Dame Mathematical Lectures No. 2 (2nd edition), Notre Dame, Indiana, 1944.

Faith, C., Algebra: Categories, Rings and Modules, Springer-Verlag, Berlin, 1973.

Herstein, I., Noncommutative Rings, Mathematics Association of America, distributed by J. Wiley, 1968.

Herstein, I., Topics in Algebra, Blaisdell Publishing Company, Waltham, Mass., 1964.

Jans, J., Rings and Homology, Holt, Rinehart and Winston, Inc., New York, 1964.

Janusz, G., Algebraic Number Fields, Academic Press, New York, 1973.

Hartshorne, R., Algebraic Geometry, Springer-Verlag, New York, 1977.

Kaplansky, I., Commutative Rings, Allyn and Bacon, Inc., Boston, 1970.

Kaplansky, I., Fields and Rings (2nd edition), University of Chicago Press, Chicago, 1972.

Kaplansky, I., Infinite Abelian Groups (2nd edition), University of Michigan Press, Ann Arbor, Michigan, 1969.

Lambek, J., Lectures on Rings and Modules, Blaisdell Publishing Company, Waltham, Mass., 1966.

Passman, D., The Algebraic Structure of Group Rings, Wiley-Interscience, New York, 1977.

Rotman, J., The Theory of Groups (2nd edition), Allyn and Bacon, Inc., Boston, 1973.

Scott, W., Group Theory, Prentice-Hall, Inc., Englewood Cliffs, N.J., 1964.

Stewart, I., Galois Theory, Chapman Hall, Ltd., London, 1973.

Index